LASERS – APPLICATIONS IN SCIENCE AND INDUSTRY

Edited by **Krzysztof Jakubczak**

Lasers – Applications in Science and Industry
Edited by Krzysztof Jakubczak

Published by InTech
Janeza Trdine 9, 51000 Rijeka, Croatia

Publishing Process Manager Petra Nenadic

Technical Editor Teodora Smiljanic

Cover Designer InTech Design Team

Additional hard copies can be obtained from orders@intechopen.com

Lasers – Applications in Science and Industry, Edited by Krzysztof Jakubczak
p. cm.
ISBN 978-953-307-755-0

For EU product safety concerns:
IN TECH d.o.o., Prolaz Marije Krucifikse Kozulić 3, 51000 Rijeka, Croatia,
info@intechopen.com or visit our website at intechopen.com.

Meet the editor

Dr. Krzysztof Jakubczak is a graduate from Military University of Technology, Warsaw, Poland. He achieved his education in the fields of laser physics, optics, infrared technology, and light modulation and detection. He earned his MSc degree for research on laser-plasma source for lithography and micro- and nano-structures production with the lobster-eye optics assembly. He continued his education at Czech Technical University in Prague. He earned his PhD for construction, examination and applications of high-order harmonics source using titanium-sapphire femtosecond laser system and noble gases. Amongst others he expanded his professional skills to such areas likę vacuum technology, spectroscopy, plasma physics, ultrafast laser systems and optics. He started his professional career as R&D Engineer with the Academy of Sciences of the Czech Republic. Then he got involved in HiLASE Project as Deputy Project Manager and led the R&D team Research Program Leader working on development of thin-disk lasers. Finally, he decided to pursue industrial career in medical devices (intraocular lenses). He works as Head of Engineering Department with Croma Polska sp. z o.o. and is now based in Warsaw, Poland. Krzysztof Jakubczak is a member of SPIE and Photonics Society of Poland.

Contents

Preface

Recent advents in laser technology and discoveries in laser physics have enabled their very new and exciting applications. Some of them cover production of new materials: nano-particles, periodic structures in nano-scale, and thin films. Others allowed better understanding of laser-matter interaction when sample is subjected to intense laser pulses of various time duration, shape in space and time domains as well as different spectral components contained in a pulse. This allowed control of molecules behavior in electromagnetic field, control of chemical reactions or even direct examination of intra-atomic processes with femtosecond or even attosecond resolution.

Lasers are also a perfect tool for medicine: since there is no contact of a tool with the tissue they are naturally aseptic; there is now tool wear and potential bio-contamination is minimized leading to minimum after-treatment trauma. Their development allowed appearing of new metrology: optical coherence tomography (OCT) which is now a routinely used device in vast majority of the hospitals. Finally, they found their use is very sophisticated applications like in-vitro fertilization or varicose veins treatment with a fiber laser with extremely short required patient stay in the hospital.

Dr. Krzysztof Jakubczak
Croma Polska Sp. z o.o.
Warsaw
Poland

Part 1

Thin Films and Nanostructures

1

Nanoparticles and Nanostructures Fabricated Using Femtosecond Laser Pulses

Chih Wei Luo
Department of Electrophysics, National Chiao Tung University, Taiwan Republic of China

1. Introduction

Recently, the processing of materials by femtosecond (fs) laser pulses has attracted a great deal of attention, because fs pulse energy can be precisely and rapidly transferred to the materials without thermal effects (Stuart et al., 1995). In particularly, periodic microstructures can be produced in almost any materials using fs pulses directly and without the need for masks or chemical photoresists to relieve the environmental concerns. For instance, nanoripples (Hsu et al., 2007; Luo et al., 2008; Sakabe et al., 2009; Jia et al., 2010; Yang et al., 2010; Bonse & Krüger, 2010; Okamuro et al., 2010; Huang et al., 2009), nanoparticles (Jia et al., 2006; Luo et al., 2008; Teng et al., 2010), nanocones (Nayak et al., 2008), and nanospikes (Zhao et al., 2007b) have been induced in various materials using single-beam fs laser pulses in air. In addition, fs laser ablation for metals and semiconductors in a vacuum environment (Amoruso et al., 2004; Liu et al., 2007a) and in liquid (Tsuji et al., 2003) have also been extensively investigated. These results are a strong indicator of the application potential of fs laser pulses in science and industry.

In this chapter, we demonstrate the generation of nanoparticles and nanostructures (including ripples and dots) using fs laser pulses. Initially, we selected the II-VI semiconductor ZnSe to demonstrate the fabrication of nanoparticles. Following the irradiation of fs laser pulses at a wavelength of 800 nm and pulse duration of 80 fs, many hexagonal-phase ZnSe nanoparticles formed on the surface of an undoped (100) cubic ZnSe single-crystal wafer. The interesting phase transition from the cubic structure of ZnSe single-crystal wafer to the hexagonal structure of ZnSe nanoparticles may have been caused by the ultra-high ablation pressure at the local area due to the sudden injection of high-energy leading to solid-solid transition. This chapter discusses the details of the mechanisms underlying this process.

In the second part of this chapter, we introduce controllable nanoripple and nanodot structures to high-T_c superconducting $YBa_2Cu_3O_7$ (YBCO) thin films. We also introduce the surface morphology of YBCO thin films under single-beam and dual-beam fs laser irradiation. The generation of periodic ripple and dot structures is determined by the application of laser fluence, the number of pulses, polarization and the incident angles of the laser beam. The period and orientation of ripples and even the size and density of dots can be controlled by these parameters.

2. Fabrication of hexagonal-phase ZnSe nanoparticles

Zinc selenide (ZnSe) has been studied extensively since the 1970s for implementation in II-VI semiconductors, due to its promising opto-electrical and electrical properties of direct wide band gap 2.7 eV at 300 K (Tawara et al., 1999; Dinger et al., 2000; Xiang et al., 2003). Over the last decade, the development of nanotechnologies has had a tremendous impact on industry and basic scientific research. The nanostructures of ZnSe, in particular, have attracted considerable attention recently (Tawara et al., 1999; Sarigiannis et al., 2002). Generally, crystalline ZnSe exhibits two structural phases, cubic and hexagonal. In ambient environments, the cubic phase is most often studies because the hexagonal structure is thermodynamically unstable (Sarigiannis et al., 2002; Che et al., 2004). In this section, we demonstrate the fabrication of hexagonal-phase ZnSe nanoparticles using femtosecond laser pulses and characterize their properties.

2.1 Experimental setup and procedure

In this study, the laser source plays an important role causing materials to undergo various changes. To reach the nonlinear region, a light source with high pulse energy is required. The seed pulses at 800 nm were produced using a mode-locked Ti:sapphire laser (Coherent-Micra10) pumped by a diode pump solid state laser (Coherent-Verdi). After being stretched to ~200 ps, these pulses were synchronously injected into a Ti:sapphire regenerative amplifier (Coherent-Legend) pumped by a 5-kHz Nd:YLF laser and the amplified pulses (pulse energy ~0.4 mJ) were recompressed to $\tau_p \sim 80$ fs at sample surface.
Figure 1 shows the experimental setup used for the fabrication of ZnSe nanoparticles. A plano-convex fused silica cylindrical lens with the focal length of 100 mm was used to focus the femtosecond laser pulses into a line spot (2270 μm×54 μm). The (100) cubic ZnSe single-crystal wafers were mounted on a motorized X-Y-Z translation stage in air and scanned using the focused laser spot at the scanning speed of 100 μm/s as shown in Fig. 1. The pulse energy was varied using metallic neutral density filters with OD0.1-OD2 (Thorlabs ND series).
Following femtosecond laser irradiation, a white-yellow powder [as shown in the inset of Fig. 2(b)], i.e. ZnSe nanoparticles, was observed on the surface of a ZnSe single-crystal wafer. Depending on the experimental objectives, these ZnSe nanoparticles could be dissolved in ethanol with ultrasonic waves or picked up with Scotch tape. After removing

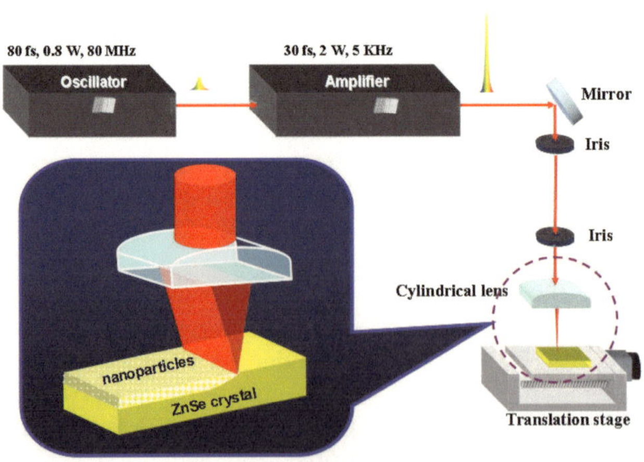

Fig. 1. Experimental setup for the fabrication of ZnSe nanoparticles.

the ZnSe nanoparticles from the surface of a ZnSe single-crystal wafer, many sub-wavelength ripples were observed on the surface, as shown in Fig. 2(b). These ripples appeared perpendicular to the scanning direction of the laser beam and the polarization of laser pulses, which are presented by the dashed and solid arrows, respectively, in Fig. 2(b).

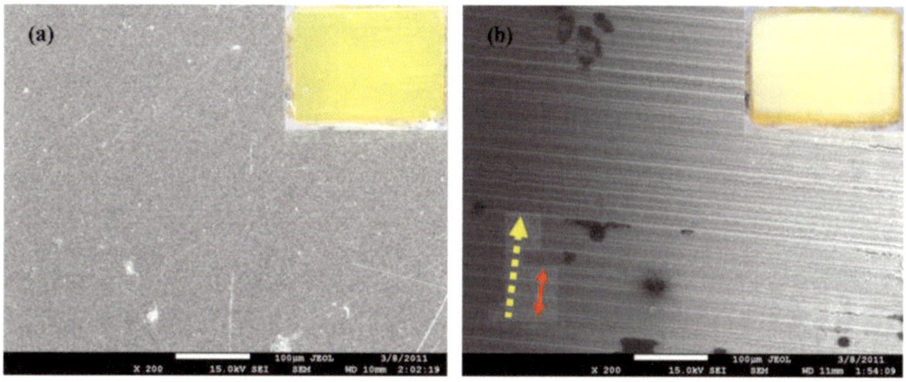

Fig. 2. SEM images of ZnSe single-crystal wafers; (a) before; and (b) after femtosecond laser pulse irradiations. Inset: OM images of ZnSe single-crystal wafer; (a) before; and (b) after femtosecond laser pulse irradiations. The dashed arrow indicates the scanning direction of a laser beam. The solid arrow indicates the polarization of laser pulses.

2.2 Characteristics of ZnSe nanoparticles

Figure 3(a) shows X-ray diffraction patterns of ZnSe nanoparticles fabricated at various fluences, which can be indexed by the hexagonal structure according to the JCPDS card no.80-0008 for ZnSe ($a = b = 3.974$ Å, $c = 6.506$ Å). It can be clearly seen that the cubic phase of the ZnSe single-crystal wafers has been transferred to the hexagonal phase in the ZnSe nanoparticles. Because hexagonal ZnSe is a metastable phase under ambient conditions, it can only be fabricated under the very strict growth conditions (Jiang et al., 2004; Liu et al., 2007b). However, hexagonal ZnSe nanoparticles can be easily and reliably achieved using femtosecond laser ablation as demonstrated in this study. Additionally, Figure 3(b) shows the room-temperature Raman scattering spectra of the ZnSe wafer, before and after the laser irradiation, and fabricated nanoparticles. The Raman peak at 252 cm^{-1} can be assigned to the longitudinal optical (LO) phonon mode of the cubic structure observed both in the ZnSe wafer before and after laser processing. For ZnSe nanoparticles, a strong peak appears at 234 cm^{-1} which is the so-called surface phonon mode (Shan et al., 2006). Typically, this surface phonon mode is a characteristic feature of nanostructures due to their large surface to volume ratio. Besides, no LO phonon mode of cubic structure is observed in ZnSe nanoparticles indicating that the crystal structure of ZnSe nanoparticles is pure hexagonal phase which is in accord with the X-ray diffraction patterns shown in Fig. 3(a).

Figure 4(a) shows a typical TEM image of ZnSe nanoparticles with the smooth spherical shape. A high-resolution TEM image at the atomic scale for one ZnSe nanoparticle is presented in the inset of Fig. 4(b). Furthermore, the six-fold electron diffraction pattern can be clearly observed in Fig. 4(b). Through the analysis of distance and angles between the nearest diffraction points and the center (biggest) point, the crystal structure of ZnSe nanoparticles was identified as a hexagonal and the orientation of each diffraction point is marked in Fig. 4(b), which consists with the results of XRD in Fig. 3(a). The energy dispersive spectroscopy (EDS) spectrum in the inset of Fig. 4(a) illustrates the composition of these ZnSe nanoparticles, comprising only two elements of Zn and Se. This reveals that the high purity of hexagonal ZnSe nanoparticles can be reliably and simply fabricated using femtosecond laser pulses.

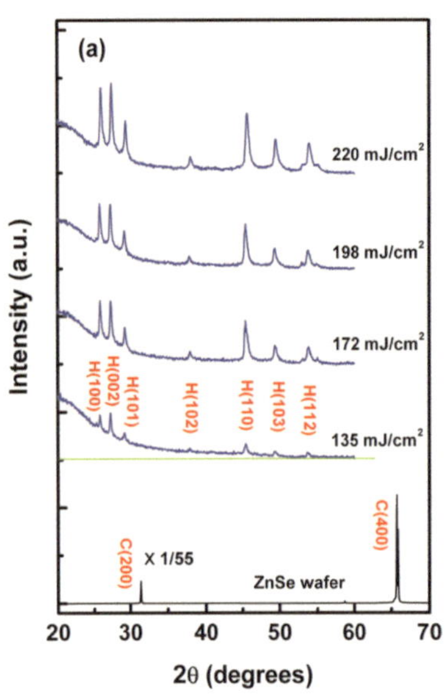

Fig. 3. (a) X-ray diffraction patterns of ZnSe wafer and ZnSe nanoparticles fabricated at various laser fluences. H: Hexagonal. C: Cubic. (b) Raman spectra of ZnSe wafer and ZnSe nanoparticles fabricated at the fluence of 220 mJ/cm^2. The 632.8 nm line of laser with 0.33 mW was used as the excitation light.

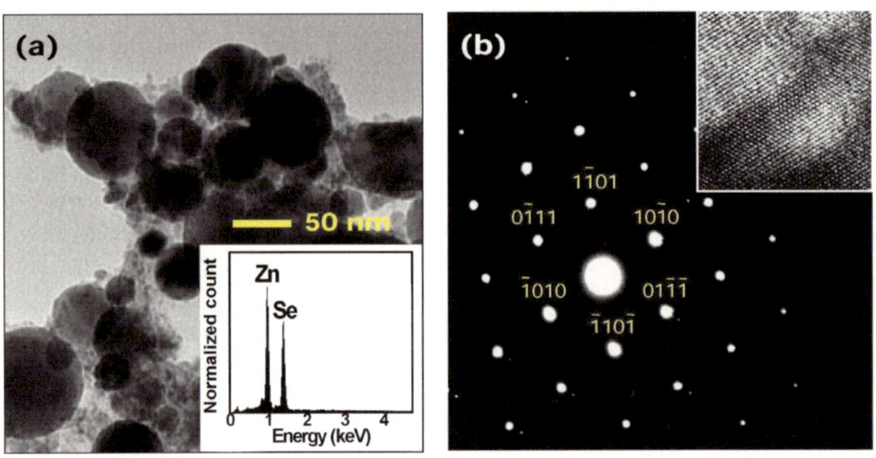

Fig. 4. (a) TEM images of ZnSe nanoparticles fabricated by the fluence of 220 mJ/cm^2. (b) TEM diffraction patterns of ZnSe nanoparticles in (a). Insets: (a) The EDS spectrum shows the composition of ZnSe nanoparticles; (b) High-resolution TEM image at the atomic scale.

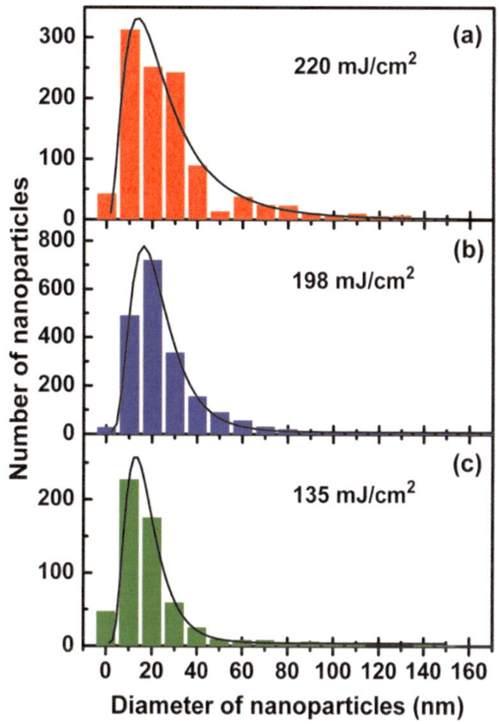

Fig. 5. Size distribution of ZnSe nanoparticles at various laser fluences corresponding to the TEM images in Fig. 4(a) with an area of 3.2 μm × 2.6 μm. The solid lines are the log-normal fitting.

The size distribution of ZnSe nanoparticles fabricated at various fluences was analyzed in Fig. 5. By the fitting of the log-normal function, we determined that the average diameter of ZnSe nanoparticles was approximately 16 nm in the case of 135 mJ/cm^2. With an increase in the laser fluence to 198 mJ/cm^2 and 220 mJ/cm^2, the average size of the ZnSe nanoparticles increased to 20 nm and 22 nm, respectively. This indicates that the size of ZnSe nanoparticles can be controlled by laser fluence. Furthermore, the generation rate of ZnSe nanoparticles using fs laser pulses is approximately 3.63×10^{10} s^{-1} (or 7.26×10^{6} per pulse) with a fluence of 135 mJ/cm^2. For the higher fluence of 220 mJ/cm^2, the generation rate of ZnSe nanoparticles increased by one order of magnitude to 3.63×10^{11} s^{-1} (or 7.26×10^{7} per pulse).

2.3 Mechanism underlying the formation of hexagonal ZnSe nanoparticles

During femtosecond laser irradiation, a large amount of energy is transferred to the specimens thereby inducing dense plasma on the surface of the sample. However, the duration of energy transfer (~80 fs) is too short for the lattice and the energy is only absorbed by the electrons within the extremely short interaction time. The ablated plume is confined within the laser focused position of the laser by the surrounding air. Thus, the rapidly cooling leads to the formation of nanoparticles on the surface of samples within the ablated plume to avoid a reaction with the air. That is the reason for the lack of impurities in the ZnSe nanoparticles fabricated by fs laser pulses in a study.

According to the early research, ZnSe transforms from a cubic structure to the hexagonal structure when the temperature is above the transition temperature (T_{tr}) of 1698 K (Rudolph et al., 1995). When ZnSe crystals are irradiated by the femtosecond laser pulses, the temperature of the ZnSe crystals increases. In the case of pulse lasers, an increases in the transient temperature ΔT in materials can be estimated according to the relationship of $\Delta T = W/(C\times V)$, where W is the pulse energy, C is the heat capacity, and V is the illuminated volume. For ZnSe at 300 K, C is 1.89×10^{6} J/m^{3}K (Martienssen & Warlimont, 2005), V is 2.29×10^{-13} m^3 [absorption depth ~1.87 μm estimated from the nonlinear absorption coefficient β (Tseng et al., 1996)], and W is on the order of 0.243 mJ (which is assumed to be totally absorbed by ZnSe). Thus, the ΔT is approximately 560 K, which is far below the structural transition temperature of 1698 K. Therefore, a structural transition could not be induced by the increase in temperature. To identify the mechanism underlying the phase transition of ZnSe from cubic to hexagonal, we further analyzed the influence of "ablation pressure" (Batani et al., 2003), which has been studied from various perspectives over the past few decades (Key et al., 1980; Groot et al., 1992). When solids are irradiated by laser pulses, high-density plasma is formed on the surface of the samples. The compressed plasma in laser driven implosions has been characterized as the ablating or exploding pusher according to the surface ablation pressure and bulk pressure due to the preheating through electrons.

In 2003, Batani et al. (Batani et al., 2003) derived the shock pressure with the laser and target parameters expressed as

$$P(\text{Mbar}) = 11.6\left(\frac{I}{10^{14}}\right)^{\frac{3}{4}} \lambda^{-\frac{1}{4}} \left(\frac{A}{2Z}\right)^{\frac{7}{16}} \left(\frac{Z\times t}{3.5}\right)^{-\frac{1}{8}} \qquad (1)$$

where I is the laser intensity on target with the unit of W/cm^2, λ is the laser wavelength in μm, A and Z are, respectively, the mass number and the atomic number of the target, and t is the time in ns. Figure 6 shows the effective pressure in the irradiated region with the laser

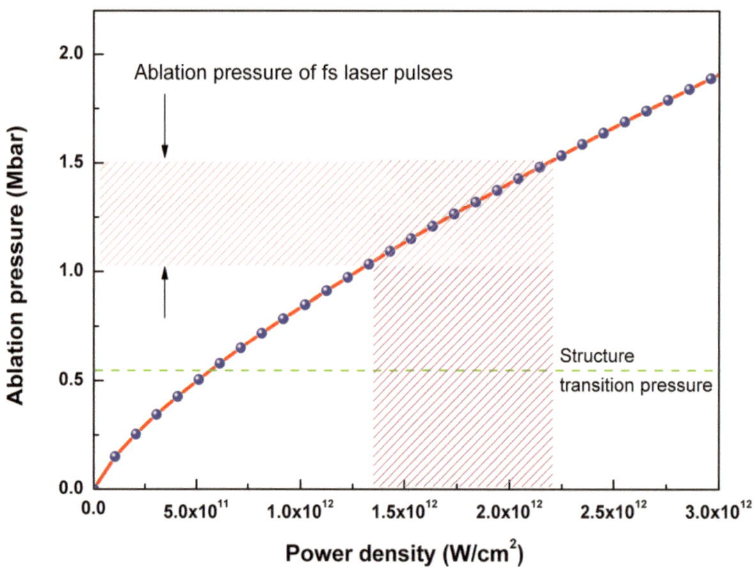

Fig. 6. Simulated ablation pressure as a function of the laser peak power density according to the Eq. (1). The shadow area indicates the range of laser peak power density in this study and corresponding ablation pressure. The dashed line represents the pressure of cubic-hexagonal phase transition, was obtained from ref. (Greene et al., 1995).

peak power density of the laser of 0 ~ 3.0×10^{12} W/cm^2. In this study, the maximum pressure induced by the laser reached approximately 1.5 Mbar. According to the studies of Greene et al. in II-VI compounds (Greene et al., 1995), the solid-solid transition point, i.e. the cubic-hexagonal phase transition, of ZnSe is approximately 0.55 Mbar. In our experiments, the ablation pressure induced by the femtosecond laser pulses on the ZnSe single crystals was in the range of 1.0 Mbar to 1.5 Mbar as shown in the shadow area of Fig. 6. This exceeds the solid-solid transition pressure 0.55 Mbar (the dashed line in Fig. 6). Therefore, the hexagonal-phase ZnSe nanoparticles transferred from the cubic phase may be caused by high ablation pressure resulting from the femtosecond laser pulses, and the accompanied increase in surface to volume ratio in the nanoparticles.

3. Generation of nanoripples and nanodots on YBCO

Issues related to energy have gradually gained in value and attracted attention around world and, the high-T_c superconducting $YBa_2Cu_3O_7$ (YBCO) has potential as an alternative material for green energy applications, e.g. electric power cables, transformers, motors, electric power generators, magnetic levitation systems, due to its high critical current of 77 K. For commercialization, critical current is the key parameter, and fs laser pulses may provide a new avenue to enhance the critical current of YBCO thin films. In this section, we demonstrate the formation of laser-induced subwavelength periodic surface structures (LIPSS), such as ripples and dots, on YBCO thin films using femtosecond laser and characterize their properties.

3.1 Preparation of YBCO thin films

The YBCO thin films used in this study were prepared by pulse laser deposition (PLD) with a KrF excimer laser operating at a repetition rate of 3-8 Hz with an energy density of 2-4 J/cm^2 as shown in the inset of Fig. 7(a). The oxygen partial pressure during deposition was maintained at 0.25 Torr, and the substrate temperature was maintained at 780-790 ^{o}C. After completion of the deposition process, the film was cooled to room temperature under 600 Torr of oxygen with the heater off. The thickness of the film was approximately 200 nm. As shown in the X-ray diffraction (XRD) pattern in Fig. 7(b), the YBCO films were (001)-oriented normal

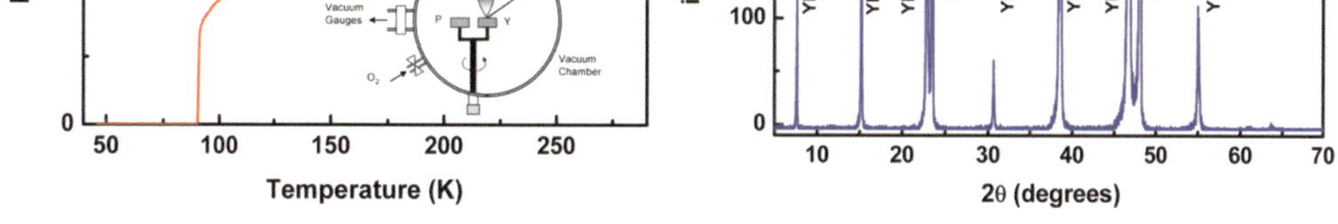

Fig. 7. (a) Resistance versus temperature curve measured on an as-deposited YBCO thin film. Inset: schematic illustration of the pulse laser deposition (PLD) system; (b) X-ray diffraction pattern of an as-deposited YBCO thin film.

to the (100) $LaAlO_3$ (LAO) substrate. The temperature-dependent resistance of an (001)-oriented YBCO thin film was measured using the standard four-probe configuration as shown in Fig. 7(a). The resistance decreased linearly with temperature in the normal state and then dropped sharply to a zero-resistance superconducting state at 90.1 K. Both features are consistent with the XRD results, indicating the high quality of the YBCO films.

3.2 Generation of YBCO ripple structures

Figure 8 shows the optical system for generating ripple structures on the YBCO thin films. A commercial regenerative amplified Ti:sapphire laser (Legend USP, Coherent) with an 800-nm wavelength, 30-fs pulse duration, ~0.5-mJ pulse energy, and 5-kHz repetition rate was used as the irradiation source. After passing through a variable neutral density (ND) filter, the normal incident laser beam was focused on the surface of the sample forming a spot of ~200 μm by means of a convex lens with a focal length of 50-mm. The number of pulses or irradiation time was precisely controlled by the electric shutter.

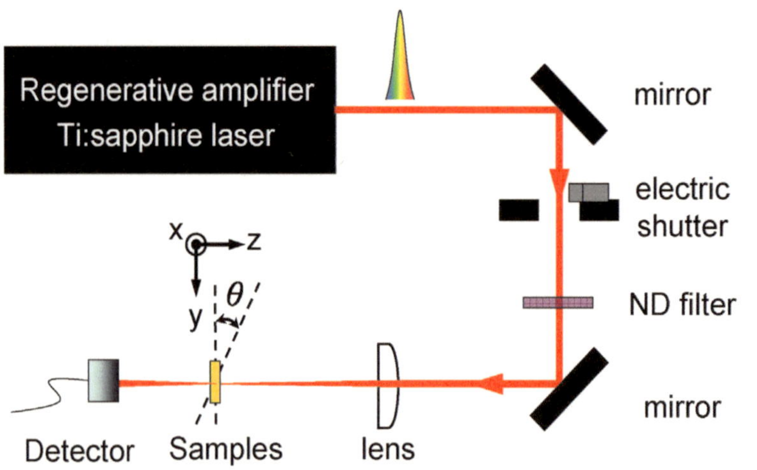

Fig. 8. Experimental setup for the generation of ripple structures on YBCO thin films.

SEM analysis (Fig. 9) indicates that the morphology of fs laser-induced surface structures depend strongly on the laser fluence. Figures 9(a)-9(f) show the evolution of the ripple structure on YBCO thin films irradiated by a single-beam fs laser with various laser fluences (*F*) and a fixed number of pulses (*N*=600,000). With an increase in laser fluence, the ripple structure becomes clear in SEM images, as evidenced by the appearance of satellite peaks in the 2D Fourier spectra in the insets of Figs. 9(c)-9(f) [there are no satellite peaks in the inset of Fig. 9(b) for the case of low laser fluence]. The spatial period Λ of ripples, estimated from the position of a satellite peak in the 2D Fourier spectra, was dependent on the laser fluence, as shown in Fig. 11(a). Once the laser fluences ≧154 mJ/cm^2, the ripple period remained at approximately 517 nm. Furthermore, the "periodicity" of the ripple-like structures was approximately 500 nm, which is much smaller than either the spot size or the wavelength of the femtosecond laser, indicating that the pattern was not formed by simple plow-and-deposit processes.

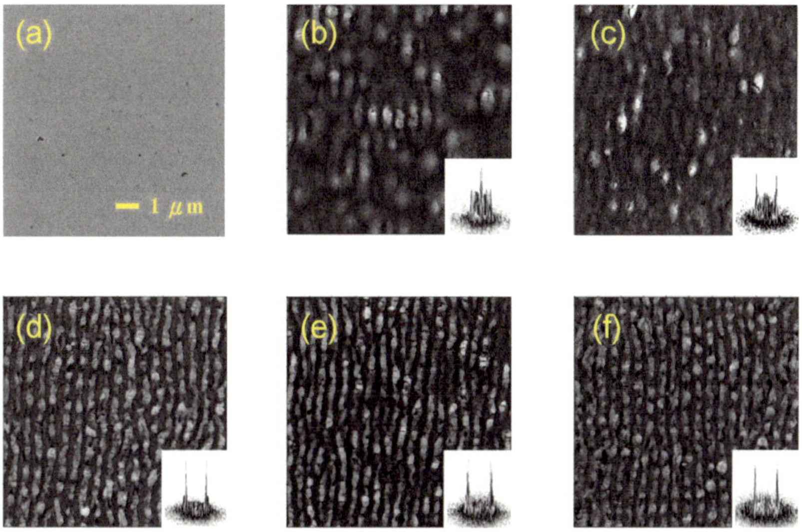

Fig. 9. Morphological evolution of structures on YBCO thin films induced by linear polarized fs laser with fixed number of pulses N=600,000 and various fluences (a) F = 0 mJ/cm^2, (b) F = 43 mJ/cm^2, (c) F = 59 mJ/cm^2, (d) F = 79 mJ/cm^2, (e) F = 154 mJ/cm^2, (f) F = 319 mJ/cm^2. Inset: 2D Fourier spectra transferred from their corresponding SEM images (10 μm×10 μm with pixel resolution of ~0.04 nm). The scale bar is applied to all pictures.

Figures 10(a)-10(c) show the evolution of the ripple structure on YBCO thin films irradiated by a single-beam fs laser with various numbers of pulses (N) and the fixed laser fluence F = 79 mJ/cm^2. With an increase in the number of pulses, the ripple structure became increasingly clear in SEM images, as evidenced by the appearance of satellite peaks in the 2D Fourier spectra in the insets of Figs. 10(b) and 10(c) [there are no satellite peaks in the inset of Fig. 10(a) for an as-deposited YBCO thin film]. The spatial period of ripples, estimated from the position of satellite peaks in the 2D Fourier spectra, is independent of the number of pulses or irradiation time, as shown in Fig. 11(b). Once the number of pulses ≧ 50,000, i.e. the sample surface was irradiated by the 75 mJ/cm^2 laser pulses for ≧10 s, ripples can be clearly observed on the surface of the sample. In addition, the real-time evolution of the ripple structure appears in the transmission measurements in Fig. 10(d). In the case of F = 154 mJ/cm^2, the transmission power of the laser beam dramatically increased to within 2 s and then saturated after ~10 s. Some specific points were marked at 79 mJ/cm^2 of Fig. 10(d), and corresponding SEM images are shown in Figs. 10(a), 10(b), and 10(c), respectively. At 0.1 s [i.e. N=500 in Fig. 10(a)], there are almost no structures on the surface of YBCO thin films. However, the rippled structure can be observed at 10 s [i.e. N=50,000 in Fig. 10(b)]; meanwhile, the transmission power dramatically increased due to the thinning of YBCO films inside the grooves. For an extended irradiation time of 30 s [i.e. N=150,000 in Fig. 10(c)], the ripple structure does not change from that of Fig. 10(b), e.g. the spatial period of ripple as shown in Fig. 11(b), except for the contrast of grooves causing slight rise in transmission power in Fig. 10(d). Furthermore, the characteristics of changes in transmission power in Fig. 10(d) are independent of laser fluence. This indicates that the formation of ripple structures is very rapid, with only ~2 s needed, and the formation processes is independent of laser fluence. Laser fluence only affects the spatial period of ripple structures, as shown in Fig. 11(a).

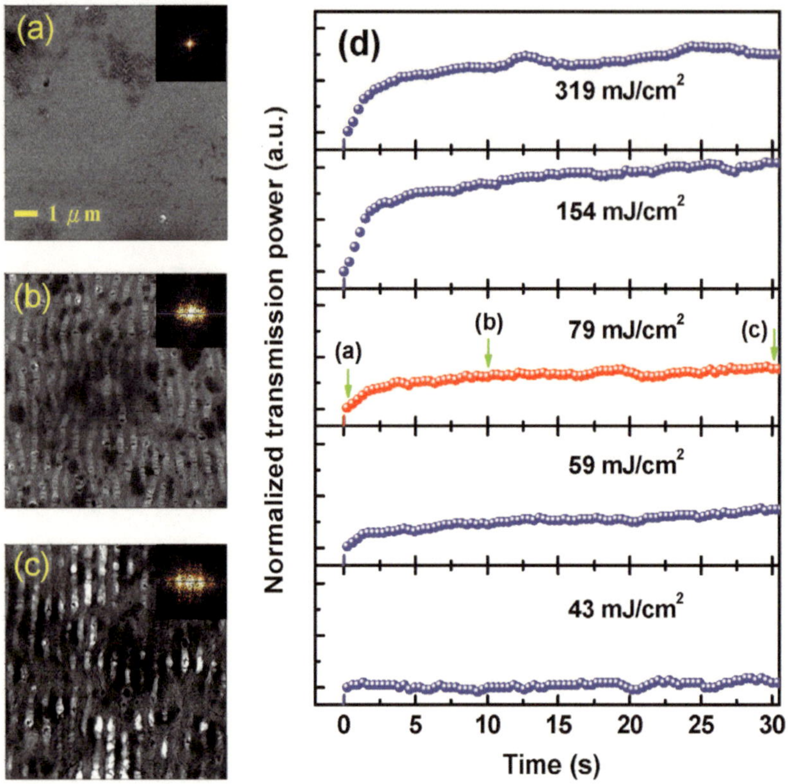

Fig. 10. Morphological evolution of structures on YBCO thin films induced by linear polarized fs laser with fixed laser fluence $F = 79$ mJ/cm² and various numbers of pulses (a) $N = 500$, (b) $N = 50{,}000$, (c) $N = 150{,}000$. (d) The transmission power of laser pulses as a function of irradiating time, *i.e.* pulse number N. Inset: 2D Fourier spectra which were transferred from their corresponding SEM images (10 μm×10 μm with pixel resolution of ~0.04 nm). The scale bar is applied to all pictures.

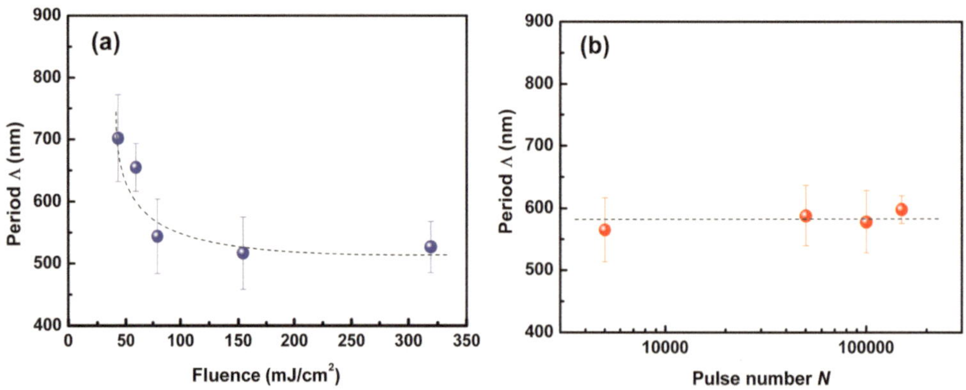

Fig. 11. (a) Dependence of the ripple period on the fluence. (b) Dependence of the ripple period on the number of pulses. The dashed lines are a guide to the eyes.

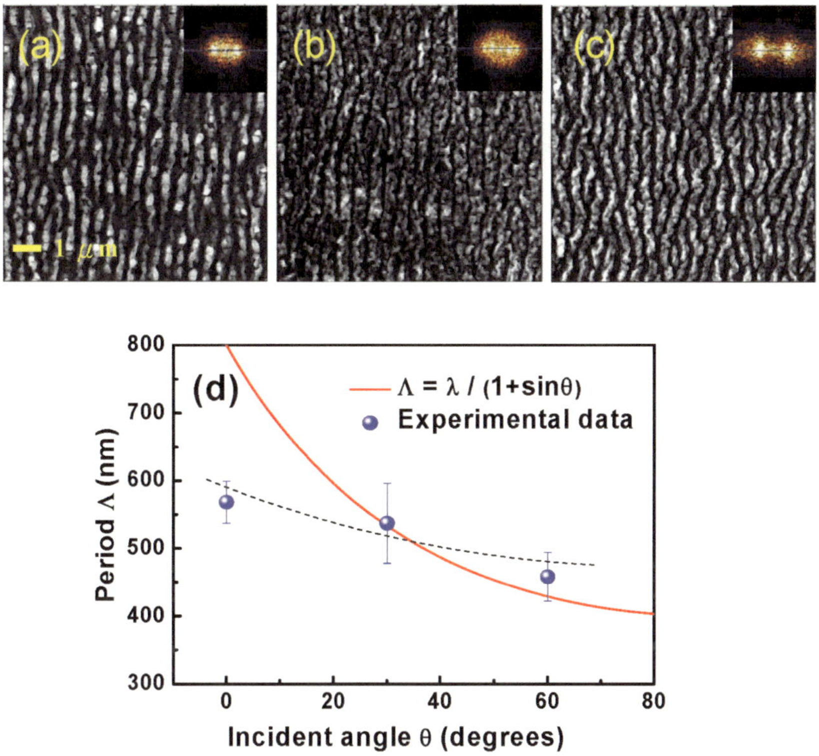

Fig. 12. Morphological evolution of ripple structures on YBCO thin films induced by linear polarized fs laser with F = 300 mJ/cm², N=150,000, and various incident angles (a) $\theta = 0°$, (b) $\theta = 30°$, (c) $\theta = 60°$. (d) Dependence of the ripple period on the incident angle of laser pulses. The dashed lines are a guide to the eyes. All SEM images are 10 μm×10 μm with pixel resolution of ~0.04 nm.

On the other hand, with the fluence and pulse number fixed at ~300 mJ/cm² and 150,000, respectively, we found that the spatial period decreased with an increase in the incident angle (θ) [see Fig. 12(d)]. However, the observed period of ripple at θ = 0° was significantly smaller than the prediction of $\Lambda=\lambda/(1+\sin\theta)$ (Zhou et al., 1982). In addition, the incident angle-dependent period of ripples on YBCO thin films cannot be described using this simplified scattering model [the solid line in Fig. 12(d)]. Therefore, the influence of surface electromagnetic waves, i.e. surface plasmons (SPs) should be taken into account in the formation of subwavelength ripples (Sakabe et al., 2009; Huang et al., 2009). According to Shimotsuma's et al. results (Shimotsuma et al.; 2003), femtosecond incident light easily excites plasmons on the surface of various materials. As shown in Fig. 13(c), once the momentum conservation condition for the wave vectors of the linear polarized laser light ($\mathbf{K}_i$), the plasma wave ($\mathbf{K}_p$), and the laser-induced subwavelength periodic surface structures (LIPSS, $\mathbf{K}_L$) is satisfied, such plasmons could couple with the incident light. The interference between the plasmons and the incident light would generate a periodically modulated electron density causing nonuniform melting. After irradiation with a femtosecond laser, the interference ripple was inscribed on the surface of the YBCO thin film.

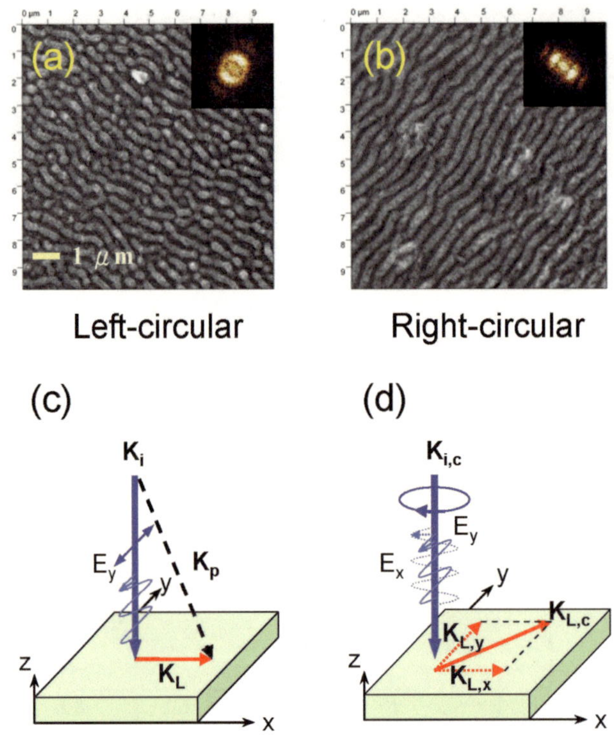

Fig. 13. SEM images (10 μm×10 μm with pixel resolution of ~0.04 nm) of fs LIPSS induced by (a) the left- and (b) right-circularly polarized beams; (c) Schematic of the momentum conservation condition of wave vectors of linear polarized laser light ($\mathbf{K}_i$), plasma wave ($\mathbf{K}_P$), and LIPSS ($\mathbf{K}_L$); (d) Schematic processes of the LIPSS by circularly polarized laser light ($\mathbf{K}_{i,C}$). The scale bar is applied to all pictures.

Interestingly, when we used a circularly polarized beam, the rippled structures were still produced, as shown in Figs. 13(a) and 13(b). The orientation of the ripples was set at -45° and +45° for left and right circularly polarized beams, respectively, with respect to the incident plane of the beam. In both cases, the spatial period was 491 nm, as produced by fs laser pulses with a fluence of 185 mJ/cm^2 and number of pulses set to 150,000. These results show the orientation of rippled structures strongly depend on the polarization-state of incident fs pulses. These results are consistent with the results of Zhao et al. on tungsten (Zhao et al., 2007a, 2007b). In principle, circularly polarized light ($\mathbf{K}_{i,c}$) can be decomposed to two perpendicular linear-polarization lights (E_x and E_y) through retardation of $\lambda/4$ in phase, as shown in Fig. 13(d). Linearly polarized light E_x and E_y can induce the LIPSS $\mathbf{K}_{L,x}$ and $\mathbf{K}_{L,y}$, respectively, as long as the momentum conservation condition in Fig. 13(c) is satisfied. Thus, both $\mathbf{K}_{L,x}$ and $\mathbf{K}_{L,y}$ with phase coherent further cause the $\mathbf{K}_{L,c}$ according to the momentum conservation condition of $\mathbf{K}_{L,c} = \mathbf{K}_{L,x} + \mathbf{K}_{L,y}$. The 45° wave vector of LIPSS, $\mathbf{K}_{L,c}$, is completely consistent with the direction of the satellite peaks in the 2D Fourier spectra [the inset of Fig. 13(a)]. Namely, the orientation of ripples is -45° for left-circularly polarized beams with respect to the incident plane of the beam. Similarly, right-circularly polarized beams induce a +45° orientation of LIPSS, $\mathbf{K}_{L,c}$, according to the momentum conservation condition of $\mathbf{K}_{L,c} = -\mathbf{K}_{L,x} + \mathbf{K}_{L,y}$ consistent with the results in Fig. 13(b).

3.3 Generation of YBCO dot structures

To produce dot structures on YBCO thin films, we adopted a dual-beam scheme using the modified Michelson interferometer shown in Fig. 14. The polarization of both beams was individually controlled by two quarter-wave plates before the reflection mirrors in both arms of the dual-beam setup. After the beam splitter in the dual-beam setup, both beams were collinearly and simultaneously focused on the surface of the sample using a convex lens with a focal length of 50-mm. Before generating the YBCO dot structures, we measured the interference patterns between two beams to check the temporal overlap of the two pulses. In the inset of Fig. 14, the interference pattern between the two pulses with parallel polarization can be clearly observed after adjusting the delay in one of the two pulses. The polarization of two pulses was set perpendicularly to each other to eliminate interference patterns and generate the YBCO dot structures. All experiments were performed in air under atmospheric pressure.

As shown in Figs. 15(a1)-15(d1), it is surprising that many dots rather than regular ripples appeared on the surface of YBCO thin films using a dual-beam setup with perpendicularly linear polarization. In the case of the dual-beam setup, the $\mathbf{K}_{L,x}$ and $\mathbf{K}_{L,y}$ without coherence in phase induced by random phase and perpendicularly linear-polarization beams (E_x and E_y), respectively, would not satisfy the conservation of momentum of $\mathbf{K}_{L,c} = \pm\mathbf{K}_{L,x} + \mathbf{K}_{L,y}$ and be unable to create ±45° wave vector of LIPSS, $\mathbf{K}_{L,c}$ as shown in Fig. 13(d). Therefore, the $\mathbf{K}_{L,x}$ and $\mathbf{K}_{L,y}$ which are perpendicular to each other would lead 2D nonuniform melting and further aggregation to form randomly distributed dots [see the 2D Fourier spectra in the inset of Figs. 15(a2)-15(d2)] due to surface tension. In the case of N = 25,000, the average diameter of dots was approximately 632 nm estimated by the log-normal fitting presented in Fig. 15(a2). An increase in the number of pulses resulted in a marked broadening in the size distribution, although the average size only slightly increased from 632 nm to 844 nm [see Figs. 15(a2)-15(d2)]. For N = 300,000, the size of a part of dots was on the order of micrometers. However, larger dots influence the dot density on the surface of YBCO thin films. For instance, the density of dots increases with the number of pulses ≦150,000. Once the dots grow too large to merge with the nearest neighbors, or even next nearest neighbors, the density of the dots significantly shrank, as shown in Fig. 15(c1). In this manner, the size and density of YBCO dots can be controlled by the numbers of pulses from the fs laser.

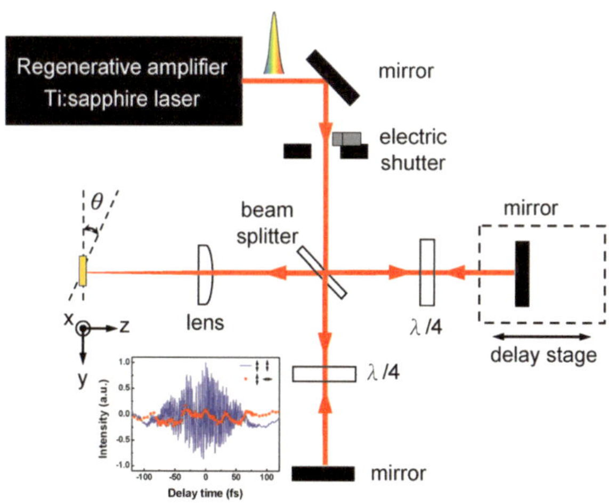

Fig. 14. Experimental setup for the generation of nanodots on YBCO thin films.

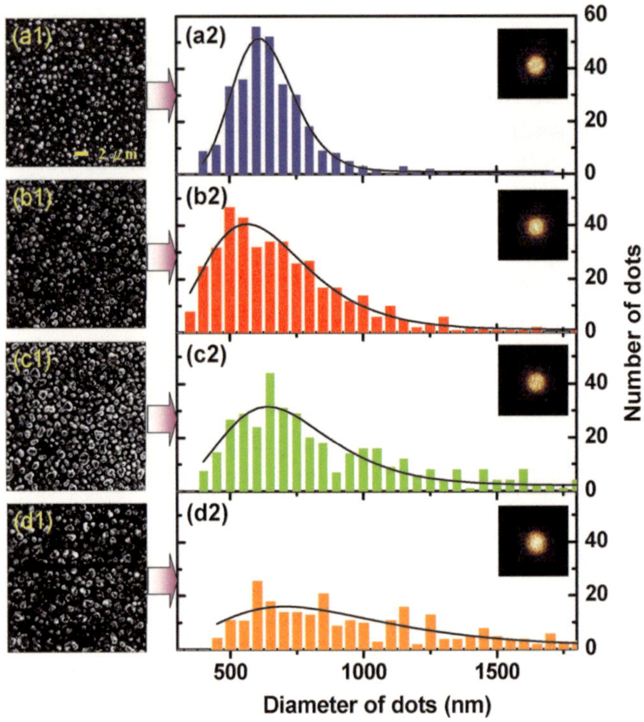

Fig. 15. Dot structures on YBCO thin films induced by a dual-beam setup with fluence = 87 mJ/cm² and various numbers of pulses (a1) N =25,000, (b1) N =50,000, (c1) N =150,000, (d1) N =300,000. (a2)-(d2) The size distribution corresponds to the SEM images (10 μm×10 μm with pixel resolution of ~0.04 nm) (a1)-(d1), respectively. Solid lines are the log-normal fitting. Inset: 2D Fourier spectra which were transferred from their corresponding SEM images (a1)-(d1), respectively. The scale bar is applied to all pictures.

3.4 Characteristics of YBCO nanostructures

To characterize the superconductivity of the ripple structures on YBCO thin films, the area of the ripple structure must be large enough to measure. Thus, the scanning scheme shown in Fig. 8 was adopted to prepare the large-area ripple structures on YBCO thin films. After passing through a variable neutral density filter, the beam was two-dimensionally scanned using a pair of galvanic mirrors with a speed of 7.6 cm/s. The laser beam was focused on the surface of the sample with a spot size of 220 μm using an f-theta lens. All experiments were performed in air under atmospheric pressure.

It is evident from Fig. 16(g) that the quality of the crystalline structure of the YBCO films remained high after irradiation by the femtosecond laser with fluence up to 260 mJ/cm². However, the quality deteriorated considerably with a further increase in laser fluences. For instance, with an irradiation fluence of 530 mJ/cm², the intensity of the characteristic X-ray diffraction peaks diminished considerably. As shown in Fig. 17, while the superconductivity of the YBCO films remained nearly unchanged under low fluence irradiation, it began degrading at irradiation levels of 320 mJ/cm² and disappeared at 530 mJ/cm², indicating structural and compositional changes with higher irradiation fluence.

As mentioned above, the crystalline structure of these YBCO nanodots induced by the laser irradiation (260 mJ/cm^2) remained oriented with the c-axis, with sharp diamagnetic Meissner effect characteristics at 89.7 K (Fig. 17), indicating that even after the dramatic morphological reconstruction, the obtained nanodots maintained most of their intrinsic properties. Indeed, as indicated by the energy dispersive spectroscopy (EDS) spectrum displayed in Fig. 16(h), which was taken on one of the nanodots [marked as area 1 in Fig. 16(e)], the composition of the nanodot had not changed from that of the original YBCO films. EDS results taken in the area between the dots [marked as area 2 in Fig. 16(e)] indicates no signal of Ba. Instead, traces of Al, presumably from the LAO substrate, were detected [see the second spectrum from the top in Fig. 16(h)]. This indicates that the composition of the area between any two nanodots has severely deviated from the stoichiometric composition of the original YBCO. The question is, how does this occur?

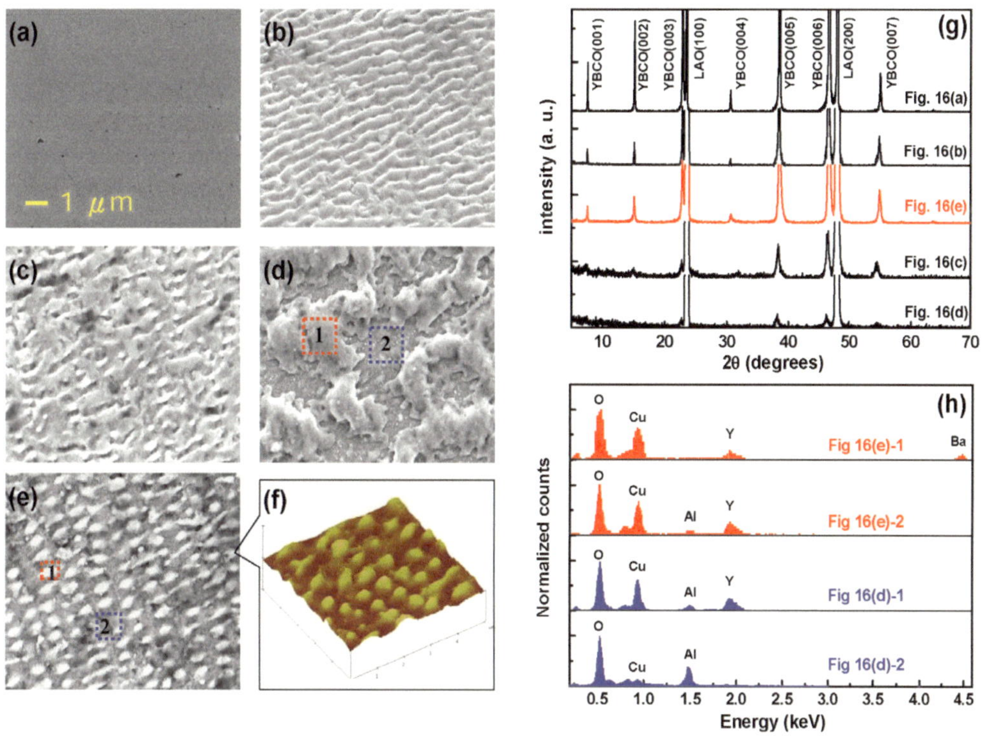

Fig. 16. (a) SEM images show the surface morphology of YBCO thin films at various laser fluences (a) $F = 0$ mJ/cm^2, (b) $F = 210$ mJ/cm^2, (c) $F = 320$ mJ/cm^2, (d) $F = 530$ mJ/cm^2, (e) $F = 260$ mJ/cm^2. (f) AFM image of (e). (g) X-ray diffraction patterns of YBCO thin films at various laser fluences corresponding to (a)-(e). (h) EDS spectra show the composition of area 1 and area 2 in (d) and (e).

Due to the laser pulses, the transient increase in temperature, ΔT, can be estimated using the following relation $\Delta T = W / CV$, where W is the pulse energy, C is the heat capacity, and V is the illuminated volume. For YBCO at 300 K using $C = 2.86\times10^6$ J/m^{3}K [derived from the Debye heat capacity and the Debye temperature of YBCO was obtained from ref. (Stupp &

Ginsberg, 1989)], $V = 1.14\times10^{-14}$ m^3 (the absorption length ~ 300 nm), and W on the order of 0.1 mJ (which is assumed to be totally absorbed by YBCO). ΔT is approximately 3000 K. This increase in temperature, in principle, will lead to massive global melting of a thin layer beneath the surface of YBCO thin films. Thus, a more random pattern would be expected when re-solidified. However, due to the interference induced by the inhomogeneous input energy, the YBCO in melted phase initially forms ripples according to the interference pattern which pushes the YBCO to the line of destructive interference. This interference pattern also leads to a periodic distribution of the fluctuations in temperature, ΔT, which happen to be higher than the boiling point of Ba [1897 K (Thompson & Vaughan, 2001)] along the line of constructive interference and lower than the boiling point of Ba [1897 K (Thompson & Vaughan, 2001)] along the line of destructive interference. As a result, in the regions of the constructive interference most Ba was vaporized, while in the destructive regions the Ba remained. Moreover, due to the surface tension and heterogeneous nucleation on the surface of the substrate, the melted YBCO along the lines of destructive interference aggregates to form nanodots in a periodic fashion, as shown in Fig. 16(b), 16 (e), and 16(f). These results suggest that, by using single-beam femtosecond laser irradiation, it is possible to fabricate a self-organized array of YBCO nanodots with most of the crystallinity and superconducting properties remaining intact, provided proper control of irradiation fluence is practiced. This technique could potentially be applied to the fabrication of microwave filter devices with array structure or the weak-link Josephson junction arrays.

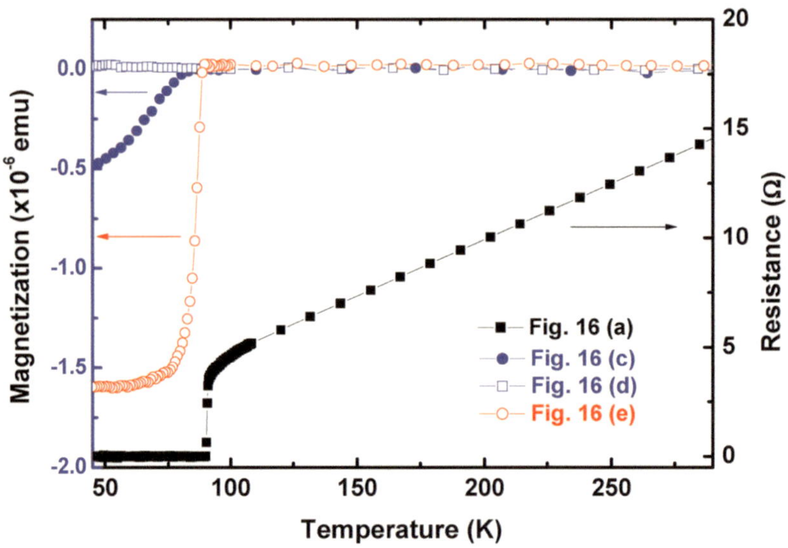

Fig. 17. Resistance versus temperature curve measured prior to femtosecond laser irradiation (F = 0 mJ/cm^2) and the magnetization versus temperature curve measured at 10 Oe after femtosecond laser irradiation, with various fluences corresponding to the Fig. 16 (c), 16(d), and 16(e), respectively.

Finally, as the fluence reached ≧ 320 mJ/cm^2, irregular, disordered patterns were observed on the surface of the LAO substrate, as shown in Fig. 16(c) and Fig. 16(d). The characteristic XRD peaks of the (001)-YBCO films deteriorated significantly [Fig. 16(g)], indicating that the crystalline structure of YBCO had been destroyed by the higher laser fluence. EDS analysis [Fig. 16(h)] also shows that Ba was absent in both area 1 and area 2, marked in Fig. 16(d). In area 2, even the composition of Y is absent in the EDS spectrum. Using the previous estimation with $W \geqq 0.12$ mJ (fluence ≧ 320 mJ/cm^2), $\Delta T \geqq 3700$ K was obtained, which is higher than the boiling point of Ba [1897 K (Thompson & Vaughan, 2001)] at the positions of both constructive and destructive interference, but only higher than the boiling point of Y [3345 K (Thompson & Vaughan, 2001)] at the position of constructive interference. In this case, the aggregation of melted YBCO becomes more disordered and the stoichiometric composition is more severely influenced, leading to the loss of crystalline integrity and superconductivity in the remaining residue of the original YBCO film.

4. Conclusions

In this chapter, we demonstrated a simple, rapid means to obtain the hexagonal ZnSe nanoparticles, YBCO ripples, and dot structures. In the fabrication of ZnSe nanoparticles, while femtosecond laser pulses were focused on the surface of ZnSe wafers in air and the ablated plume cannot expand as rapidly as plumes would in a vacuum chamber which causes an instantaneous high-energy, high-pressure region around the focal point of the laser; meanwhile, a large amount of spherical-shape ZnSe nanoparticles with an average diameter of 16-22 nm (depending on the laser fluence) forms on the surface of the wafer. During the formation of ZnSe nanoparticles, the structural phase further changes from cubic to metastable hexagonal phase due to the ultrahigh localized ablation pressure caused by the rapid injection of high laser energy within a femtosecond time scale.

For the generation of ripple and dot structures, we have systematically studied the surface morphology of YBCO thin films under a single-beam and a dual-beam fs laser irradiation. The generation of ripple and dot periodic structures was determined by the applied laser fluence, number of pulses, and polarization of the laser. The period and orientation of ripples, and even the size and density of dots can be controlled by these parameters. With lower laser fluence, the (001)-YBCO film turns into (001)-ripple or dot arrays with superconductivity remaining nearly intact. These rippled (or dotted) structures and superconductivity, however, were rapidly destroyed with higher fluence. These results may be applied to enhance the critical current of YBCO thin films and the fabrication of the microwave filter devices with array structures or the weak-link Josephson junction arrays.

The present results clearly demonstrate that the femtosecond laser, in addition to its crucial role in studying the ultrafast dynamics of matter, they can also serve as a new avenue for engineering materials and structures into their surfaces at a nanometer scale.

5. Acknowledgments

The author would like to express his sincere appreciation and gratitude to his collaborators and colleagues, Ms. H. I. Wang, Mr. W. T. Tang, Ms. C. C. Lee, and Mr. L. W. Liao, Profs. T. Kobayashi, K. H. Wu, J. Y. Juang, J.-Y. Lin, T. M. Uen, C. S. Yang. This work was supported by the MOE-ATU program at NCTU and National Science Council of Taiwan, under Grant No. NSC 98-2112-M-009-008-MY3.

6. References

Amoruso, S.; Bruzzese, R.; Spinelli, N.; Velotta, R.; Vitiello, M.; Wang, X.; Ausanio, G.; Lannotti, V. & Lanotte, L. (2004). Generation of Silicon Nanoparticles via Femtosecond Laser Ablation in Vacuum. *Applied Physics Letters*, Vol.84, No.22, (May 2004) pp. 4502-4504, ISSN 0003-6951

Batani, D.; Stabile, H.; Ravasio, A.; Lucchini, G.; Strati, F.; Desai, T.; Ullschmied, J.; Krousky, E.; Skala, J.; Juha, L.; Kralikova, B.; Pfeifer, M.; Kadlec, Ch.; Mocek, T.; Präg, A.; Nishimura, H. & Ochi, Y. (2003). Ablation pressure scaling at short laser wavelength. *Physical Review E*, Vol.68, No.6, (December 2003) pp. 067403, ISSN 1539-3755

Bonse, J. & Krüger, J. (2010). Pulse Number Dependence of Laser-Induced Periodic Surface Structures for Femtosecond Laser Irradiation of Silicon. *Journal of Applied Physics*, Vol.108, No.3, (August 2010) pp. 034903, ISSN 0021-8979

Che, J.; Yao, X.; Jian, H. & Wang, M. (2004). Application and preparation of ZnSe nanometer powder by reduction process. *Ceramics International*, Vol.30, No.7, (July 2004) pp. 1935-1938, ISSN 0272-8842

Dinger, A.; Becker, R.; Goppert, M.; Petillon S.; Grun, M.; Klingshirm, C.; Liang, J.; Wagner, V. & Geurts, J. (2000). Lattice dynamical properties of cubic CdS/ZnSe strained-layer superlattices. *Journal of Crystal Growth*, Vol.214, No.2, (June 2000) pp. 676-679, ISSN 0022-0248

Groot, J. S. De; Estabrook, K. G.; Kruer, W. L.; Drake, R. P.; Mizuno, K. & Cameron, S. M. (1992). Distributed absorption model for moderate to high laser powers. *Physics of Fluids B*, Vol.4, No.3, (March 1992) pp. 701-707, ISSN 0899-8221

Greene, R. G.; Luo, H. & Ruoff, A. L. (1995). High pressure x-ray and raman study of ZnSe. *Journal of Physics and Chemistry of Solids*, Vol.56, No.3/4, (March-April 1995) pp. 521-524, ISSN 0022-3697

Hsu, E. M.; Crawford, T. H. R.; Tiedje, H. F. & Haugen, H. K. (2007). Periodic Surface Structures on Gallium Phosphide after Irradiation with 150 fs–7 ns Laser Pulses at 800 nm. *Applied Physics Letters*, Vol.91, No.11, (September 2007) pp. 111102, ISSN 0003-6951

Huang, M.; Zhao, F.; Cheng, Y.; Xu, N. & Xu, Z. (2009). Origin of Laser-Induced Near-Subwavelength Ripples: Interference between Surface Plasmons and Incident Laser. *ACS Nano*, Vol.3, No.12, (November 2009) pp. 4062-4070, ISSN 1936-0851

Jiang, Y.; Meng, X. M.; Yiu, W. C.; Liu, J.; Ding, J. X.; Lee, C. S. & Lee, S. T. (2004). Zinc Selenide Nanoribbons and Nanowires. *The Journal of Physical Chemistry B*, Vol.108, No.9, (March 2004) pp. 2784-2787, ISSN 1520-6106

Jia, T. Q.; Zhao, F. L.; Huang, M.; Chen, H. X.; Qiu, J. R.; Li, R. X.; Xu, Z. Z. & Kuroda, H. (2006). Alignment of Nanoparticles Formed on the Surface of 6H-SiC Crystals Irradiated by Two Collinear Femtosecond Laser Beams. *Applied Physics Letters*, Vol.88, No.11, (March 2006) pp. 111117, ISSN 0003-6951

Jia, X.; Jia, T. Q.; Zhang, Y,; Xiong, P. X.; Feng, D. H.; Sun, Z. R.; Qiu, J. R. & Xu, Z. Z. (2010). Periodic Nanoripples in the Surface and Subsurface Layers in ZnO Irradiated by Femtosecond Laser Pulses. *Optics Letters*, Vol.35, No.8, (April 2010) pp. 1248-1250, ISSN 0146-9592

Key, M. H.; Rumsby, P. T.; Evans, R. G.; Lewis, C. L. S.; Ward, J. M. & Cooke, R. L. (1980). Study of Ablatively Imploded Spherical Shells. *Physical Review Letters*, Vol.45, No.22, (December 1980) pp. 1801-1804, ISSN 0031-9007

Liu, B.; Hu, Z.; Che, Y.; Chen, Y. & Pan, X. (2007a). Nanoparticle generation in ultrafast pulsed laser ablation of nickel. *Applied Physics Letters*, Vol.90, No. 4, (January 2007) pp. 44103, ISSN 0003-6951

Liu, S. Y.; Choy, W. C. H.; Jin, L.; Leung, Y. P.; Zheng, G. P.; Wang, J. & Soh, A. K. (2007b). Triple-Crystal Zinc Selenide Nanobelts. *The Journal of Physical Chemistry C*, Vol.111, No.26, (July 2007) pp. 9055-9059, ISSN 1932-7447

Liu, S. Y.; Choy, W. C. H.; Jin, L.; Leung, Y. P.; Zheng, G. P.; Wang, J. & Soh, A. K. (2007). Triple-Crystal Zinc Selenide Nanobelts. *The Journal of Physical Chemistry C*, Vol.111, No.26, (July 2007) pp. 9055-9059, ISSN 1932-7447

Luo, C. W.; Lee, C. C.; Li, C. H.; Shih, H. C.; Chen, Y.-J.; Hsieh, C. C.; Su, C. H., Tzeng, W. Y.; Wu, K. H.; Uen, T. M.; Juang, J. Y.; Chen, S. P.; Lin, J.-Y. & Kobayashi, T. (2008). Ordered YBCO Sub-Micron Array Structures Induced by Pulsed Femtosecond Laser Irradiation. *Optics Express*, Vol.16, No.25, (December 2008) pp. 20610-20616, ISSN 1094-4087

Martienssen, W. & Warlimont, H. (2005). *Springer Handbook of Condensed Matter and Materials Data*, Heidelberg: Springer-Verlag, Berlin

Nayak, B. K.; Gupta, M. C. & Kolasinski, K. W. (2008). Formation of Nano-Textured Conical Microstructures in Titanium Metal Surface by Femtosecond Laser Irradiation. *Applied Physics A: Materials Science & Processing*, Vol.90, No.3, (December 2007) pp. 399-402, ISSN 0947-8396

Okamuro, K.; Hashida, M.; Miyasaka, Y.; Ikuta, Y.; Tokita, S. & Sakabe, S. (2010). Laser Fluence Dependence of Periodic Grating Structures Formed on Metal Surfaces under Femtosecond Laser Pulse Irradiation. *Physical Review B*, Vol.82, No.16, (October 2010) pp. 165417, ISSN 1098-0121

Rudolph, P.; Schäfer, N. & Fukuda, T. (1995). Crystal growth of ZnSe from the melt. *Materials Science and Engineering: R: Reports*, Vol.15, No.3, (September 1995) pp. 85-133, ISSN 0927-796X

Stupp, S. E. & Ginsberg, D. M. (1989). A review of the linear term in the low temperature specific heat of $YBa_2Cu_3O_{7-\delta}$. *Physica C* Vol.158, No.3, (May 1989) pp. 299-310, ISSN 0921-4534

Stuart, B. C.; Feit, M. D.; Rubenchik, A. M.; Shore, B. W. & Perry, M. D. (1995). Laser-Induced Damage in Dielectrics with Nanosecond to Subpicosecond Pulses. *Physical Review Letters*, Vol.74, No.12, (March 1995) pp. 2248-2251, ISSN 0031-9007

Sarigiannis, D.; Peck, J. D., Kioseoglou, G.; Petrou, A. & Mountziaris, T. J. (2002). Characterization of Vapor-Phase-Grown ZnSe nanoparticles. *Applied Physics Letters*, Vol.80, No.21, (May 2002) pp. 4024-4026, ISSN 0003-6951

Shimotsuma, Y.; Kazansky, P. G.; Qiu, J. & Hirao, K. (2003). Self-Organized Nanogratings in Glass Irradiated by Ultrashort Light Pulses. *Physical Review Letters*, Vol.91, No.24, (December 2003) pp. 247405, ISSN 0031-9007

Shan, C. X.; Liu, Z.; Zhang, X. T.; Wong, C. C. & Hark, S. K. (2006). Wurtzite ZnSe nanowires: growth, photoluminescence, and single-wire Raman properties. *Nanotechnology*, Vol.17, No.22, (November 2006) pp. 5561-5564, ISSN 0957-4484

Sakabe, S.; Hashida, M.; Tokita, S.; Namba, S. & Okamuro, K. (2009). Mechanism for Self-Formation of Periodic Grating Structures on a Metal Surface by a Femtosecond Laser Pulse. *Physical Review B*, Vol.79, No.3, (January 2009) pp. 033409, ISSN 1098-0121

Tseng, K. Y.; Wong, K. S. & Wong, G. K. L. (1996). Femtosecond Time-Resolved Z-scan Investigations of optical nonolinearities in ZnSe. *Optics Letters*, Vol.21, No.3, (February 1996) pp. 180-182, ISSN 0146-9592

Tawara, T.; Tanaka, S.; Kumano, H. & Suemune, I. (1999). Growth and luminescence properties of self-organized ZnSe quantum dots. *Applied Physics Letters*, Vol.75, No.2, (July 1999) pp. 235-237, ISSN 0003-6951

Thompson, A. C. & Vaughan, D. (2001). *X-ray data booklet*, Lawrence Berkeley National Laboratory, California, USA

Tsuji, T.; Kakita, T. & Tsuji, M. (2003). Preparation of Nano-Size Particles of Silver with Femtosecond Laser Ablation in Water. *Applied Surface Science*, Vol.206, No.1-4, (February 2003) pp. 314-320, ISSN 0169-4332

Teng, Y.; Zhou, J.; Luo, F.; Ma, Z.; Lin, G. & Qiu, J. (2010). Shape- and Size-Controllable Microstructure on Glass Surface Induced by Femtosecond Laser Irradiation. *Optics Letters*, Vol.35, No.13, (July 2010) pp. 2299-2301, ISSN 0146-9592

Xiang, B.; Zhang, H. Z.; Li, G. H.; Yang, F. H.; Su, G. H.; Wang, R. M.; Xu, J.; Lu, G. W.; Sun, X. C.; Zhao, Q. & Yu, D. P. (2003). Green-light-emitting ZnSe nanowires fabricated via vapor phase growth. *Applied Physics Letters*, Vol.82, No.19, (May 2003) pp. 3330-3332, ISSN 0003-6951

Yang, Y.; Yang, J.; Xue, L. & Guo, Y. (2010). Surface Patterning on Periodicity of Femtosecond Laser-Induced Ripples. *Applied Physics Letters*, Vol.97, No.14, (October 2010) pp. 141101, ISSN 0003-6951

Zhou, G.; Fauchet, P. M. & Siegman, A. E. (1982). Growth of spontaneous periodic surface structures on solids during laser illumination. *Physical Review B*, Vol.26, No.10, (November 1982) pp. 5366-5381, ISSN 1098-0121

Zhao, Q. Z.; Malzer, S. & Wang, L. J. (2007a). Formation of subwavelength periodic structures on tungsten induced by ultrashort laser pulses. *Optics Letters*, Vol.32, No.13, (July 2007) pp. 1932-1934, ISSN 0146-9592

Zhao, Q. Z.; Malzer, S. & Wang, L. J. (2007b). Self-Organized Tungsten Nanospikes Grown on Subwavelength Ripples Induced by Femtosecond Laser Pulses. *Optics Express*, Vol.15, No.24, (November 2007) pp. 15741-15746, ISSN 1094-4087

2

Production of Optical Coatings Resistant to Damage by Petawatt Class Laser Pulses

John Bellum[1], Patrick Rambo, Jens Schwarz, Ian Smith, Mark Kimmel, Damon Kletecka[1] and Briggs Atherton
Sandia National Laboratories, Albuquerque, NM
USA

1. Introduction

There are a number of ultra-high intensity lasers in operation around the world that produce petawatt (PW) class pulses. The Z-Backlighter lasers at Sandia National Laboratories belong to the class of these lasers whose laser beams are large (tens of cm) in diameter and whose beam trains require large, meter-class, optics. This chapter provides an in-depth overview of the production of state-of-the-art high laser-induced damage threshold (LIDT) optical coatings for PW class laser pulses, with emphasis on depositing such coatings on meter-class optics.

We begin with a review of ultra-high intensity laser pulses and the various approaches to creating them, in order to establish the context and issues relating to high LIDT optical coatings for such pulses. We next describe Sandia's PW Z-Backlighter lasers as a specific example of the class of large-scale lasers that generate PW pulses. Then we go into details of the Sandia Large Optics Coating Operation, describing the features of the large optics coating chamber in its Class 100 clean room environment, the coating process controls, and the challenges in the production of high LIDT coatings on large dimension optical substrates. The coatings consist of hafnia/silica layer pairs deposited by electron beam evaporation with temperature control of the optical substrate and with ion assisted deposition (IAD) for some coatings as a means of mitigating stress mismatch between the coating and substrate. We continue with details of preparation of large optics for coating, including the polishing and washing and cleaning of the substrate surfaces, in ways that insure the highest LIDTs of coatings on those surfaces. We turn next to LIDT tests with nanosecond and sub-picosecond class laser pulses while emphasizing the need, when interpreting LIDT test results, to take into account the differences between the test laser pulses and the pulses of the actual PW laser system. We present a comprehensive summary of results of LIDT tests on Sandia coatings for PW pulses.

Two sections of the chapter present specific coating case studies, one for designs of a high reflection (HR) coating with challenging performance specifications and one for the anti-reflection (AR) coatings of a diagnostic beamsplitter. The coatings are for non-normal angle

---

[1] Contract Associate to Sandia (JB with Sandia Staffing Alliance; DK with LMATA Government Services)

of incidence (AOI), and the designs take into account behaviors of both S and P polarization (Spol and Ppol) electric field intensities resulting from interference of forward and backward propagating fields during reflection and transmission by the coatings. For the HR coating, a 68 layer design and a 50 layer design both meet the stringent reflectivity requirements (> 99.6% reflectivity of PW pulses in both Ppol and Spol over AOIs from 24° to 47° within ~ 1% bandwidth at both 527 nm and 1054 nm), but the 68 layer coating's LIDT is 5 times less than that of the 50 layer coating because the electric field exhibits high intensity peaks deep within the former coating, but exhibits peaks of moderate intensity that quench rapidly into the latter coating. The study of the AR coatings features measurements of their reflectivities, and of their uniformity over the 92 cm dimension of test optics in the coating chamber. The final section of the chapter presents a conclusion.

2. Ultra-high intensity laser pulses and approaches to creating them

Many ultra-high intensity laser facilities are in operation or under development around the world. Information on these facilities has been compiled by The International Committee on Ultra-High Intensity Lasers (ICUIL) and is available on its website, www.icuil.org. Such high intensity lasers are opening up an ever widening scope of research into laser-matter interactions beyond linear and non-linear optical phenomena at the level of molecular electronic structure and excitation to production of high energy density plasmas, energetic x-rays, inertial confinement fusion and laser induced acceleration of electrons and ions up to relativistic speeds (Perry & Mourou, 1994; Mourou & Umstadter, 2002; Tajima et al., 2010; Mourou & Tajima, 2011). Ultra-high intensity lasers depend on methods of creating laser pulses either of large energy per pulse, or of short pulse duration, or both. By large pulse energies we mean in the range from J to MJ but typically in the kJ regime for a single laser beam train; and by short pulse durations we mean in the ns, ps, fs or shorter regimes. Actually, in the world of ultra-high intensity lasers, reference to "long" in terms of pulse duration means ns class pulses; and "short" means sub-ns class pulses. The resulting intensities of these laser pulses are typically terawatt (TW) to PW and even higher. Focusing of the beams leads to corresponding fluences of 10^{16} W/cm^2 to 10^{19} W/cm^2 and beyond, approaching 10^{22} W/cm^2, depending on the particular laser system and on the achievable minimum focal spot size. Aberrations prevent focusing in the diffraction limit, so minimizing beam train aberrations is critical to achieving the highest fluences at focus. On the other hand, defocusing the beam in a controlled way is sometimes useful as a means of lowering the fluence to some specific level within a focal spot larger than the minimum achievable one.

Regardless of a laser's pulse duration/energy combination, its practical and optimal operation is feasible only to the extent that the laser pulses can traverse the beam train without causing damage or aberrations to its components (windows, mirrors, lenses, gain media, etc.) or their optical coatings. Such laser-induced damage has been the focus of extensive research (Wood, 1990, 2003). It can result from any linear or non-linear laser-matter interaction and is characterized by its LIDT, the laser fluence at or above which it occurs. Optical coatings are our particular concern, and we will deal with both HR and AR types in this chapter. HR and AR coatings are, like optical coatings in general, specific to their use wavelengths, which are the wavelengths of the ultra-high intensity lasers in this context. AR coatings consist of a few (usually < 10) alternating high and low index of refraction thin film layers while HR coatings consist of typically a few tens (< 40) of such layers. They serve the crucial role of reducing loss of energy of the laser pulses in the beam

train; in the case of AR coatings, by minimizing reflection losses at the surfaces of transmissive optics (i. e., windows or lenses) through which the pulses propagate; and, in the case of HR coatings, by minimizing transmission losses (i.e., by providing excellent reflectivities) at the surfaces of mirrors that reflect the pulses. In any case, unless these coatings as well as the optics of a laser beam train can resist damage and aberrations induced by the laser's pulses, the high energy, high intensity pulses of light will not arrive at their final focal volume efficiently enough to reach the fluence levels that produce the ultra-high energy density laser-matter interactions of interest.
The main approaches in creating ultra-high intensity laser light are as follows.

2.1 Laser systems with beam trains of large dimension and cross section

These lasers, owing to the distribution of the pulse energy over large beam cross sectional areas, can generate and handle pulses of large energies at fluences below the LIDTs of the laser optics and coatings. Such lasers, of which there about 15 around the world according to the ICUIL website, www.icuil.org, depend on major government support to provide the large facilities and infrastructure they require. They face the challenges and costs of fabricating and coating large dimension optics to high optical precision. The costs start becoming prohibitive at optic dimensions approaching a meter and beyond, especially for parabolic or other non-planar, non-spherical polished surfaces. But, because energy capacity per pulse increases linearly with beam cross sectional area, up to 4 orders of magnitude increase in pulse energies are possible in going from table top lasers with cm class beam trains to large scale lasers with meter class beam trains. Meter class laser beam trains can support kJ class energies per pulse. Perhaps the most well known of this class of lasers are the National Ignition Facility (NIF) laser system, comprised of 192 laser beam trains, at Lawrence Livermore National Laboratory (LLNL) in the United States (https://lasers.llnl.gov/), and the Laser MegaJoule (LMJ) laser system, comprised of 240 laser beam trains, at the Commissariat a l'Energie Atomique in France (http://www-lmj.cea.fr/).

2.2 Implementation of gain media, optics and coatings with superior resistance to laser-induced damage or aberrations

High LIDT gain media, optics and optical coatings are the focus of important, on-going research. Gains in energy capacity per pulse of a given laser system due to improvements in the LIDTs of optics and coatings can be significant, amounting to factors of 2 or more, but usually less than 10. As mentioned, laser-induced aberrations within gain media and optics undermine the achievement of ultra-high intensities by causing distortion of the beam's wave front and corresponding decrease of its fluence at focus. This latter effect can easily spoil the focal fluence by 1 or 2 orders of magnitude. Most ultra-high intensity lasers utilize optics and gain media with the highest fluence thresholds for laser-induced aberrations and operate at energies per pulse up to but not beyond those thresholds. They then use spatial filtering to restore the wave front of the high energy beam back closer to what it was at lower pulse energy. But, regardless of the optical medium, as laser intensities become higher and higher, the laser-induced aberrations eventually lead to local run-away self-focusing and catastrophic damage along fine, filament-like pathways (Perry & Mourou, 1994). This is due to an accumulation (referred to as the B integral) of laser-induced non-linear optical phase distortions along the propagation path, and correlates especially with intensity hot spots that are not uncommon in the cross section of high intensity laser beams. Fused silica and BK7 are among the most laser damage resistant optical grade glasses (Wood, 2003), and Nd:Phosphate Glass and Ti:Sapphire are laser gain media that also exhibit high fluence thresholds for laser-induced damage (Wood, 2003) and at the same time afford some of the highest energy storage capacities (Perry & Mourou, 1994), at the optimal wavelengths of 1054 nm in the former case and 800 nm in the latter case. Ti:Sapphire can, however, also provide reasonable energy storage and lasing over a broad spectral range. As to thin film optical coatings, LIDTs depend not only on the coating materials but also on the coating design, on the techniques of preparing the optics for coating, and on the coating process itself. We will treat issues of coating design in more detail in this chapter. Regarding the polishing and preparing of optics for coating, we have demonstrated in the case of an AR coating that using one combination of polishing compound and wash preparation for the substrate prior to coating over another can lead to an improvement by a factor of 2 in the laser damage threshold of the coating, and hence the energy capacity per pulse of the laser (Bellum et al., 2010).

2.3 Methods of generating laser pulses of ever shorter duration

For a given energy per pulse, the intensity of the laser light varies inversely with pulse duration. So, techniques such as Q-switching or mode locking to produce short laser pulses, of ns, ps, fs, or even shorter durations, without appreciably reducing the energy per pulse, can lead to orders of magnitude increases in laser intensities.

All ultra-high intensity laser systems involve trade-offs between the above 3 approaches. Avoiding self-focusing is a major factor in any laser design. It not only limits the thicknesses of gain media and optics for given laser pulse energies and durations, but also prevents sub-ns class laser pulses produced by means of laser cavity based techniques such as Q-switching and mode locking from being able to undergo effective amplification in high energy capacity solid state gain media like Ti:Sapphire and Nd:Phosphate Glass. The reason for this latter limitation is that sub-ns pulses, as they increase in energy per pulse, reach the fluence levels resulting in self-focusing before they reach the saturation fluences necessary for efficient extraction of stored energy in the gain medium (Perry & Mourou, 1994). Due to this, the successful ultra-high intensity laser systems developed during the first few decades after the advent of the laser in the 1960s were based on approaches 2.1 and 2.2 above featuring ns class pulses. These were large laser systems using solid state gain media and generating kJ per pulse class laser beams of large, meter class dimensions, and were the predecessors of the NIF and LMJ class of lasers.

The advent of chirped pulse amplification (CPA) in the mid 1980s was a major breakthrough in opening up the realm of sub-ns ultra-high intensity laser pulses (Perry & Mourou, 1994; Strickland & Mourou, 1985; Maine et al., 1988). CPA technology uses optical gratings or other optical techniques to "stretch" a low energy sub-ps class laser pulse of sufficient bandwidth into a ps to ns class pulse, which can then undergo efficient amplification without the self-focusing problems that would occur for the sub-ps class pulse. A reverse version of the "stretching" process then recompresses the amplified ps to ns class pulse into a high energy, sub-ps class pulse. Focusing of these high energy laser pulses is the final step in achieving the ultra-high fluences of coherent light and their associated electric and magnetic optical fields that in turn lead to the high energy density laser-matter interactions. CPA with ps and fs class pulses has permitted the development of ultra-high intensity table top lasers, but is also a technique that has become more and more common in the context of the large, meter-class, ultra-high intensity laser systems, taking them from ns pulses at TW intensity levels with 10^{18} J/cm^2 to sub-ps pulses at PW intensity levels with > 10^{21} J/cm^2.

3. The Sandia TW and PW Z-Backlighter lasers

The Z-Backlighter lasers at Sandia National Laboratories are part of the Pulsed Power Sciences program (http://www.sandia.gov/pulsedpower/) in support of the Z-Accelerator, which produces extremely high energy density conditions by means of a magnetic pinch along the vertical (Z) direction, and is the most powerful source of x-rays in the world. There are two basic Z-Backlighter lasers, Z-Beamlet (Rambo et al., 2005) with TW, ns class pulses and Z-Petawatt (Schwarz et al., 2008) with 100 TW up to PW, sub-ps class pulses. These pulses, after propagating nearly 200 feet from the Z-Backlighter Laser Facility to the Z-Accelerator, undergo focusing onto target foils near the Z pinch. Their focused fluences, ranging from 10^{16} to 10^{20} W/cm^2, produce highly energetic x-rays that back-light the magnetic pinch with enough energy to penetrate its high energy density core and, in this way, provide a diagnostic of the pinch as it occurs (Sinars et al., 2003).

The ns class Z-Beamlet laser pulses undergo multi-pass power amplification in Xe flashlamp pumped Nd:Phosphate Glass amplifier slabs at 1054 nm laser wavelength corresponding to the fundamental laser frequency of Nd:Phosphate Glass. Z-Beamlet then converts these amplified pulses by means of frequency doubling in a large dimension KDP crystal to the second harmonic at 527 nm. Its pulses are of duration in the range 0.3 - 8 ns, but the most common operation is with 1 - 2 ns pulses, and pulse energies of up to ~ 2 kJ at 527 nm in a beam of about 900 cm^2 cross sectional area. The sub-ps class Z-Petawatt laser uses optical parametric chirped pulse amplification (OPCPA). A Ti:Sapphire laser operating at 1054 nm provides 100 fs pulses at low (nJ) energies. A double-pass grating stretcher temporally expands these pulses to ~ 2 ns duration. The stretched pulses then undergo optical parametric amplification (OPA) in three stages, by means of a BBO crystal in each stage pumped by amplified, ~ 2 ns pulses at 532 nm of a frequency doubled Nd:YAG laser. After amplification in double-pass rod amplifiers, the OPA output pulses undergo final double-pass amplification in the main amplifier consisting of 10 Xe-flashlamp pumped Nd:Phosphate Glass slabs (44.8 cm X 78.8 cm X 4.0 cm). The output pulses from the main amplifier then are temporally compressed to ~ 500 fs by means of large, meter class gratings. The Z-Petawatt output pulses can range in duration down to ~ 500 fs and the energies per pulse can extend up to ~ 420 J in the current configuration that uses gratings produced on gold coated meter-class fused silica substrates. New gratings have now been produced for Sandia by Plymouth Grating Laboratory (www.plymouthgrating.com) by means of a laser based nano-ruler process (Smith et al., 2008) on large (94 cm X 42 cm X 9 cm) fused silica substrates which, prior to the nano-ruler process, were coated by Sandia with a multi-layer dielectric (MLD) coating. These new MLD gratings will permit energies per sub-ps pulse approaching 1 kJ due to their superior resistance to laser damage as compared to that of the gratings on the gold coated substrates. The expanded Z-Beamlet laser beam can present 2.5 - 10 J/cm^2 in a 1 ns pulse of 527 nm light over its cross section. In the case of the Z-Petawatt laser, the beam can present 1 - 2 J/cm^2 in a 700 fs pulse of 1054 nm light over its cross section. Our goal in large optics coatings is that their LIDTs exceed these fluences, and preferably by factors of ~ 2 in order to handle hot spots in the beams.

4. Depositing high LIDT coatings at Sandia's large optics coating operation

Coating large optics goes hand in hand with large vacuum coating chambers. In Sandia's case, the coating chamber is 2.3 m x 2.3 m x 1.8 m in size and opens to a Class 100 clean room equipped for handling and cleaning the large optics for coating (see Fig. 1). Such a highly clean environment, with downward laminar air flow into a perforated raised floor to enhance the laminar quality, is critically important to the production of optical coatings exhibiting the highest possible LIDTs. This is due to the fact that even nano-scale particulates on an optical surface prior to coating become initiation sites for laser damage of the coated surface to occur at lower LIDTs (Stolz & Genin, 2003). A major issue with particulates is that, when the coating chamber is not under vacuum and its door is open, coating material on the chamber walls tends to flake off, violating Class 100 conditions inside and in front of the chamber. This calls for measures to prevent these particulates from contaminating the surfaces of product optics prior to coating. One such measure is the use of clean room curtains, as shown in Fig. 1, to separate the area in front of the coater from the rest of the Class 100 area, shown in Fig. 2, in which optics undergo cleaning and preparation for coating. Another such measure is to handle optics in preparation for coating and to load them into the chamber using special tooling and techniques that protect the surfaces undergoing coating from exposure to the non-Class 100 conditions in front of and inside the open chamber. Once the chamber door is closed, the downward laminar flow of Class 100 air quickly restores the area in front of the chamber to Class 100 status; and the risks of particle contamination inside the chamber are negligible when it is under vacuum.

Fig. 1. The Sandia large optics coating chamber and process control console.

Among deposition methods that produce high quality coatings, conventional electron beam (e-beam) evaporation of thin film materials is the most suitable for coating large optical substrates. This is because of the high levels of uniformity of the coating over large substrate areas that are achievable with e-beam deposition due, in part, to the relatively large cone angles of the plumes of e-beam evaporated coating molecules. In addition, motion of the substrates in planetary fixtures as well as masks with special design and placement between the thin film material sources and the substrates are necessary as a means of controlling and averaging out the deposition to insure uniform thin film layer thicknesses. In Sandia's 3-planet configuration, as shown in Fig. 3, each planetary fixture can hold optical substrates up to 94 cm in diameter. The planet fixtures of a 2-planet, counter-rotating option, can hold substrates up to 1.2 m in one dimension and 80 cm in the other. The coater has three e-beam sources (see Fig. 3) for evaporation of the thin film materials. Hafnia and silica are, respectively, the high and low index of refraction layers of choice for high LIDT coatings, due to their high resistance to laser damage by visible and near infra-red light (Fournet et al., 1995; Stolz & Genin, 2003; Stolz et al., 2008). Crystal sensors in locations on the bottom sides of the masks, which are near the plane of the optical surfaces undergoing coating, serve to monitor the coating process by detecting the amount and rate at which they accumulate coating material during deposition. The Sandia chamber also can accommodate optical monitoring of the coating deposition process. An RF ion source (see Fig. 3) provides the option of IAD. The base pressure of the coating chamber needs to be ~ 1 - 2 X 10^{-6} Torr in order to insure contamination free conditions for the deposition process.

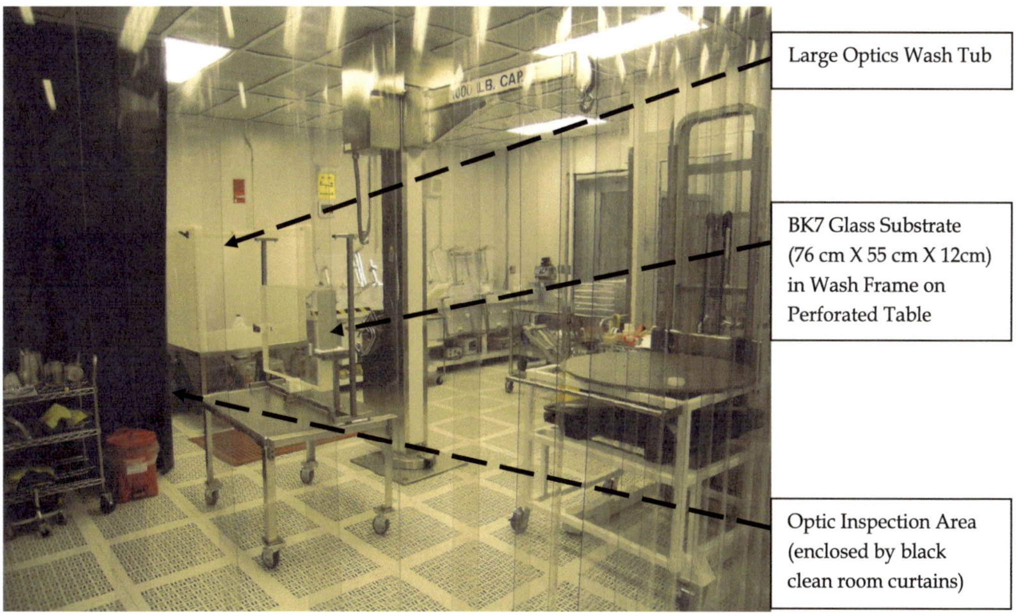

Fig. 2. Sandia's Class 100 clean room for washing and preparing large optics for coating.

Achieving high LIDT coatings depends not only on use of coating materials with high resistance to laser damage, but also on the methods of preparing the substrate surfaces for coating and on the deposition processes and process control, as we mentioned above, and, as we will see later in the chapter, on the coating design. Direct e-beam evaporation of silica, because it occurs at moderate e-beam current and voltage, leads to generally defect-free thin film layers. This is, however, not the case for hafnia because it requires much higher e-beam current and voltage to evaporate, which in turn increases the risk of the evaporation process producing hafnia particulates along with hafnia molecules. Such particulates that attach to the coating as it is forming become defect sites that can initiate laser damage. To avoid this, we use direct e-beam evaporation of hafnium metal in combination with a back pressure of oxygen at ~ 10^{-4} Torr that is sufficient to insure that all of the evaporated hafnium atoms react with oxygen to form hafnia molecules that then form the hafnia coating layer. This occurs in a defect-free way because evaporation of hafnium metal occurs at more moderate e-beam current and voltage than evaporation of hafnia, with correspondingly lower risk of producing particulates in the evaporation process.

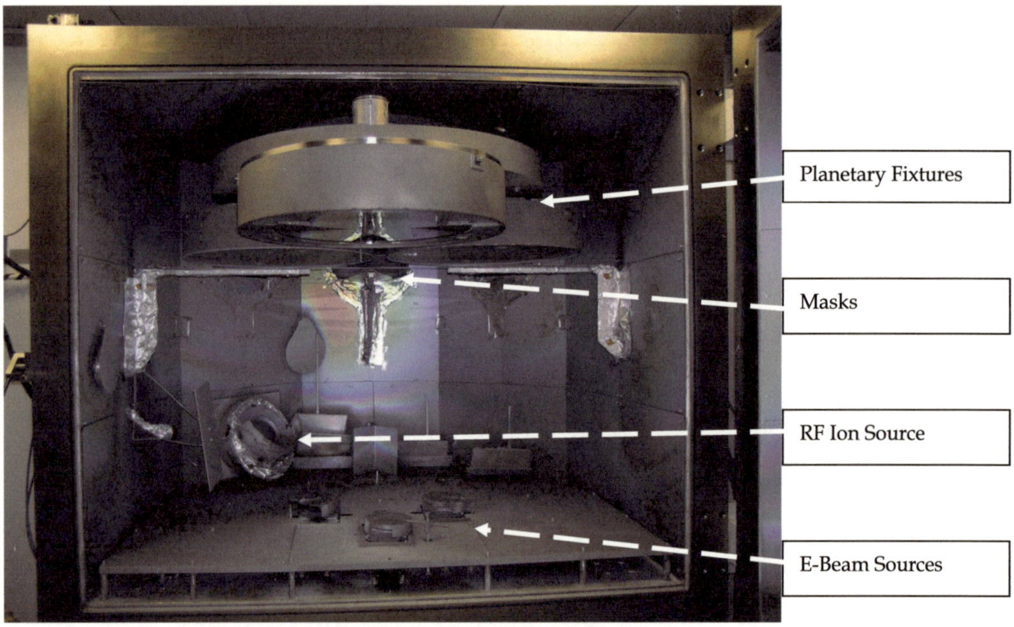

Fig. 3. Interior of the Sandia large optics coating chamber.

A feature of the Sandia large optics coater is the control of the substrate temperature - that is, the temperature within the coating chamber - during deposition. The temperature governs the energy of molecular motion, both of the coating molecules as they assemble to form a coating layer and of the substrate molecules in their phonon degrees of freedom. Thus, lowering or raising the temperature can change the dynamics at the molecular level by which coatings form. In particular, coating at an elevated temperature of ~ 200 °C can promote formation of coatings with mechanical stress (Strauss, 2003) that matches or is close to that of the substrate. This is important because stress differences between a coating and substrate increase the risk of the coating delaminating from the substrate. The case of HR coatings on BK7 optical glass is a good example of how deposition at ~ 200 °C results in low stress differences between coating and substrate. With IAD, ions from the ion source bombard the coating layer as it forms, thus modifying how the coating molecules assemble into a layer. Such IAD coatings are usually denser with a higher level of surface roughness, and have less stress mismatch with the substrate, than do non-IAD e-beam deposited coatings, and their LIDTs tend to be as high as or somewhat higher than those of non-IAD coatings. The increase in surface roughness leads to diffuse reflection, detracting from the specular reflection that an HR coating could otherwise provide. We have investigated techniques of reducing the surface roughness of IAD HR coatings based on using an elevated chamber temperature during the coating run and on turning the ion beam off during the pause between layers in the deposition process (Bellum et al., 2009).

The risks of system or process failures in a coating run increase with the number of coating layers being deposited whether the coating system is large or small, and process control measures constitute the primary means of mitigating these risks. There are, however, additional risks and challenges when it comes to coating large optics. The amounts of thin film material that must be evaporated by the e-beam process increase with the size of the coating chamber to the extent that depletion of coating materials starts becoming a problem in a large optics coating run after ~ 20 coating layers. Related to material depletion is the problem that the topology of the depleted material's surface melt or glaze becomes irregular, and this can cause random steering of the plume of e-beam evaporated material and lead to degradation of coating uniformity. This is especially the case in the deposition of silica in that more silica must undergo evaporation to form a layer of a given optical thickness because of silica's lower index of refraction and thin film density compared to hafnia. For this reason, we use two e-beam sources for silica so that material depletion is less for each source since it needs to provide for only half the number of silica layers in a coating run. An associated challenge is achieving layer pair thickness accuracy. Though layer pair thickness errors tend to be random, the overall effect of the errors increases with number of layers. This is not so critical for standard quarter-wave layer coatings because for each layer that is a bit thinner than a quarter of a wave there is likely to be one that is a bit thicker, and the errors tend to cancel out. It is, however, critical for non-quarter-wave coatings of more than ~ 20 layers in which layer pair thickness accuracy is important especially in the outer (last deposited) layers. Figure 4 summarizes these large optics coating production challenges. Successful production of coatings on large optical substrates requires ongoing efforts to find ways of meeting and mitigating these challenges through coating process control measures.

5. Preparation of large optics for coating – polishing, washing and cleaning

Because of their size, large optical substrates usually undergo single-sided pitch polishing. For optics with optically flat side 1 and side 2 surfaces, double-sided polishing is very effective, but cannot yet handle optics of dimension more than ~ 0.6 m. Polishing large optics to scratch/dig (American National Standards Institute, 2006, 2008) surface qualities of 30/10 and surface figures of $1/10^{th}$ wave peak-to-valley is achievable, but at significant costs and lead times (often more than a year) for the fabrication and polishing processes. Going beyond these optical surface properties moves fabrication and polishing costs and lead times from significant to daunting.

The polishing compound itself influences the laser damage properties of an optically polished substrate, whether coated or uncoated, because residual amounts of it remain to some extent embedded in the microstructure of the polished surface. Alumina, ceria and zirconia are some of the most laser damage resistant polishing compounds, and this correlates in part to their sizable energy thresholds for electronic excitation and ionization. But laser damage also correlates to the degree to which trace levels of polishing compound remain in the microstructure of a polished surface, which in turn depends on the hardness and size of the polishing compound particles. In any case, the achievement of the highest possible laser damage threshold for a coated optic depends on techniques of washing and cleaning the optical surface prior to coating in a way that removes as much surface contamination as possible, including residual polishing compound.

At Sandia, washing of meter-class optics is by hand in the large optics wash tub (see Fig. 2) following the wash protocol of Table 1. Inspection of the cleaned surfaces is by eye in the dark inspection area (see Fig. 2) using bright light emerging from a fiber optic bundle within a small cone angle to illuminate the optic surfaces. For large optics, such manual washing and inspection are most common, although hands-off, automated wash and inspection processes offer advantages and are becoming available (Menapace, 2010). The first 8 steps of Table 1 include an alumina slurry wash step along with mild detergent wash and clear water rinse steps. This protocol relies on copious flow of highly de-ionized (DI) water (resistivity > 17.5 MΩ) and on washing using ultra-low particulate hydro-entangled polyester/cellulose Texwipes. The mild detergent is Micro-90 diluted with DI water. The alumina slurry is Baikalox (also under the name, Rhodax) ultra pure, agglomerate free, 0.05

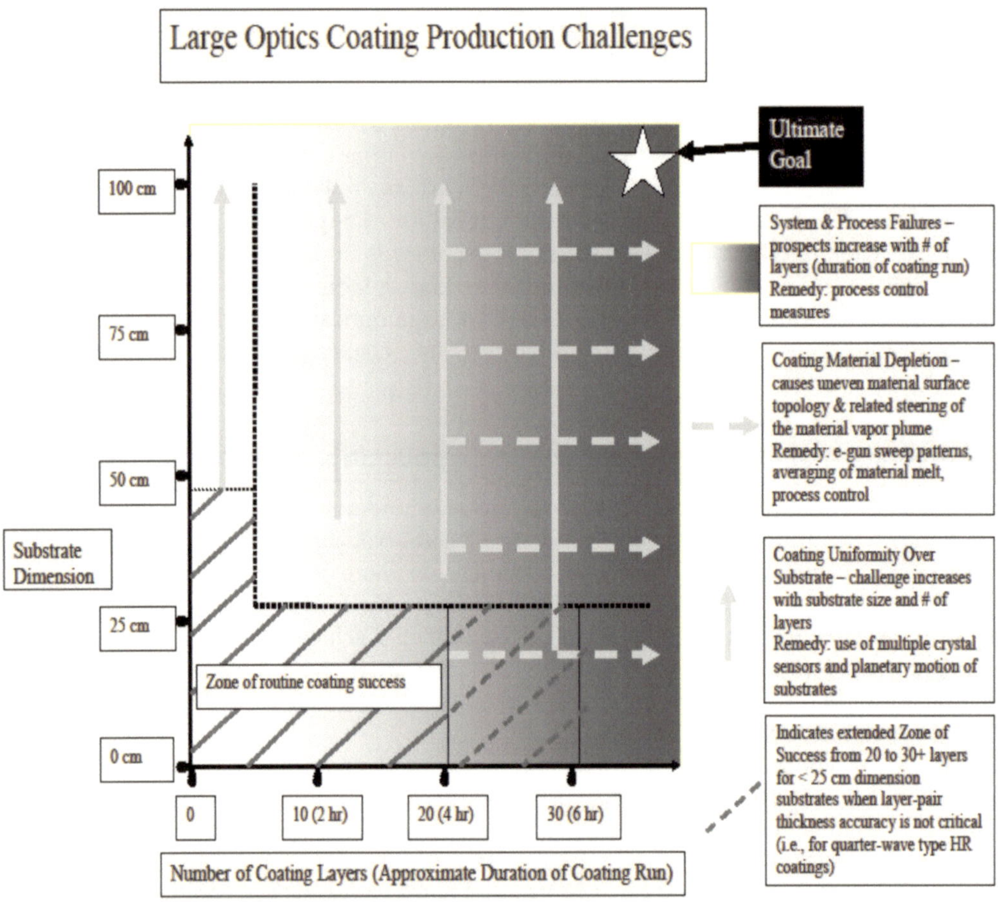

Fig. 4. Summary of large optics coating production challenges.

CR alumina polishing liquid, which is a suspension of alumina particles with nominal size of 0.05 μm. Washing using the slurry with its extremely fine alumina particles serves to remove, at least partially, the residual polishing compound embedded in the microstructure of the optical surface, and does so without degrading the optically polished surface's scratch and dig properties. This is important because polishing compounds are usually less resistant to laser damage than are the optical surfaces or the coatings, so removing residual polishing compound can enhance the LIDT of the coated surface. Our recent study on this (Bellum et al., 2010) found that LIDTs of an AR coating on fused silica substrates polished with ceria or zirconia polishing compounds were ~ 2 times higher for the substrates we washed with compared to without the alumina wash step, confirming that the alumina slurry wash step significantly reduces residual polishing compound on the optic surface and leads to improved LIDTs of coatings on those surfaces.

The steps of Table 1 proceed with repetition as necessary until Step 9, the Class 100 laminar air flow drying, occurs with the optic surface properly sheeting off excess DI water and being free of any cleaning residue or particles as verified by Step 10. In Step 11, the optic either passes inspection or fails, in which case we return to Step 1. An optic that passes inspection should, within hours the same day, be loaded into the chamber for coating. Otherwise it must undergo the wash process again because the risks of particulates attaching to its surface become unacceptably high even after a few hours in the Class 100 environment. In Step 9, the washed substrate rests in its wash frame, as shown for the BK7 substrate in Fig. 2, such that the laminar air flow occurs along the washed surfaces. Use of a perforated table, like that of Fig. 2, on which to place the washed optics helps maintain the laminar quality of this downward air flow at the high level required to prevent particulates from attaching to the optical surface to be coated. As we mentioned earlier, keeping the surface free of particulates is necessary to achieving the highest laser damage resistance of the eventual coating on the surface, since such particulates serve as likely sites for initiation of laser damage.

Step 1.	Clear water rinse/wipe
Step 2.	Vigorous mild detergent wash
Step 3.	Clear water flow rinse
Step 4.	Vigorous alumina slurry wash
Step 5.	Clear water flow rinse
Step 6.	Vigorous mild detergent wash
Step 7.	Vigorous clear water wash/rinse
Step 8.	Thorough clear water flow and/or spray rinse
Step 9.	Class 100 laminar air flow drying
Step 10.	Inspection of washed optic
Step 11.	Optic passes – or return to Step 1

Table 1. Large Optics Wash Protocol

6. LIDT tests

Laser-induced damage to optics and their optical coatings varies greatly as to the mechanisms by which it occurs (Wood, 1009, 2003), as to whether it does or does not grow or propagate in physical size, and as to how deleterious its effects are to the operation of a laser. These variations depend on factors such as the frequency (i.e., wavelength) of the laser light, its transverse and longitudinal mode structure, the duration and temporal behavior of the laser pulse, and the laser fluence. The LIDT refers to the maximum laser fluence, usually expressed in J/cm^2, that a coated optic in a given laser beam train can tolerate before it suffers damage to an extent that prevents satisfactory operation of the laser. LIDT tests should ideally take place with the actual optic in the actual laser of interest which, in the present context, is a PW class laser with meter-class optics. This is, however, not practical. Instead, LIDT tests are commonly done on small damage test optics using table top high energy lasers whose laser wavelength, transverse and longitudinal mode structure, and pulse duration and temporal behavior are similar to those of the ultra high intensity laser of interest. Such damage test lasers need only be capable of producing moderately high intensity laser pulses whose fluences can, with focusing if necessary, range up to and beyond those expected in the transverse beam cross section of the ultra high intensity laser. For the LIDT tests to be as valid and informative as possible, the damage test optic must match the large, meter-class laser optic in type of optical glass, in polishing compound and process, in washing and cleaning prior to coating, and in optical coating, including that both the test optic and the meter-class optic be coated in the same coating run. Even so, because of differences between the test and use lasers, results of LIDT tests require careful interpretation in determining how they relate and apply to the design and performance of a given PW class laser.

By convention, LIDTs are the fluences as measured in the laser beam cross section regardless of whether or not the AOI of the laser is normal to the coated optical surface. Thus, the measured LIDT fluence projects in its entirety onto the optic surface only for LIDT tests at normal AOI. For LIDT tests with the laser beam at a non-normal AOI, the measured LIDT fluence projects only partially onto the optic surface, with the corresponding projected fluence on the surface being less than the measured LIDT by the geometric projection factor of cosine of the AOI. Even though this can be confusing, it is important to keep in mind. For LIDT tests to be valid for optical coatings whose designs are for specific non-normal AOIs and Spol or Ppol, the AOIs and polarization of the test laser beams must match those of the coating designs. This is especially important because of the differences in boundary conditions satisfied by Spol and Ppol components of the optical electric fields at interfaces between optical media (Born & Wolf, 1980). For coatings, these interfaces are those between the coating and the substrate, the coating and the incident medium, and the coating layers. These boundary condition differences at non-normal AOIs can lead to significant differences between Spol and Ppol LIDTs, as we have shown for various 4-layer AR coatings (Bellum et al., 2011).

The Z-Backlighter lasers operate with two pulse types: single longitudinal mode, ns class pulses at 1054 nm and 527 nm in the case of the Z-Beamlet TW class laser; and mode-locked, sub-ps class pulses at 1054 nm in the case of the 100 TW and PW class lasers. The lasers fire on a single shot basis, usually with hours between shots. Their laser beams all exhibit single transverse mode intensities resulting from spatial filtering, and also exhibit intensity hot spots across the beam cross section. LIDT tests on coatings of the Z-Backlighter laser optics are also with single transverse mode laser pulses, but with differing longitudinal mode properties. The tests at or near the 1054 nm wavelength are with multi longitudinal mode, ns class pulses or with mode-locked, sub-ps class pulses; and the tests at or near the 527 nm wavelength are with multi or single longitudinal mode, ns class pulses. Multi longitudinal mode pulses exhibit intensity spikes due to random mode beating and may for this reason be more effective in causing laser damage at a given fluence than single longitudinal mode or mode-locked pulses, which tend to exhibit temporally smooth intensity behavior [see, for example, (Do & Smith, 2009)]. The enhancement of laser damage associated with intensity spiking in LIDT tests with multi longitudinal mode pulses tends, however, to make these tests realistic in that it is a counterpart to (though different from) actual enhancement of laser damage that occurs in the Z-Backlighter laser beam trains due to beam hot spots.

LIDT tests of Z-Backlighter laser coatings are of several types. First is an important type of long pulse test which is performed by Spica Technologies Inc. (www.spicatech.com) using 3.5 ns, multi longitudinal mode Nd:YAG laser pulses at 1064 nm or frequency doubled at 532 nm. These wavelengths are close enough to the 1054 nm or 527 nm Z-Backlighter wavelengths that LIDTs measured at 1064 nm or 532 nm reliably match those at 1054 nm or 527 nm. The pulses are incident one shot at a time per site of a 1 cm X 1 cm grid of ~ 2500 such sites on the coating. This testing protocol originated out of the NIF laser program (National Ignition Facility, 2005) and we refer to it as the NIF–MEL protocol. In the raster scans, the laser spot overlaps itself from one grid site to the next at its 90% peak intensity radius. In our tests, the fluence in the cross section of the laser beam usually starts at 1 J/cm^2 for the first raster scan and increases in increments of 3 J/cm^2 for each successive scan. This procedure amounts to performing a so-called N:1 LIDT test (Stolz & Genin, 2003) at each of the ~ 2500 raster scan sites over the 1 cm^2 area, conducted by means of raster scan iterations with the fluence increasing iteration to iteration. At each fluence level, the test monitors the number of new laser induced damage sites, of which there are two basic types; those that are non-propagating in that they form but then do not grow in size as the laser fluence increases, and those that are propagating in that they form and then continue growing in size as the laser fluence increases. The NIF-MEL protocol specifies the LIDT as the lowest between the two fluence thresholds, the propagating damage threshold for which at least one propagating damage site occurs, or the non-propagating damage threshold for which the number of non-propagating damage sites accumulates to at least 25, corresponding to non-propagating damage over ~ 1% of the 1 cm^2 scan area (~ 1% of the ~ 2500 scan sites). This LIDT protocol indicates the damage behavior we can realistically expect of a coating when it is in the laser beam train exposed daily to Z-Backlighter laser shots. The propagating damage threshold specifies the fluences at which we can avoid catastrophic coating failure resulting from one or more propagating damage sites. Such propagating damage typically grows into large damage craters and definitely constitutes an unacceptable degradation to the coating's optical performance. The non-propagating damage threshold, on the other hand, specifies the fluences at which we can keep the area coverage of non-propagating damage to the coating at ~ 1% or less of the area of the coating exposed to the laser beam. This 1% gauge is based on an estimate of when non-propagating damage becomes unacceptable. As the area coverage of non-propagating damage increases to the 1% level, we expect based solely on geometry that the optical losses due to scattering of light by the non-propagating damage sites become appreciable compared to 1% of the laser beam intensity. This approaches a level of loss that we try hard to avoid. For example, by means of AR coatings on transmissive optics we try to keep surface reflection losses below 0.5%. So, the non-propagating damage threshold is indeed a reasonable gauge for assessing the laser fluence beyond which the degradation of a coating's optical performance due to non-propagating damage is no longer acceptable.

Next are our in-house LIDT tests, which are in the short pulse regime with 350 fs, mode locked pulses at 1054 nm on a single shot basis, and in the long pulse regime with 7 ns, single or multi longitudinal mode pulses at 532 nm on a single shot basis, and also on a multi shot basis (10 shots at 10 Hz pulse repetition frequency) but only in the case of multi longitudinal mode pulses. Our recent papers provide a detailed description of the test set-up and formats for the 350 fs pulses at 1054 nm (Kimmel et al., 2009) and the 7 ns pulses at 532 nm (Kimmel et al., 2010). For the latter in-house tests at 532 nm, the single longitudinal mode condition is achieved by injection seeding of the laser with the output of a single longitudinal mode seed laser. Within the overall long pulse regime, the pulse duration

		NIF-MEL Tests		**Sandia In-House Tests**	
	AOI	1064 nm (3.5 ns pulses)	532 nm (3.5 ns pulses)	1054 nm (350 fs pulses)	532 nm (7 ns pulses)
AR coatings					
for 1054 nm	0 deg	18, 18, 19, 19, 21, 25, 25, 27, (33)		(1.8)	
for 1054 nm	32 deg	Spol: (37); Ppol: (34)			
for 1054 nm	45 deg	Spol: 47; Ppol: 19			
for 527 & 1054 nm	0 deg	(25), ((19)), [23], [[29]], 19, 22	(9), ((6)), [8], [[13]]	[[~ 2]]	[[38]], [[38]]; 10 shot: [[28]]
for 527 & 1054 nm	22.5 deg	Spol: (38), ((46)); Ppol: (38), ((55))	Spol: (12), ((11)); Ppol: (12), ((13))		
HR coatings (quarter-wave type)					
for 1054 nm	0 deg	IAD: 37, 56, 75; Non-IAD: 82			
for 1054 nm	32 deg	Spol: (79), ((82)); Ppol: (88), ((79)), 70, 91			
for 1054 nm	45 deg	Spol: (82), ((88)), [88]; Ppol: (73), ((75)), [88], 58, 79, 88, 88, 91, 91, 97			
for 527 & 1054 nm	30 deg			Ppol: (1.32), (1.71)	Ppol: 70

Table 2. Measured LIDTs (in J/cm^2) of Sandia AR and HR coatings. For each listed coating, values in similar brackets are for the same coating run.

differences (7 ns pulses of our in-house 532 nm tests, 3.5 ns pulses of the NIF-MEL tests, and ~ 1 ns pulses of the Z-Backlighter lasers) lead to corresponding differences in LIDTs, with the longer pulses affording higher LIDTs at a given fluence than those with the shorter pulses. Finally, concerning LIDTs, the NIF-MEL criteria [see above and (Bellum et al., 2009, 2010; National Ignition Facility, 2005)] involves each raster scan site on the coating receiving multi longitudinal mode laser shots one at a time, with minutes between shots, over and over at increasing fluence until damage (non-propagating or propagating) occurs. For our in-house tests, by contrast, each new site on the coating receives either a single laser shot or 10 laser shots (at 10 Hz) at a given fluence with the next new site similarly receiving one shot or 10 shots at a higher fluence, etc., until damage occurs (Kimmel et al., 2009, 2010). In addition, the NIF-MEL laser damage test protocol, with its 2500 raster scan sites in a 1cm X 1cm area, samples an appreciable area of the coating. On the other hand, our in-house testing is at tens of specific sites on the coating with one level of laser fluence at each site, and so affords a more limited sampling of the coating. The important point is that interpretation of the various LIDT tests requires taking into account their differing conditions and relating these conditions to those of the PW laser. Table 2 summarizes results from our previous reports of these LIDT tests on Sandia coatings (Bellum et al., 2009, 2010, 2011; Kimmel et al., 2009, 2010). The LIDTs are all reasonably high and adequate to insure that the coatings will stand up to the laser fluence levels of the PW class pulses in the Z-Backlighter beam trains.

7. HR coating case study: Electric field intensity behaviors favorable to high LIDTs

A key optic in the next generation Z-backlighter laser beam train is the PW Final Optics Assembly (FOA) steering mirror. It has very challenging coating performance specifications, well beyond what we normally face, and provides an instructive coating design case study. We included an initial report on this mirror and its coating in a recent paper (Bellum, 2009). The mirror's fused silica substrate, shown in Fig. 5, is 75 cm in diameter with a sculpted back surface and corresponding thickness ranging from ~ 3 cm at the edge to a maximum of ~ 15 cm in an annular zone centered about the optic axis. It weighs ~ 100 kg, and serves as the final optic steering the Z-Backlighter laser beams to focus. Its use environment is in vacuum so its coating needs to be IAD, as we explained in the recent paper (Bellum, 2009). The Z-Backlighter reflectivity performance requirements of its HR coating are very demanding: R for Ppol and Spol > 99.6 % for AOIs from 24° to 47° and for both the Nd:Phosphate Glass fundamental and second harmonic wavelengths with extended bandwidths; that is for 1054 nm +/- 6 nm and for 527 nm +/- 3 nm. Furthermore, the coating's LIDT must allow it to handle the ns as well as sub-ps pulses of the Z-Backlighter lasers; namely, LIDT > 2 J/cm^2 for the sub-ps Z-Petawatt laser pulses at 1054 nm, and LIDT > 10 J/cm^2 for the ns Z-Beamlet laser pulses at 527 nm.

We begin this case study by reviewing the considerations that influence the process of designing an optical coating consisting of alternating layers of high and low index of refraction materials. Perhaps the most basic one is that of determining the layer thicknesses of the coating such that it reflects or transmits light according to design specifications for the wavelengths, AOIs and polarization of the incident light. This in turn depends on how the incident light divides up into forward and backward propagating components due to partial transmission and/or reflection at each boundary between coating layers, and on how these

Fig. 5. The PW FOA steering mirror substrate, held by the large optics loading tool.

forward and backward propagating components interfere with one another. The perplexity of this design step is that different combinations of layer thicknesses (i.e., of interfering forward and backward propagating components of light) can lead to similar overall transmission or reflection. In other words, there is not a unique optical coating design for a given set of transmission and reflection performance criteria. Excellent coating design software codes are available. They rely on various design algorithms based on minimizing differences between design criteria and the calculated performance of the coating. The minimization procedures depend on the starting choice of layers and their thicknesses and lead to local minima. A better minimum may be achievable with a better, or just different, choice of starting layers or with a different choice of design algorithm. In the end, these software codes serve as useful tools for exploring coating design options, and the best coatings result from judicious assessment and exploration of theoretical designs by the designer based on his or her knowledge and experience with coating deposition and performance. Our design process relies on the OptiLayer thin film software (www.optilayer.com), which has proven to be a very effective tool for exploring coating design options. Other coating design considerations include how feasible it is to produce the coating on the intended product optic with the available coating deposition system and, for coatings for ultra-high intensity lasers, whether the design provides the required transmission or reflection properties with the highest possible LIDT.

Coating designs that meet the PW FOA steering mirror's daunting, dual-wavelength, and wide ranging AOI HR performance requirements will differ from standard quarter-wave type coatings, like those we reported before (Bellum et al., 2009), that are suitable for HR at a single wavelength and AOI. Our first design attempt for the PW FOA steering mirror coating was based only on meeting the challenging HR performance goals, and resulted in a 68 layer coating about 9 μm thick. Figure 6 shows the calculated Ppol reflection spectra of this coating in spectral regions near the dual design wavelengths of 1054 nm and 527 nm for a sample of 5 AOIs, 25°, 30°, 35°, 40°, and 45°, within the coating's 24° to 47° performance range of AOIs. These calculated reflectivities confirm that the coating should very successfully meet these stringent HR performance specifications.

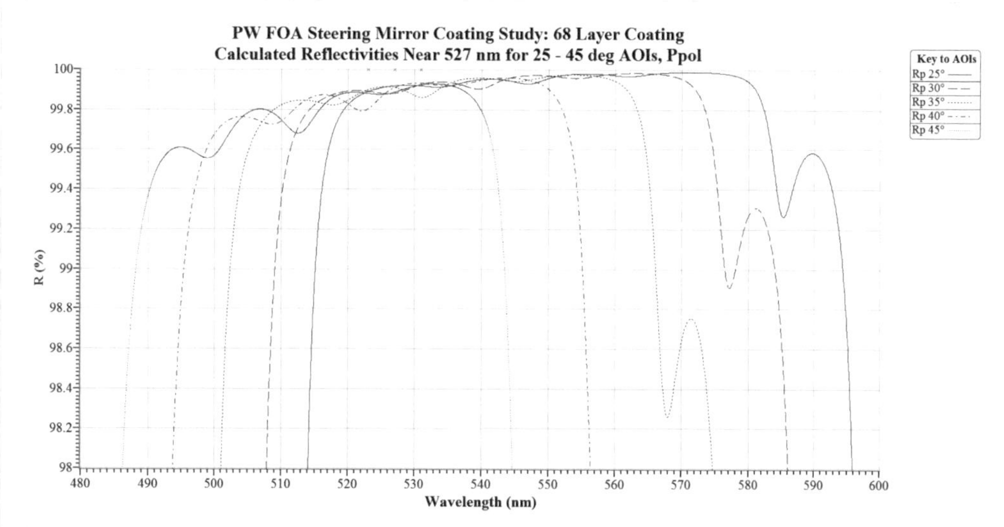

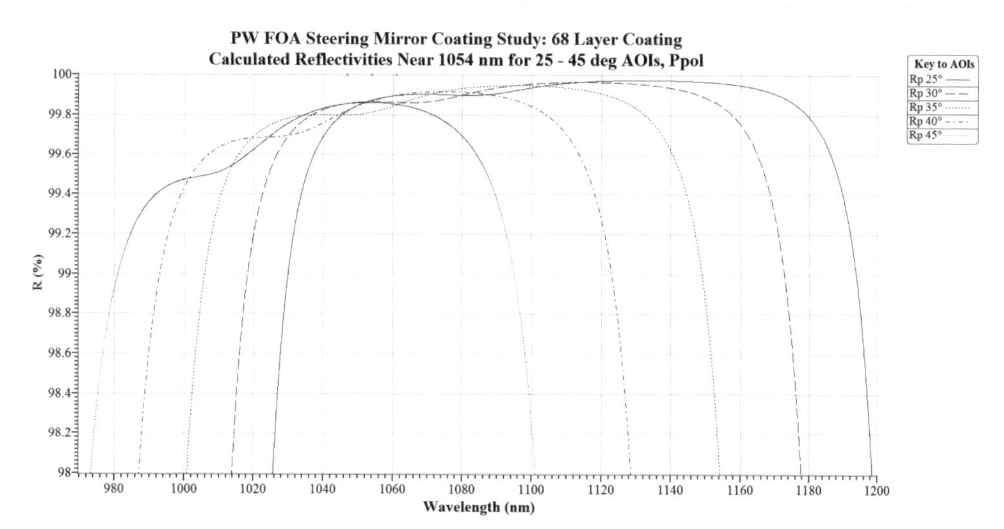

Fig. 6. Calculated reflectivities for Ppol at 25°, 30°, 35°, 40° and 45° AOIs and wavelengths near 527 nm (top figure) and 1054 nm (bottom figure) according to the 68 layer coating design for the PW FOA steering mirror.

The reflectivities of Fig. 6 indicated this 68 layer design would be a good one to use despite the risks we explained above of unforeseen coating process problems that tend to increase with the number of coating layers and process time, which is about 8 hours for this coating. But, LIDTs measured in the NIF-MEL protocol at 25°, 30° and 35° AOIs, Ppol, for this coating are all similar and proved to be disappointing at 532 nm, though excellent at 1064 nm. Figure 7 shows these LIDT results for the case of 35° AOI. The figure displays the cumulative number of non-propagating damage sites versus laser fluence and indicates by a horizontal dashed line the fail threshold of 25 non-propagating damage sites. At 1064 nm, the number of non-propagating damage sites accumulates to only 5 (with no propagating damage sites) as the laser fluence increases to 79 J/cm² (which was the highest fluence the test laser could produce in this particular test configuration). We conclude that the LIDT at 1064 nm in this case is > 79 J/cm²; which is to say that since, at 79 J/cm², neither has the number of non-propagating damage sites exceeded 25 nor has propagating damage occurred, the former will exceed 25, or the latter will occur, only at a fluence > 79 J/cm². This is a very adequate LIDT for ns class Z-Backlighter laser pulses at 1054 nm. At 532 nm, on the other hand, the non-propagating damage sites accumulate to 93, well in excess of 25, at a laser fluence of only 2.5 J/cm². This, then, is the NIF-MEL LIDT in this case, and it is well below the > 10 J/cm² required for the ns class Z-Backlighter laser pulses at 527 nm. The corresponding LIDT results at 25° and 30° AOIs are, respectively, 2.5 J/cm² and 4 J/cm² at 532 nm and, respectively, 76 J/cm² and 79 J/cm² at 1064 nm, completely consistent with their 35° AOI counterparts.

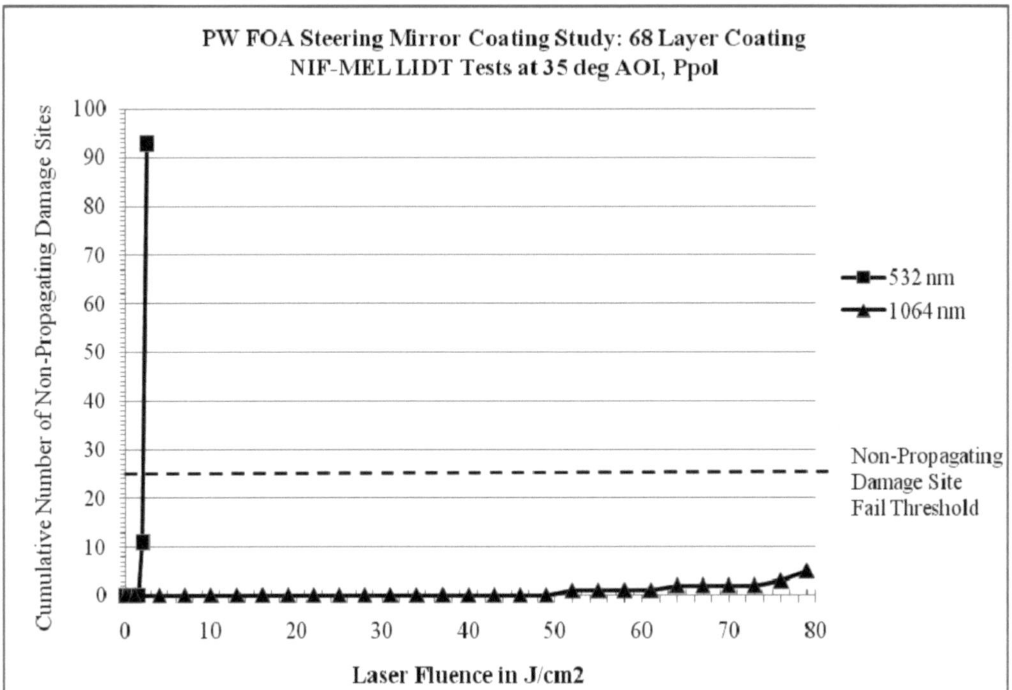

Fig. 7. NIF-MEL LIDT test results at 532 nm and 1064 nm, and 35° AOI, Ppol, for the 68 layer PW FOA steering mirror coating.

We discovered the reason for these disappointing LIDTs at 532 nm by looking at the behavior of the optical electric field intensities for this 68 layer coating. Figure 8 shows the 527 nm field intensities for the 35° AOI, Ppol case. These intensities exhibit significant ringing, with many intensity peaks over 200% of the incident intensity within ~ 34 layers into the coating, and with the highest peak at 340% of the incident intensity. The 527 nm, Ppol intensities for 25°, 30°, 40° and 45° AOIs are all similar to those of Fig. 8. This explains why this 68 layer coating suffered laser damage so readily. Its design is a set of coating layers that provide excellent reflectivities for 527 nm over the 24° to 47° range of AOIs (Fig. 6), but in a way in which highly constructive interference of the forward and backward propagating components of light occurs within the first 34 layers of the coating. This interference becomes destructive, with rapid quenching of the intensity, only within layers 34 to 46 (see Fig. 8), which is where the reflection of the 527 nm light actually takes place within the coating. This means that the 527 nm light must propagate more than half way into the coating before it reaches the layers that reflect it. And in this process, the reflected light interferes constructively with the incoming light within the first 34 layers, leading to the strong intensity peaks that in turn make the coating more susceptible to laser damage at the lower fluences.

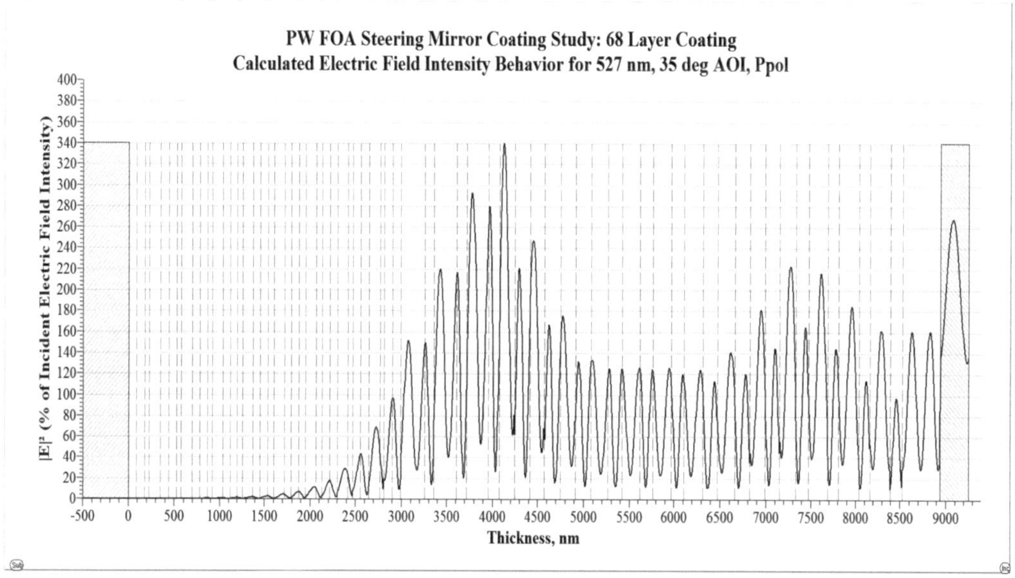

Fig. 8. Calculated electric field intensity at 527 nm for the 68 layer PW FOA steering mirror coating for 35° AOI, Ppol. Shaded areas denote the substrate (left), which is fused silica, and incident medium (right), which is air or vacuum. Vertical dashed lines mark the boundaries of the coating layers.

A very different behavior of electric field intensity is exhibited by 1054 nm light incident on this 68 layer coating, as Fig. 9 shows for 35° AOI, Ppol. The optical electric field intensity peaks quench rapidly into the coating, progressing from ~ 160% of the incident intensity in the outermost silica layer to ~ 100% by the 3rd layer and on down to < 10% beyond the 12th layer. Thus, reflection at 1054 nm is based primarily on interference between forward and backward propagating components of light within the first 12 to 15 layers of the coating, and this interference leads to intensity peaks well below the incident intensity except in the outer silica layer where the peak is moderate, at ~ 160 % of the incident intensity. This type of electric field behavior is favorable to high LIDTs (Stolz & Genin, 2003; Bellum et al., 2009), as is confirmed by the high 1054 nm LIDTs of this 68 layer coating. The thicker outermost silica layer of this 68 layer coating is a feature of its design that enhances this type of electric field pattern for 1054 nm light favorable to high LIDTs (Stolz & Genin, 2003; Bellum et al., 2009).

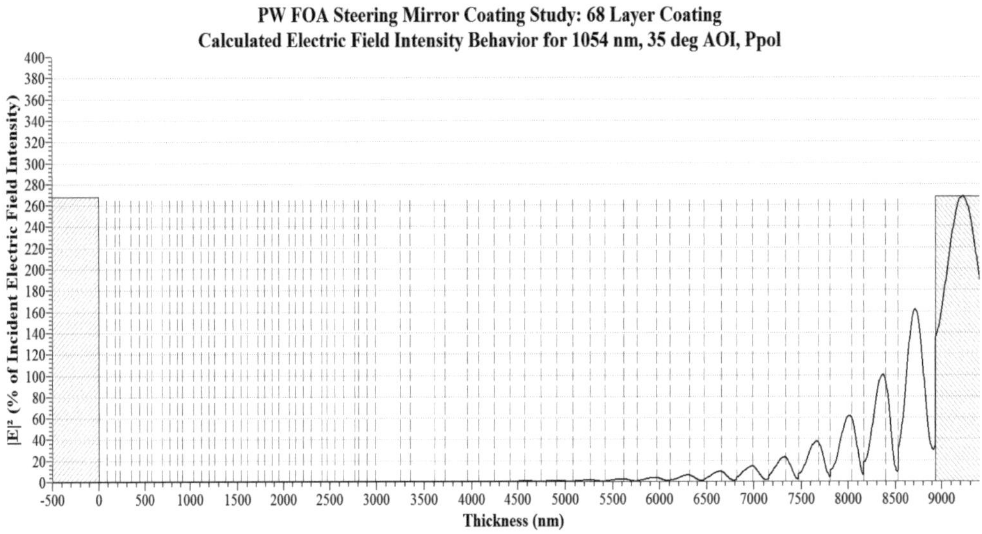

Fig. 9. Calculated electric field intensity at 1054 nm for the 68 layer PW FOA steering mirror coating for 35° AOI, Ppol. Left and right shaded areas and dashed vertical lines identify optical media, as in Fig. 8.

We returned to the design process, looking for design options based not only on meeting the HR requirements but also on meeting the requirement that the optical electric field intensity behavior within the coating show moderate intensity peaks that rapidly quench within the first ~ 15 coating layers for 527 nm as well as for 1054 nm. The result was a suitable 50 layer design for a coating about 8 μm thick that meets both of these requirements. Figure 10 shows its 25°, 30°, 35°, 40°, and 45° AOI, Ppol reflection spectra near 527 nm and 1054 nm, confirming the PW FOA HR performance specifications (R > 99.6% for 527 nm +/- 3 nm and 1054 nm +/- 6 nm), but now over narrower ranges of wavelengths (R > 99.6% for 523 nm - 533 nm and 1048 nm - 1065 nm) as compared to the 68 layer coating (see Fig. 6; R > 99.6% for 518 nm - 541 nm and 1038 nm - 1084 nm). Meeting such an HR specification within narrower spectral range margins places increased demands on coating process control and achievement of layer pair accuracies in the deposition of the 50 layer coating. On the other hand, the risks of coating system and process failures for the 50 layer deposition are not as high as for the 68 layer deposition.

Figure 11 shows the 527 nm and 1054 nm electric field behaviors within the 50 layer coating for 35° AOI and both Ppol and Spol, and they all meet the design goal of exhibiting rapid quenching into the coating. We include the Spol intensities in Fig. 11 to contrast them with the Ppol intensities. The intensity patterns for both 527 nm and 1054 nm are similar in their moderate peaks that quickly quench within the coating. But, in each case, the Spol intensities are slightly lower than the Ppol intensities within the coating but peak much higher in the incident medium just in front of the coating. The Spol intensities also reach near zero intensity minima at the coating layer interfaces and at the interface between the coating and the incident medium, and show no intensity jumps. The Ppol intensities, on the other hand, exhibit intensity jumps at the media interfaces, particularly at the interface between the coating and the incident medium. These Spol and Ppol intensity behaviors are characteristic of HR coating designs like the 50 layer design, and their differences are due to

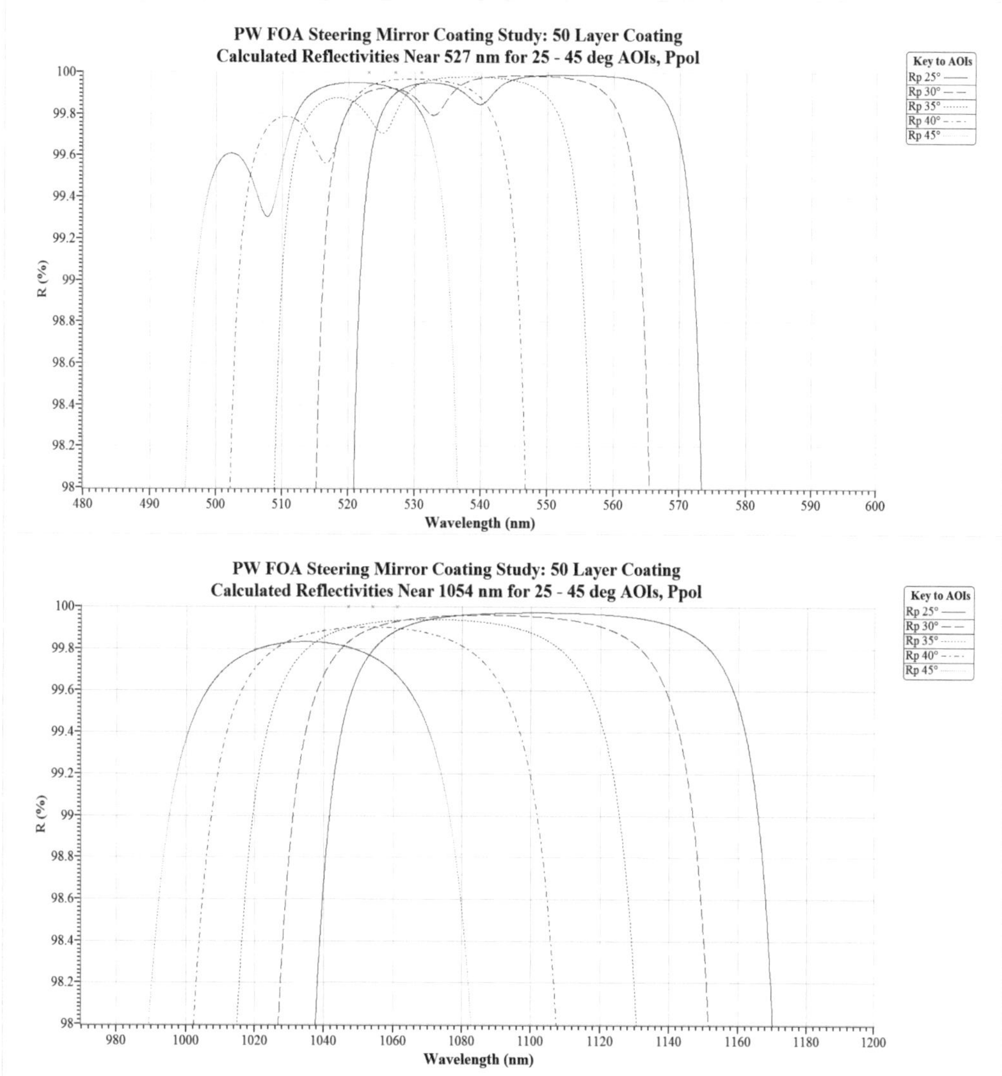

Fig. 10. Calculated reflectivities for Ppol at 25°, 30°, 35°, 40° and 45° AOIs and wavelengths near 527 nm (top figure) and 1054 nm (bottom figure) according to the 50 layer coating design for the PW FOA steering mirror.

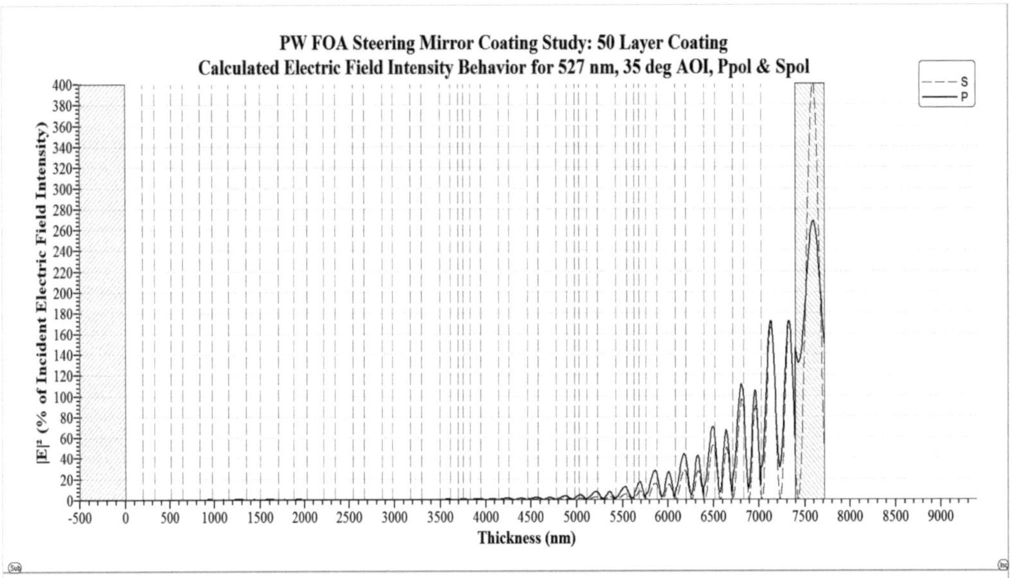

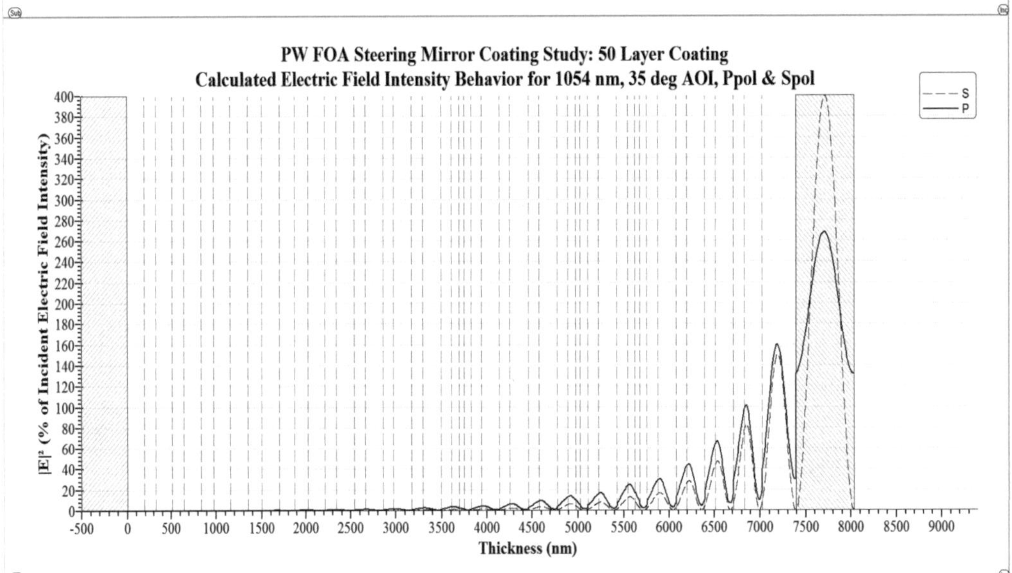

Fig. 11. Calculated electric field intensity at 527 nm (top figure) and 1054 nm (bottom figure) for the 50 layer PW FOA steering mirror coating design for 35° AOI, Ppol and Spol. Left and right shaded areas and dashed vertical lines identify optical media, as in Fig. 8.

the differences in boundary conditions satisfied by Spol and Ppol components of the optical electric field at media interfaces (Born & Wolf, 1980; Bellum, et al., 2011). In any case, because Ppol intensities exhibit jumps at media interfaces and are somewhat higher than the Spol intensities for these HR coatings, their Ppol LIDTs should be lower than their Spol counterparts. That is why our LIDT tests of HR coatings are usually with Ppol, providing a more conservative assessment of the coatings' resistance to laser damage. Another difference between Ppol and Spol behaviors for HR coatings is that the Spol reflectivities are usually higher, and remain high over a broader spectral range, than is the case for their Ppol reflectivity counterparts. Thus, the 50 layer coating will meet the stringent HR performance specifications of the PW FOA steering mirror for Spol within spectral margins near 527 nm and 1054 nm that are wider than the very narrow spectral margins (see Fig. 10) in which it meets those specifications for Ppol.

The LIDTs are indeed high at both 1064 nm and 532 nm for this 50 layer PW FOA steering mirror HR coating as confirmed by the LIDT test results of Fig. 12 for 35° AOI, Ppol, showing in this case that the 1064 nm LIDT is 76 J/cm² (based on propagating damage as opposed to non-propagating damage sites exceeding 25) and the 532 nm LIDT is ~ 12 J/cm² (based on both propagating and non-propagating damage criteria since, at 13 J/cm², non-propagating damage sites had accumulated to 43 and propagating damage had also occurred). The 50 layer coating's 1064 nm LIDT of 76 J/cm² at 35° AOI, Ppol is similar to its counterpart (>79 J/cm²) for the 68 layer coating, but its 532 nm LIDT of ~ 12 J/cm² at 35° AOI, Ppol is nearly 5 times higher than the 2.5 J/cm² LIDT of its 68 layer coating counterpart. The corresponding LIDT results at 25°, 30°, 40° and 45° AOIs are, respectively, 16 J/cm², 16 J/cm², 19 J/cm² and 19 J/cm² at 532 nm; and, respectively, 70 J/cm², 67 J/cm², 82 J/cm² and 64 J/cm² at 1064 nm. These are consistent with their 35° AOI counterparts. This is a satisfying result for the 50 layer coating, indicating that both its 1054 nm and 527 nm LIDTs meet the laser damage resistance required by ns class Z-Backlighter laser pulses over the entire 24° - 47° range of AOIs.

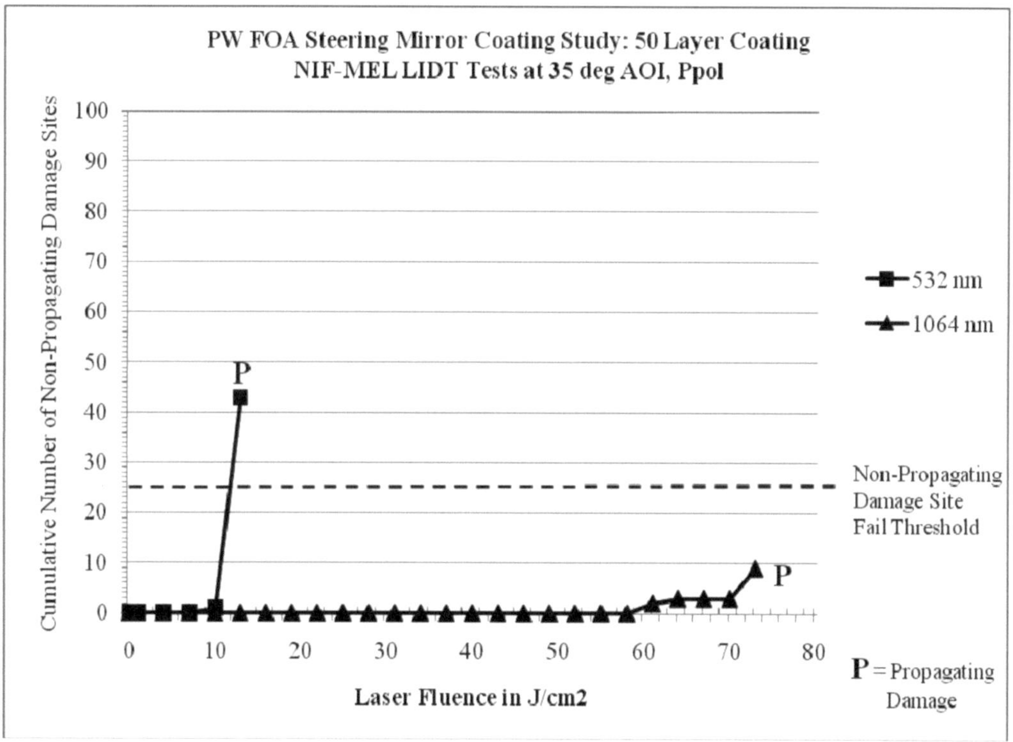

Fig. 12. NIF-MEL LIDT test results at 532 nm and 1064 nm, and 35° AOI, Ppol, for the 50 layer PW FOA steering mirror coating.

This case study for the complex and demanding PW FOA steering mirror HR coating requirements demonstrates the critical role that coating design plays in obtaining coatings that not only meet reflection or transmission specifications, but do so in terms of electric field behaviors within the coating that favor the highest achievable LIDTs. In our study of electric field intensity behaviors for AR coatings (Bellum et al., 2011), we found that the interference of forward and backward propagating components of light leads to electric field intensity behaviors quite different from those for HR coatings, consistent with AR coating design goals of transmitting rather than reflecting incident light. We also found interesting correlations for AR coatings between their LIDTs and the behaviors of the optical electric fields within them, and especially the behaviors of Ppol intensity jumps at coating layer boundaries in the case of non-normal AOIs (Bellum et al., 2011).

8. AR coating case study: Reflectivities and uniformity of AR coatings for a TW diagnostic beamsplitter

The next case study highlights aspects of reflectivity and uniformity of coatings for meter-class, high intensity laser optics in the context of the Side 1 and Side 2 AR coatings of a diagnostic beamsplitter for the TW class Z-Beamlet laser beam at 527 nm and 22.5° AOI. This beamsplitter and diagnostic of the 527 nm beam are located just beyond where it is generated by means of frequency doubling of the 1054 nm beam in a large KDP crystal. Because the frequency doubling process is about 70% efficient, the actual beam emerging from the KDP crystal consists of the 527 nm TW beam as its primary component, comprising ~ 70% of the total beam intensity and the one of interest on target, and a residual 1054 nm beam of much lower intensity whose role on target is relatively minor and inconsequential. The schematic of Fig. 13 depicts the 527 nm and 1054 nm components of the TW laser beam together with the beamsplitter, which is a 61.5 cm diameter, fused silica optic with ~ 50 cm diameter central clear aperture.

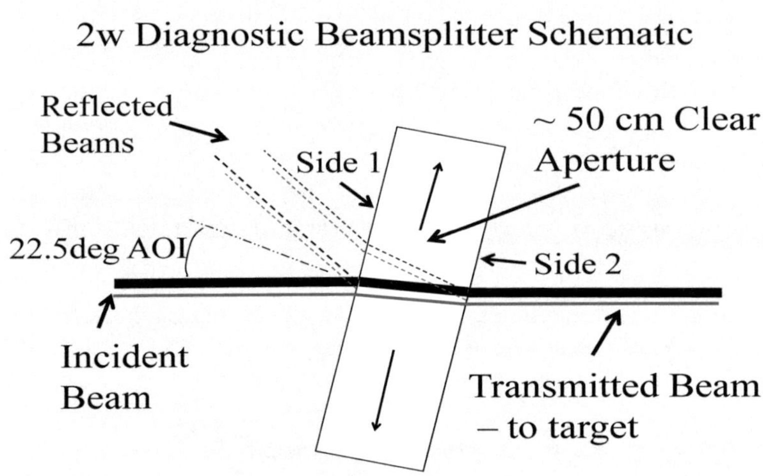

Fig. 13. Schematic diagram of the diagnostic beamsplitter. The solid and dashed lines in black represent the laser beam components at 527 nm while the solid and dashed lines in gray represent the laser beam components at 1054 nm.

The purpose of the Side 1 AR coating of the beamsplitter is to sample the 527 nm TW beam, which undergoes diagnostics of transverse intensity and phase that faithfully match those of the 527 nm TW beam to the extent that the reflectivity at 527 nm across the Side 1 clear aperture is uniform. To do this, the Side 1 coating must not only offer very uniform performance over the beamsplitter clear aperture but also must strike a balance between excellent and merely good AR performance at 527 nm. An excellent 527 nm AR, with reflectivity in the range of ~ 0.14%, would be desirable for minimizing intensity losses and delivering the 527 nm TW beam to target with maximum intensity in the target focal volume. But such low reflectivities afford insufficient intensity in the sample beam to ensure reliable diagnostics. So, in the design of this Side 1 AR coating, we had to sacrifice somewhat the excellence of the 527 nm AR performance, to a level allowing adequate sample intensity for good diagnostics at the expense of a higher loss of transmitted TW intensity than we would like. Accordingly, we set a design goal for the Side 1 AR coating to reflect 527 nm light in the range of 0.5% - 1.0%. For 1054 nm, the Side 1 coating needs to provide an excellent AR to minimize the amount of reflected light at 1054 nm co-propagating with the 527 nm sample beam and possibly interfering with the 527 nm diagnostics.

The Side 2 AR coating, unlike that of Side 1, does not provide a sample of the 527 nm TW beam for diagnostics. Rather, it should offer excellent 527 nm AR so as to add the least amount intensity loss as possible to the losses incurred by the 527 nm TW beam at Side 1. On the other hand, the amount of the 1054 nm residual component of the TW beam reaching the target is not critical and the 1054 nm AR property of the Side 2 coating is also not critical, but need only be in the range of excellent to good in order to keep 1054 nm light reflected by Side 2 at reasonably low intensities. In summary, we designed the coatings for this beamsplitter to meet AR performances at the 22.5° AOI as follows: for Side 1, 0.5% - 1.0% reflectivity at 527 nm and ~ 0.15% or less reflectivity at 1054 nm; and, for Side 2, ~ 0.15% or less reflectivity at 527 nm and 0.5% - 1.5% reflectivity at 1054 nm. We reported the actual layer thicknesses of these Side 1 and Side 2 AR coatings in our previous paper on correlations between LIDTs and electric field intensity behaviors for AR coatings (Bellum et al., 2011). A slight wedge angle between Sides 1 and 2 prevents 527 nm and 1054 nm components of light reflected by Side 2 from entering the 527 nm diagnostic beam train and possibly interfering with the diagnostics.

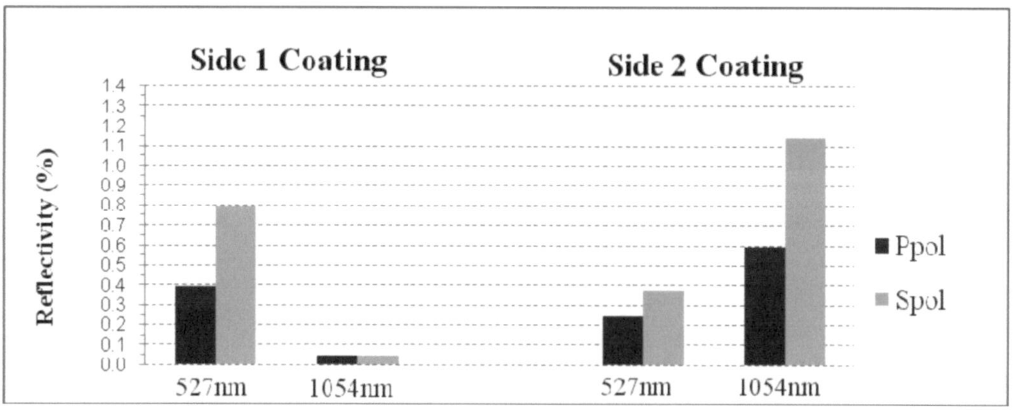

Fig. 14. Measured reflectivities at 527 nm and 1054 nm of the as-deposited Side 1 and Side 2 AR coatings of the diagnostic beamsplitter at its 22.5° use AOI for both Ppol and Spol.

Figure 14 shows the measured reflectivities at 527 nm and 1054 nm for the actual, as-deposited, Side 1 and Side 2 AR coatings at their 22.5° use AOI for both Ppol and Spol. These measurements were on small coated witness substrates using the Sandia reflectometer in a configuration that can also accommodate large, meter-class, optical substrates. We met our coating design goals for Side 1, with reflectivities of ~ 0.4% for Ppol and ~ 0.8% for Spol in the case of 527 nm, and ~ 0.045% for both Ppol and Spol in the case of 1054 nm; and for Side 2 at 1054 nm, with a reflectivity of ~ 0.6% for Ppol and ~ 1.14% for Spol. But our Side 2 reflectivity at 527 nm, of ~ 0.24% for Ppol and ~ 0.37% for Spol, while reasonably low, is about 2 times larger than our design goal of 0.15% or less, indicating that we need to improve on our Side 2 AR coating design in this respect.

Fig. 15 presents measured results for the uniformity of these two coatings. These measurements are based on broadband reflection spectra of the coatings, from roughly 400 nm to 900 nm, recorded in 2 cm intervals along a 5 cm wide uniformity witness optic spanning the full 94 cm diameter of one of the three equivalent planetary fixtures during the Side 1 and Side 2 product coating runs. Another of these planetary fixtures held the diagnostic beamsplitter product optic during these runs. We track the wavelengths of spectral peaks or valleys, which are easily identifiable features of the spectra, measuring them at each 2 cm interval along the planetary diameter according to the percent deviations from their average values. As Fig. 15 shows, the averages of these spectral peak and valley percent deviations are within +/- 0.5% over the central 60 cm of the planet diameter for both

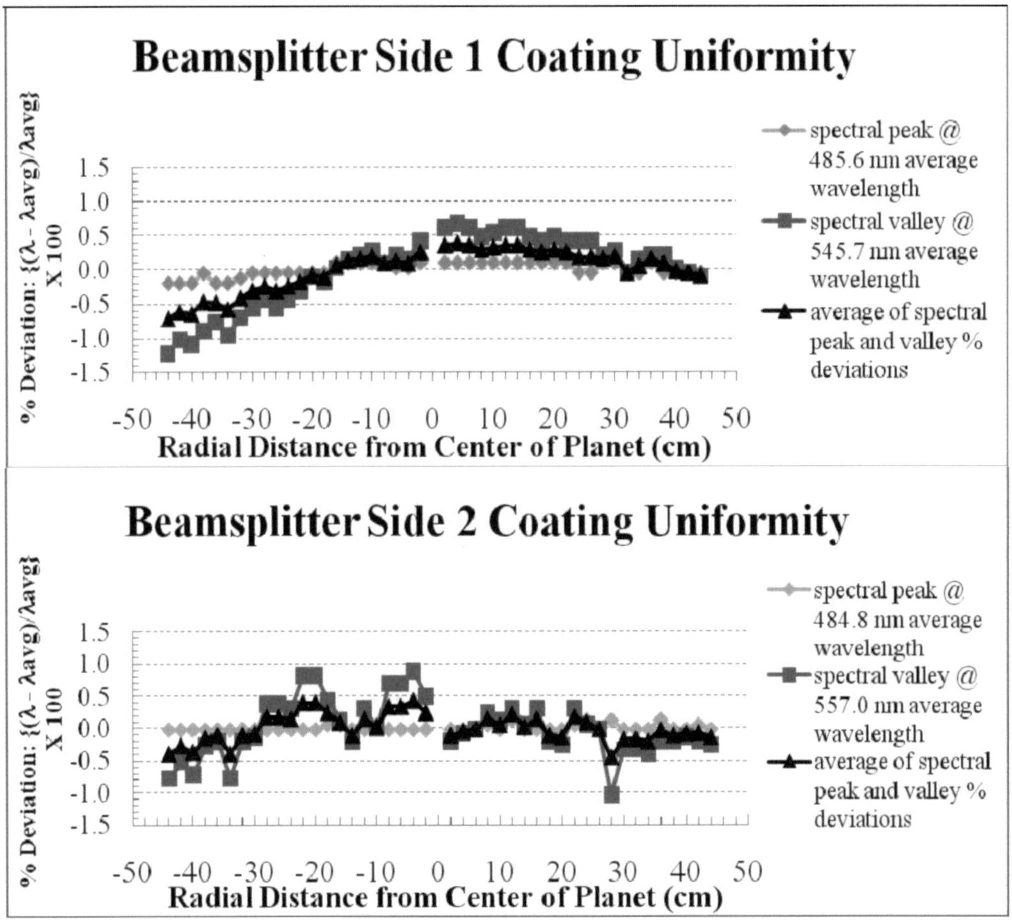

Fig. 15. Measured uniformity for the diagnostic beamsplitter Side 1 (top figure) and Side 2 (bottom figure) AR coatings. See text for details.

the Side 1 and Side 2 coatings. This high level of uniformity, which is typical for our coatings, is critical to insuring that the transverse phase and relative intensity properties of the 527 nm sample beam reflected from Side 1 closely match those of the 527 nm TW laser beam incident on the diagnostic beamsplitter. Only with such accurate matching of phase and relative intensity between the 527 nm sample and TW beams will the diagnostics of the sample beam reliably describe those of its TW counterpart. The Spol and Ppol LIDTs of these Side 1 and Side 2 diagnostic beamsplitter AR coatings measured in the NIF-MEL protocol at their 22.5^{o} use AOI are > 10 J/cm^2 at 532 nm and > 38 J/cm^2 at 1064 nm, as we reported previously (Bellum et al., 2011), and are thus adequate to protect against laser damage in the Z-Backlighter laser beam trains.

9. Conclusion

This chapter is an in-depth overview of the production of high LIDT optical coatings for PW class laser pulses. Lasers that generate such ultra-high intensity pulses use various approaches involving large energies per pulse and/or extremely short pulse durations, including the use of CPA techniques which have revolutionized ultra-high intensity laser technology. The successful operation of these lasers depends on optical coatings of the highest possible LIDTs to insure that the ultra-high intensity laser pulses, regardless of their pulse duration/energy combination, are able to propagate along the laser beam train without causing damage or aberrations. Our focus is on producing these high LIDT optical coatings on the large, meter-class optics required by the important category of ultra-high intensity lasers that use large cross section beam trains to accommodate large energies per pulse. Such large scale lasers were the earliest sources of PW class pulses and continue as important sources of PW pulses not only in the ns regime but also in the sub-ps regime by means of CPA. Sandia's Z-Backlighter TW and PW lasers, with their large cross section beam trains supporting ns pulses at 527 nm and 1054 nm and sub-ps, CPA pulses at 1054 nm, and its Large Optics Coating Operation together provide an excellent context for our overview of high LIDT coatings.

The LIDT of an optical coating depends not only on the resistance of the coating materials to laser damage but also on the design of the coating, on the techniques of keeping the optic surface free of particulates or contamination and of preparing it for coating, and on the coating process itself. Even a single particulate on an optic surface prior to coating can initiate laser damage and undermine an otherwise high LIDT of the coated surface. For this reason, a coating operation for producing high LIDT coatings must use a Class 100 or cleaner environment with excellent downward laminar flow of the clean air. In this regard, integrating the coating chamber into the Class 100 environment, with appropriate clean room curtain partitions, is also crucial. Of related importance is to transfer an optic into the coating chamber in a way that prevents the surface to be coated from exposure to particulates or contamination from the coating chamber or tooling. Proper coating process control is also important to obtaining coatings with high LIDTs. This includes deposition of hafnia by means of e-beam evaporation of hafnium metal in an oxygen back pressure, and use of IAD and temperature control of the coating chamber/substrate to tailor the molecular dynamics of coating formation as a means of fine tuning the coating's stress and density. Planetary motion of the substrates undergoing coating is necessary for obtaining good uniformity of coatings over large substrate surfaces. Coating large dimension optics poses unique challenges related to coating material depletion and the risk of system and process failures associated with producing uniform coatings in large coating chambers, and we summarize these large optics coating production challenges.

Regarding polishing, washing and cleaning of an optic prior to coating it, we point out that residual amounts of the polishing compound embedded in the microstructure of the polished surface can compromise the LIDT of the coated surface. As a result, the wash process must remove not only surface contamination but also polishing compound embedded in the microstructure of the polished optical surface. Using a wash protocol that includes an alumina slurry wash step in addition to mild detergent wash and clear water rinse steps does partially remove residual polishing compound from optic surfaces, and leads to improved LIDTs of AR coatings on those surfaces.

Useful LIDT tests are essential to the development and fielding of high LIDT optical coatings. Important here is to take into account the differences between the LIDT test laser conditions and the use laser conditions. This means that results of LIDT tests require careful interpretation in determining how they relate and apply to the design and performance of a given PW class laser. Our evaluation of the NIF-MEL LIDT tests and our in-house LIDT tests, and of how they relate to the Z-Backlighter TW and PW laser conditions, illustrates this. A comprehensive summary of the results of these LIDT tests on Sandia AR and HR coatings, for ns class pulses at 532 nm and 1064 nm and sub-ps class pulses at 1054 nm, shows that the LIDTs are high and adequate to insure that the coatings can stand up to the laser fluence levels of the PW class pulses in the Z-Backlighter beam trains.

Electric field behavior due to interference of forward and backward propagating components of light in a coating can be very different for different coating designs that meet the same reflectivity specifications, and not all field behaviors favor high LIDTs. Our first case study clearly illustrates this with the PW FOA steering mirror coating according to a 68 layer design and a 50 layer design. Both designs meet the mirror's extremely challenging reflection specifications (R > 99.6 % for 527 nm and 1054 nm for Spol and Ppol at AOIs from 24° to 47°) but the LIDTs at 527 nm are ~ 5 times larger for the 50 layer coating than for the 68 layer coating. This correlates with the moderate electric field intensity peaks at 527 nm that quench rapidly into the coating for the 50 layer design in contrast to the stronger 527 nm electric field intensity peaks, at ~ 200 % of the incident intensity and deep within the coating, for the 68 layer design. Some electric field behaviors afford higher LIDTs than others, and it is possible to design a coating that not only meets reflectivity requirements but that also is characterized by electric field intensities that enhance the LIDT of the coating.

Our second case study, of the Side 1 and Side 2 AR coatings of the diagnostic beamsplitter for the Z-Backlighter pulses at 527 nm, highlights reflectivity performance and uniformity which, though always important for large optics coatings, are particularly critical for diagnostic beamsplitter coatings since the validity of the beam diagnostics depends on them. Because partial reflection of the 527 nm laser beam by the beamsplitter produces the low intensity sample beam that undergoes the beam diagnostic tests, this partial reflection process must accurately preserve the transverse phase and relative intensity of the 527 nm laser beam over its entire cross section in order for it to be reliably described by the diagnostics of the sample beam. The Side 1 and Side 2 beamsplitter AR coatings of this case study do exhibit excellent uniformity and their designs match subtle reflectivity requirements, insuring beam diagnostics based on appropriate partial reflection with integrity of transverse phase and relative intensity. The coatings also account for secondary pulses at 1054 nm co-propagating with the primary pulses at 527 nm, a dual beam situation not uncommon for PW class lasers as a by-product of frequency doubling to produce the primary laser beam.

This chapter has covered key aspects of producing high LIDT optical coatings for PW class laser pulses. We hope it is of practical value in helping researchers in the field of ultra-high intensity lasers to navigate the design and production issues and considerations for high LIDT coatings.

10. Acknowledgement

Sandia National Laboratories is a multi-program laboratory managed and operated by Sandia Corporation, a wholly owned subsidiary of Lockheed Martin Corporation, for the U.S. Department of Energy's National Nuclear Security Administration under contract DE-AC04-94AL85000.

11. References

American National Standards Institute (2006). ANSI/OEOSC OP1.002-2006, Optics and Electro-Optical Instruments - Optical Elements and Assemblies - Appearance Imperfections. Available from ANSI eStandards Store: http://www.webstore.ansi.org/default.aspx.

American National Standards Institute (2008). ISO 10110-7:2008(E), Optics and photonics – Preparation of drawings for optical elements and systems – Part 7: Surface imperfection tolerances. Available from ANSI eStandards Store: http://www.webstore.ansi.org/default.aspx.

Bellum, J., Kletecka, D., Rambo, P., Smith, I., Kimmel, M., Schwarz, J., Geissel, M., Copeland, G., Atherton, B., Smith, D., Smith, C. & Khripin, C. (2009). Meeting thin film design and production challenges for laser damage resistant optical coatings at the Sandia Large Optics Coating Operation. *Proc. of SPIE,* Vol.7504, 75040C, ISBN 9780819478825, Boulder, Colorado, USA, September 2009.

Bellum, J., Kletecka, D., Kimmel, M., Rambo, P., Smith, I., Schwarz, J., Atherton, B., Hobbs, Z. & Smith, D. (2010). Laser damage by ns and sub-ps pulses on hafnia/silica anti-reflection coatings on fused silica double-sided polished using zirconia or ceria and washed with or without an alumina wash step. *Proc. of SPIE,* Vol.7842, 784208, ISBN 9780819483652, Boulder, Colorado, USA, September 2010.

Bellum, J., Kletecka, D., Rambo, P., Smith, I., Schwarz, J. & Atherton, B. (2011). Comparisons between laser damage and optical electric field behaviors for hafnia/silica antireflection coatings. *Appl. Opt.,* Vol.50, 9, March 2011, pp. C340-C348, ISSN 0003-6935.

Born, M. & Wolf, E. (1980). *Principals of Optics* (6th Edition), Pergamon Press Inc., ISBN 0-08-026481-6, New York.

Do, B. T. & Smith, A. V. (2009). Deterministic single shot and multiple shots bulk damage thresholds for doped and undoped, crystalline and ceramic YAG. *Proc. of SPIE.* Vol.7504, 75041O, ISBN 9780819478825, Boulder, Colorado, USA, September 2009.

Fournet, C., Pinot, B., Geenen, B., Ollivier, F., Alexandre, W., Floch, H. G., Roussel, A., Cordillot, C. & Billon, D. (1992). High damage threshold mirrors and polarizers in the ZrO_2/SiO_2 and HfO_2/SiO_2 dielectric systems. *Proc. of SPIE,* Vol.1624, pp. 282-293, ISBN 9780819407665, Boulder, Colorado, USA, September 1991.

Kimmel, M., Rambo, P., Broyles, R., Geissel, M., Schwarz, J., Bellum, J. & Atherton, B. (2009). Optical damage testing at the Z-Backlighter Facility at Sandia National Laboratories. *Proc. of SPIE,* Vol.7504, 75041G, ISBN 9780819478825, Boulder, Colorado, USA, September 2009.

Kimmel, M., Rambo, P., Schwarz, J., Bellum, J. & Atherton, B. (2010). Dual wavelength laser damage testing for high energy lasers. *Proc. of SPIE,* Vol.7842, 78421O, ISBN 9780819483652, Boulder, Colorado, USA, September 2010.

Maine, P., Strickland, D., Bado, P., Pessot, M. & Mourou, G. (1988). Generation of Ultrahigh Peak Power Pulses by Chirped Pulse Amplification. *IEEE J. Quantum Electron.,* Vol.24, 2, February 1988, pp. 398-403, ISSN 0018 9197.

Menapace, J. A. (2010). Private communication with J. A. Menapace, Lawrence Livermore National Laboratory.

Mourou, G. A. & Umstadter, D. (2002). Extreme Light. *Scientific American*, Vol.286, 5, May 2002, pp. 81-86, ISSN 0036-8733.

Mourou, G. & Tajima, T. (2011). More Intense, Shorter Pulses. *Science*, Vol.331, 6013, January 2011, pp. 41-42, ISSN 0036-8075 (print), ISSN 1095-9203 (online).

National Ignition Facility (2005). Small Optics Laser Damage Test Procedure. NIF Tech. Rep. MEL01-013-0D, Lawrence Livermore National Laboratory, Livermore, California.

Perry, M. D. & Mourou, G. (1994). Terawatt to Petawatt Subpicosecond Lasers. *Science*, Vol.264, 5161, May 1994, pp. 917-924, ISSN 0036-8075 (print), ISSN 1095-9203 (online).

Rambo, P. K., Smith, I. C., Porter Jr., J. L., Hurst, M. J., Speas, C. S., Adams, R. G., Garcia, A. J., Dawson, E., Thurston, B. D., Wakefield, C., Kellogg, J. W., Slattery, M. J., Ives III, H. C., Broyles, R. S., Caird, J. A., Erlandson, A. C., Murray, J. E., Behrendt, W. C., Neilsen, N. D. & Narduzzi, J. M. (2005). Z-Beamlet: a multikilojoule, terawatt-class laser system. *Appl. Opt.*, Vol.44, 12, April 2005, pp. 2421-2430, ISSN 0003-6935.

Schwarz, J., Rambo, P., Geissel, M., Edens, A., Smith, I., Brambrink, E., Kimmel, M. & Atherton, B. (2008). Activation of the Z-Petawatt laser at Sandia National Laboratories. *Journal of Physics: Conference Series*, Vol.112, 032020, ISSN 1742-6596, Kobe, Japan, September 2007.

Sinars, D. B., Cuneo, M. E., Bennett, G. R., Wenger, D. F., Ruggles, L. E., Vargas, M. F., Porter, J. L., Adams, R. G., Johnson, D. W., Keller, K. L., Rambo, P. K., Rovang, D. C., Seamen, H., Simpson, W. W., Smith, I. C. & Speas, S. C. (2003). Monochromatic x-ray backlighting of wire-array z-pinch plasmas using spherically bent quartz crystals. *Rev. Sci. Instr.*, Vol.74, 3, March 2003, pp. 2202-2205, ISSN 0034-6748.

Smith, D. J., McCullough, M., Smith, C., Mikami, T. & Jitsuno, T. (2008). Low stress ion-assisted coatings on fused silica substrates for large aperture laser pulse compression gratings. *Proc. of SPIE*, Vol.7132, 71320E, ISBN 9780819473660, Boulder, Colorado, USA, September 2008.

Stolz, C. J. & Genin, F. Y. (2003). Laser Resistant Coatings, In: *Optical Interference Coatings*, Kaiser, N. & Pulker, H. K. (Eds.), pp. 309-333, Springer-Verlag, ISBN 3-540-00364-9, Berlin/Heidelberg.

Stolz, C. J., Thomas, M. D. & Griffin, A. J. (2008). BDS thin film damage competition. *Proc. of SPIE*, Vol. 7132, 71320C, ISBN 9780819473660, Boulder, Colorado, USA, September 2008.

Strauss, G. N. (2003). Mechanical Stress in Optical Coatings, In: *Optical Interference Coatings*, Kaiser, N. & Pulker, H. K. (Eds.), pp. 207-229, Springer-Verlag, ISBN 3-540-00364-9, Berlin/Heidelberg.

Strickland, D. & Mourou, G. (1985). Compression of amplified chirped optical pulses. *Opt. Commun.*, Vol.56, 3, December 1985, pp.219-221, ISSN 0030-4018.

Tajima, T., Mourou, G. A. & Habs, D. (2010). Highest intensities, shortest pulses: Towards new physics with the Extreme Light Infrastructure. *Optik & Photonik*, Vol.5, 4, December 2010, pp. 24-29, ISSN 2191-1975 (online).

Wood, R. M. (Ed.). (1990). *Selected Papers on Laser Damage in Optical Materials*, SPIE Press Milestone Series Vol.MS24, ISBN 9780819405432, Bellingham, Washington.

Wood, R. M. (2003). *Laser-Induced Damage of Optical Materials*, Institute of Physics Publishing, ISBN 0 7503 0845 1, Bristol & Philadelphia.

3

Effect of Pulse Laser Duration and Shape on PLD Thin Films Morphology and Structure

Carmen Ristoscu and Ion N. Mihailescu
National Institute for Lasers, Plasma and Radiation Physics,
Lasers Department, Magurele, Ilfov
Romania

1. Introduction

Lasers are unique energy sources characterized by spectral purity, spatial and temporal coherence, which ensure the highest incident intensity on the surface of any kind of sample. Each of these characteristics stays at the origin of different applications. The study of high-intensity laser sources interaction with solid materials was started at the beginning of laser era, i.e. more than 50 years ago. This interaction was called during time as: vaporization, pulverization, desorption, etching or laser ablation (Cheung 1994). Ablation was used for the first time in connection with lasers for induction of material expulsion by infrared (IR) lasers. The primary interaction between IR photons and material takes place by transitions between vibration levels.

The plasma generated and supported under the action of high-intensity laser radiation was for long considered as a loss channel only and therefore, a strong hampering in the development of efficient laser processing of materials. In time, it was shown that the plasma controls not only the complex interaction phenomena between the laser radiation and various media, but can be used for improving laser radiation coupling and ultimately the efficient processing of materials (Mihailescu and Hermann, 2010).

The plasma generated under the action of fs laser pulses was investigated by optical emission spectroscopy (OES) and time-of-flight mass spectrometry (TOF-MS) (Ristoscu et al., 2003; Qian et al., 1999; Pronko et al., 2003; Claeyssens et al., 2002; Grojo et al., 2005; Amoruso et al., 2005a).

Lasers with ultrashort pulses have found in last years applications in precise machining, laser induced spectroscopy or biological characterization (Dausinger et al., 2004), but also for synthesis and/or transfer of a large class of materials: diamond-like carbon (DLC) (Qian et al., 1999; Banks et al., 1999; Garrelie et al., 2003), oxides (Okoshi et al., 2000; Perriere et al., 2002; Millon et al., 2002), nitrides (Zhang et al., 2000; Luculescu et al., 2002; Geretovszky et al., 2003; Ristoscu et al., 2004), carbides (Ghica et al., 2006), metals (Klini et al., 2008) or quasicrystals (Teghil et al., 2003). Femtosecond laser pulses stimulate the apparition of non-equilibrium states in the irradiated material, which lead to very fast changes and development of metastable phases. This way, the material to be ablated reaches the critical point which control the generation of nanoparticles (Eliezer et al., 2004; Amoruso et al., 2005b; Barcikowski et al., 2007; Amoruso et al., 2007).

Pulse shaping introduces the method that makes possible the production of tunable arbitrary shaped pulses. This technique has already been applied in femtochemistry (Judson and Rabitz, 1992), to the study of plasma plumes (Singha et al., 2008; Guillermin et al., 2009), controlling of two-photon photoemission (Golan et al., 2009), or coherent control experiments in the UV where many organic molecules have strong absorption bands (Parker et al., 2009). Double laser pulses were shown to be promising in laser-induced breakdown spectroscopy (Piñon et al., 2008), since they allow for the increase of both ion production and ion energy. The spatial pulse shaping is required to control the composition of the plume and to achieve the fully atomized gas phase by a single subpicosecond laser pulse (Gamaly et al., 2007).

Temporally shaping of ultrashort laser pulses by Fourier synthesis of the spectral components is an effective technique to control numerous physical and chemical processes (Assion et al., 1998), like: the control of ionization processes (Papastathopoulos et al., 2005), the improvement of high harmonic soft X-Rays emission efficiency (Bartels et al., 2000), materials processing (Stoian et al., 2003; Jegenyes et al., 2006; Ristoscu et al., 2006) and spectroscopic applications (Assion et al., 2003; Gunaratne et al., 2006).

The adaptive pulse shaping has been applied for ion ejection efficiency (Colombier et al, 2006; Dachraoui and Husinsky, 2006), generation of nanoparticles with tailored size (Hergenroder et al., 2006), applications in spectroscopy and pulse characterization (Ackermann et al., 2006; Lozovoy et al., 2008).

In materials science, pulsed laser action results in various applications such as localized melting, laser annealing, surface cleaning by desorption and ablation, surface hardening by rapid quench, and after 1988, pulsed laser deposition (PLD) technologies for synthesizing high quality nanostructured thin films (Miller 1994; Belouet 1996; Chrisey and Hubler, 1994; Von Allmen and Blatter, 1995). The laser - target interaction is a very complex physical phenomenon. Theoretical descriptions are multidisciplinary and involve equilibrium and non-equilibrium processes.

There are several consistent attempts in the literature for describing the interaction of ultrashort laser pulses with materials, especially metallic ones (Kaganov et al., 1957; Zhigilei and Garrison, 2000). Conversely, there are only a few that deal with the interaction of ultrashort pulses with wide band gap (dielectric, insulator and/or transparent) materials. Itina and Shcheblanov (Itina and Shcheblanov, 2010) recently proposed a model based on simplified rate equations instead of the Boltzmann equation to predict excitation by ultrashort laser pulses of conduction electrons in wide band gap materials, the next evolution of the surface reflectivity and the deposition rate. The analysis was extended from single to double and multipulse irradiation. They predicted that under optimum conditions the laser absorption can become smoother so that both excessive photothermal and photomechanical effects accompanying ultrashort laser interactions can be attenuated. On the other hand, temporally asymmetric pulses were shown to significantly affect the ionization process (Englert et al., 2007; Englert et al., 2008).

Implementation of PLD by using ps or sub-ps laser has been predicted to be more precise and expected to lead to a better morphology, in comparison to experiments performed with nanosecond laser pulses (Chichkov et al., 1996; Pronko et al., 1995). Clean ablation of solid targets is achieved without the evidence of the molten phase, due to the insignificant thermal conduction inside the irradiated material during the sub-ps and fs laser pulse action. Accordingly, ablation with sub-ps laser pulses was expected to produce much smoother film surfaces than those obtained by ns laser pulses (Miller and Haglund, 1998). It was shown that many parameters have to be monitored in order to get thin films with the desired quality. They are, but not limited to: the laser intensity distribution, scanning speed of the laser focal spot across the target surface, energy of the pre-pulse (in case of Ti-sapphire lasers) or post-pulse (for excimer lasers), pressure and nature of the gas in the reaction chamber, and so on.

In this chapter we review results on the effect of pulse duration upon the characteristics of nanostructures synthesized by PLD with ns, sub-ps and fs laser pulses. The materials morphology and structure can be gradually modified when applying the shaping of the ultra-short fs laser pulses into two pulses succeeding to each other under the same temporal envelope as the initial laser pulse, or temporally shaped pulse trains with picosecond separation (mono-pulses of different duration or a sequence of two pulses of different intensities).

2. Role of laser pulse duration in deposition of AlN thin films

Aluminum nitride (AlN), a wide band gap semiconductor (Eg= 6.2 eV), is of interest for key applications in crucial technological sectors, from acoustic wave devices on Si, optical coatings for spacecraft components, electroluminescent devices in the wavelength range from 215 nm to the blue end of the optical spectrum, as well as heat sinks in electronic packaging applications, where films with suitable surface finishing (roughness) are requested. The effect of laser wavelength, pulse duration, and ambient gas pressure on the composition and morphology of the AlN films prepared by PLD was investigated (Ristoscu et al., 2004). We worked with three laser sources generating pulses of 34 ns@248 nm (source A), 450 fs@248 nm (source B), and 50 fs@800 nm (source C). We have demonstrated that the duration of the laser pulse is an important parameter for the quality and performances of AlN structures.

Using PLD technique (Fig. 1), AlN thin films well oriented (Gyorgy et al., 2001) and having good piezoelectric properties can be obtained. The laser beam was focused onto the surface of a high purity (99.99%) AlN target, at an incidence angle of about 45° with respect to the target surface. The laser fluence incident onto the target surface was set at 0.1, 0.2 and 0.4 J/cm^2. For deposition of one film, we applied the laser pulses for 15 or 20 minutes.

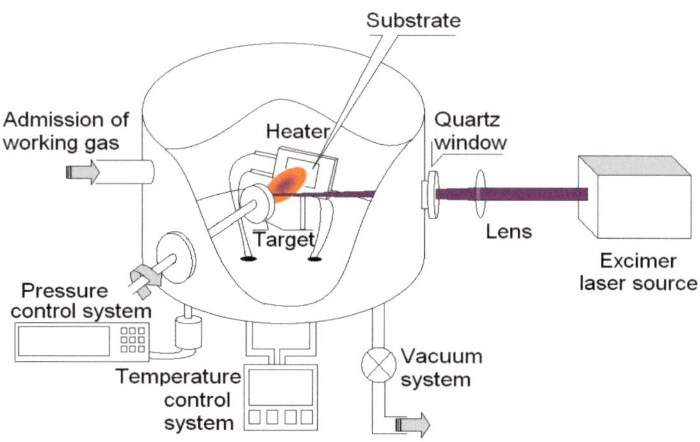

Fig. 1. PLD general setup used in the experiments reviewed in this chapter

Before each deposition the irradiation chamber was evacuated down to a residual pressure of ~ 10^{-6} Pa. The depositions have been conducted in vacuum ($5x10^{-4}$ Pa) or in very low dynamic nitrogen pressure at values in the range $(1-5)x10^{-1}$ Pa. During PLD deposition the substrates were heated up to 750 °C. The target-substrate separation distance was 4 cm. AlN thin films were deposited on various substrates: oxidized silicon wafers and oxidized silicon wafers covered with a platinum film, glass plates, suitable for various characterization techniques.
In the following, we will present detailed results for the PLD films deposited with source C. The synthesized structures were rather thin, having a thickness of 90-100 nm. The film deposited with the highest laser fluence (0.4 J/cm^2) has a thickness of about 400 nm.

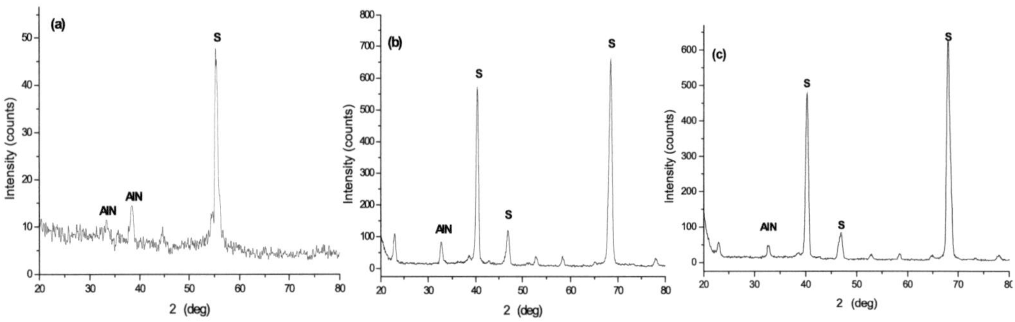

Fig. 2. XRD patterns of the films deposited from AlN target in vacuum ($5x10^{-4}$ Pa) (a), 0.1 Pa N_2 (b), and 0.5 Pa N_2 (c), respectively (Cu Kα radiation); S stands for substrate

Typical XRD patterns recorded for PLD AlN films are given in Figs. 2a-c. For the films obtained in vacuum (Fig. 2a), 0.1 Pa N_2 (Fig. 2b) as well as 0.5 Pa N_2 (Fig. 2c), a low intensity peak is present in the XRD patterns. This peak placed at 33° is assigned to AlN <100> hexagonal phase. The low intensity is due to the fact that the films are rather thin. Along with this peak, some other lines assigned to the substrate are present. Anyhow, the peaks attributed to AlN are quite large. This is indicative in our opinion for a mixture of crystalline and amorphous phases in the deposited films. This mixture was formed as an effect of the depositions temperature, 750° C. Previous depositions in which we evidenced only crystalline AlN were performed at 900° C (Gyorgy et al., 2001).
SEM investigations of the films (Figs. 3a-c) showed that the number of the particulates observed on the surface decreases with the increase of the ambient gas pressure, but their dimensions increase. The particulates present on films surface have spherical shape, with diameters in the range (100-800) nm.

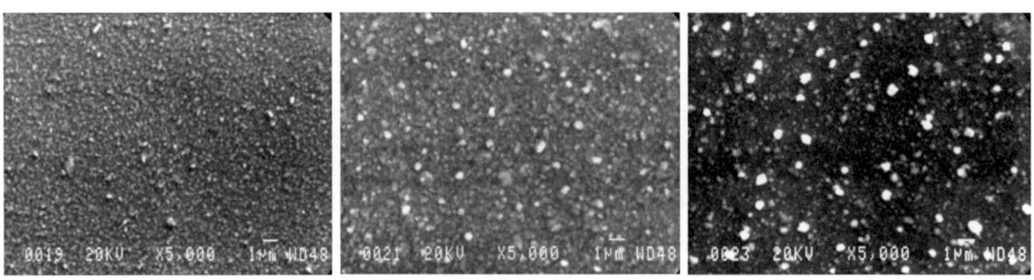

(a) (b) (c)

Fig. 3. SEM images of the films deposited from AlN target in vacuum ($5x10^{-4}$ Pa) (a), 0.1 Pa N_2 (b), and 0.5 Pa N_2 (c), respectively

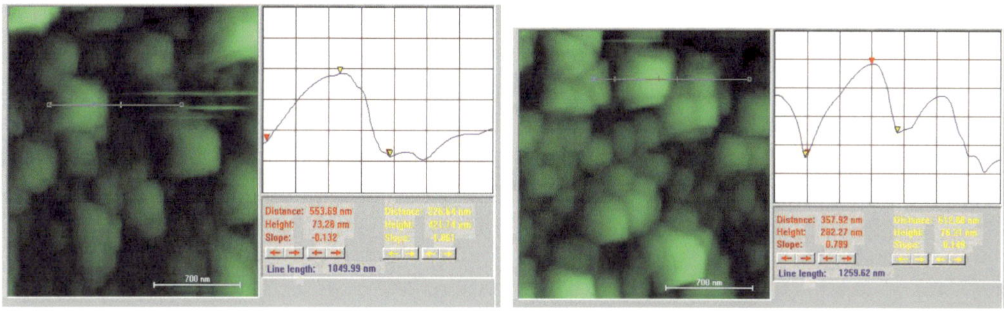

(a) (b)

Fig. 4. AFM pictures of AlN thin films obtained from AlN target in 0.1 Pa N_2 (a), and 0.5 Pa N_2 (b)

From AFM images (Figs. 4a,b), we observed that the size of grains reaches hundreds of nanometers, increasing from sample a) to sample b), in good agreement with thickness measurements and SEM investigations.

In Table 1 we summarized the characteristics of AlN thin films obtained with the three laser sources, along with the deposition rate.

Pressure	Laser wavelength	Frequency repetition rate	Pulse duration	Incident laser fluence	Phase content	Observations
Vacuum ($5x10^{-5}$ Pa)	248 nm	10 Hz	34 ns (A)	4 J / cm^2	Al(111)c, Al(200)c, Al(220)c, AlN(002)h	Microcrystallites in dendrite arrangements, 0.7 Å/pulse
	248 nm	10 Hz	450 fs (B)	4 J / cm^2	AlN(100)h	Droplets with diameters of 100 nm - 1μm, 0.05 Å/pulse
	800 nm	1 kHz	50 fs (C)	0.4J / cm^2	AlN(100)h	Droplets of less 1 μm diameter, 0.0033 Å/pulse
0.5 Pa N_2	248 nm	10 Hz	34 ns (A)	4 J / cm^2	AlN(100)h, AlN(002)h	1D low amplitude undulation 0.7 Å/pulse
	248 nm	10 Hz	450 fs (B)	4 J / cm^2	AlN(100)h	Droplets of less 1 μm diameters, 0.01 Å/pulse
	800 nm	1 kHz	50 fs (C)	0.4 J / cm^2	AlN(100)h	Lower droplets density than in vacuum, 0.0033 Å/pulse

Table 1. Main characteristics of AlN deposited films

We observed that only AlN was detected in the films obtained with laser sources B and C, while films obtained with source A contain a significant amount of metallic Al. The increase of N_2 pressure causes crystalline status perturbation for films deposited with sources B and C, but compensates N_2 loss when working with source A. The lowest density of particulates was observed for films obtained with source A. It dramatically increases (4-5 orders of magnitude) for sources B and C. The deposition rate exponentially decreases from sources A to C. These behaviors well corroborate with target examination. The crater on the surface of the target submitted to source A gets metallised in time, while the other two craters preserve the ceramic aspect. OES and TOF-MS investigations are in agreement with the studies of films, showing plasma richer in Al ions for source A (Ristoscu et al., 2003). Our studies evidenced the prevalent presence of AlN positive ions in the plasma generated under the action of sources B and C.
We deposited stoichiometric and even textured AlN thin films by PLD from AlN targets using a Ti-sapphire laser system generating pulses of 50 fs@800 nm (source C).

3. Temporal shaping of ultrashort laser pulses

Ref. (Stoian et al., 2002) demonstrated a significant improvement in the quality of ultrafast laser microstructuring of dielectrics when using temporally shaped pulse trains. Dielectric samples were irradiated with pulses from an 800 nm/1 kHz Ti:sapphire laser system delivering 90 fs pulses at 1.5 mJ. They used single sequences of identical, double and triple pulses of different separation times (0.3–1 ps) and equal fluences (Fig. 5). The use of shaped pulses enlarges the processing window allowing the application of higher fluences and number of sequences per site while keeping fracturing at a reduced level. For brittle materials with strong electron-phonon coupling, the heating control represents an advantage. The sequential energy delivery induced a material softening during the initial steps of excitation, changing the energy coupling for the subsequent steps. This leaded to cleaner structures with lower stress. Temporally shaped femtosecond laser pulses would thus allow exploitation of the dynamic processes and control thermal effects to improve structuring.

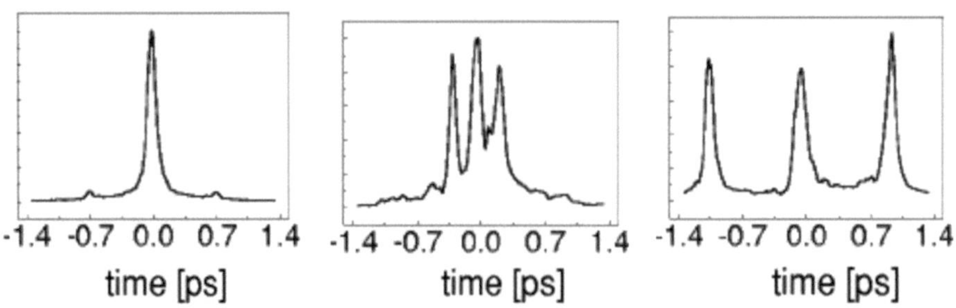

Fig. 5. Single pulses and triple-pulse sequences with different separation times (0.3–1 ps) and equal fluences (Stoian et al., 2002)

Ref. (Guillermin et al., 2009) reports on the possibility of tailoring the plasma plume by adaptive temporal shaping. The outcome has potential interest for thin films elaboration or nanoparticles synthesis. A Ti:saphirre laser beam (centered at 800 nm) with 150 fs pulse duration was used in their experiments. The pulses from the femtosecond oscillator are spectrally dispersed in a zero-dispersion unit and the spatially-separated frequency components pass through a pixellated liquid crystal array acting as a Spatial Light Modulator (SLM). The device allows relative retardation of spectral components, tailoring in turn the temporal shape of the pulse. They applied an adaptive optimization loop to lock up temporal shapes fulfilling user-designed constraints on plasma optical emission. The pulses with a temporal form expanding on several ps improved the ionic vs. neutral emission and allowed an enhancement of the global emission of the plasma plume.

Temporally shaped femtosecond laser pulses have been used for controlling the size and the morphology of micron-sized metallic structures obtained by using the Laser Induced Forward Transfer (LIFT) technique. Ref. (Klini et al., 2008) presents the effect of pulse shaping on the size and morphology of the deposited structures of Au, Zn, Cr. The double pulses of variable intensities with separation time Δt (from 0 to 10 ps) were generated by using a liquid crystal SLM (Fig. 6).

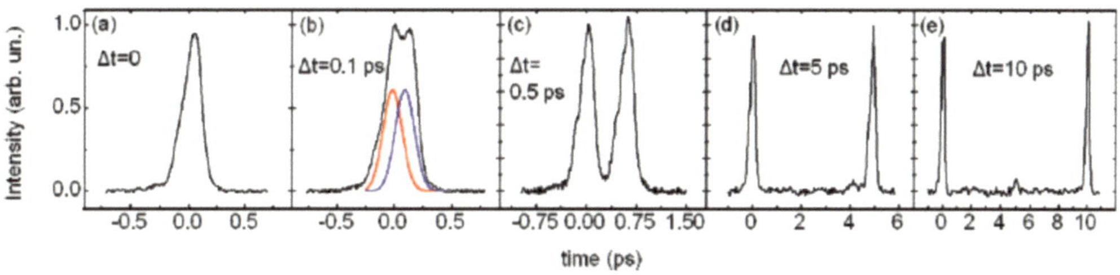

Fig. 6. Temporal pulse profiles generated with the method described in the text. Red and blue profiles in (b) are a guide to the eye to represent the underlying double pulses (Klini et al., 2008)

The laser source used for the pump-probe experiments was a Ti:Sapphire oscillator delivering 100 fs long pulses at 800 nm and with a 80 MHz repetition rate.

The temporal shape of the excitation pulse and the time scales of the ultrafast early stage processes occurring in the material can influence the morphology and the size of the LIFT dots. For Cr and Zn the electron-phonon coupling is relatively strong, and the morphology of the transferred films is determined by the electron-phonon scattering rate, i.e. very fast and within the pulse duration for Cr, and in the few picoseconds time scale for Zn. For Au the electron-phonon coupling is weak but the fast ballistic transport of electrons is very efficient. The numerous collisions of electrons with the film's surfaces determine the morphology. The internal electron thermalization rate which controls the electron-lattice coupling strength may determine the films' sizes.

The observed differences in size and morphology are correlated with the conclusion of pump-probe experiments for the study of electron-phonon scattering dynamics and subsequent energy transfer processes to the bulk. (Klini et al., 2008) proposed that in metals with weak electron-lattice coupling, the electron ballistic motion and the resulting fast electron scattering at the film surface, as well as the internal electron thermalization process are crucial to the morphology and size of the transferred material. Therefore, temporal shaping within the corresponding time scales of these processes may be used for tailoring the features of the metallic structures obtained by LIFT.

We mention here other approaches to obtain shaped pulses. Refs. (Hu et al., 2007) and (Singha et al., 2008) used an amplified Ti:sapphire laser (Spectra Physics Tsunami oscillator and Spitfire amplifier), which delivers 800 nm, 45 fs pulses with a maximum pulse energy of 2 mJ at a 1 kHz repetition rate and a Michelson interferometer to generate double pulses with a controllable delay of up to 110 ps. An autocorrelation measurement showed that the pulse is stretched by the subsequent optics to 80 fs. Ref. (Golan et al., 2009) introduced the output from the frequency doubled mode-locked Ti-sapphire laser (60 fs pulses at 430 nm, having energy of about 0.4 nJ per pulse) into a programmable pulse shaper composed of a pair of diffraction gratings and a pair of cylindrical lenses. A pair of one-dimensional programmable liquid-crystal SLM arrays is placed at the Fourier plane of the shaper. These arrays are used as a dynamic filter for spectral phase manipulation of the pulses. Using a pair of SLM arrays provides an additional degree of freedom and therefore allows some control over the polarization of the pulse. Ref. (Parker et al., 2009) uses a reflective mode, folded, pulse shaping assembly employing SLM shapes femtosecond pulses in the visible region of the spectrum. The shaped visible light pulses are frequency doubled to generate phase- and amplitude-shaped, ultra-short light pulses in the deep ultraviolet.

4. Temporally shaped vs. unshaped ultrashort laser pulses applied in PLD of SiC

Semiconductor electronic devices and circuits based on silicon carbide (SiC) were developed for the use in high-temperature, high-power, and/or high-radiation conditions under which devices made from conventional semiconductors cannot adequately perform. The ability of SiC-based devices to function under such extreme conditions is expected to enable significant improvements in a variety of applications and systems. These include greatly improved high-voltage switching for saving energy in electric power distribution and electric motor drives, more powerful microwave electronic circuits for radar and communications, sensors and controllers for cleaner burning, more fuel-efficient jet aircraft and automobile engines (http://www.nasatech.com/Briefs/Feb04/LEW17186.html).
The excellent physical and electrical properties of silicon carbide, such as wide band gap (between 2.2 and 3.3 eV), high thermal conductivity (three times larger than that of Si), high breakdown electric field, high saturated electron drift velocity and resistance to chemical attack, defines it as a promising material for high-temperature, high-power and high-frequency electronic devices (Muller et al., 1994; Brown et al., 1996), as well as for opto-electronic applications (Palmour et al., 1993; Sheng et al., 1997).
In Ref. (Ristoscu et al., 2006) it was tested eventual effects of interactions of the time shaping of the ultra-short fs laser pulses into two pulses succeeding to each other under the same temporal envelope as the initial laser pulse. This proposal was different from that used in Ref. (Gamaly et al., 2004) in case of spatial pulse shaping. The spatial Gaussian shape of the laser pulses was preserved. As known (Gyorgy et al., 2004) and demonstrated in the section 2 of this chapter, high intensity fs laser ablation deposition produces mainly amorphous structures with a prevalent content of nanoparticulates. This seems to be the consequence of coupling features of "normal" fs laser pulses to solid targets. We tried to test the effect of detaching from the "main" pulse a first signal with intensity in excess of plasma ignition threshold (Fig. 7).
The ablation is then initiated by the first pre-pulse and the expulsed material is further heated under the action of the second, longer and more energetic pulse. One expects that by proper choice of temporal delay, the second pulse intercepts and overheats the particulates generated by the pre-pulse causing their gradual boiling and elimination. Ultimately, the

deposition of a film becomes possible with a lower particulates density (till complete elimination) and with a highly improved crystalline status.

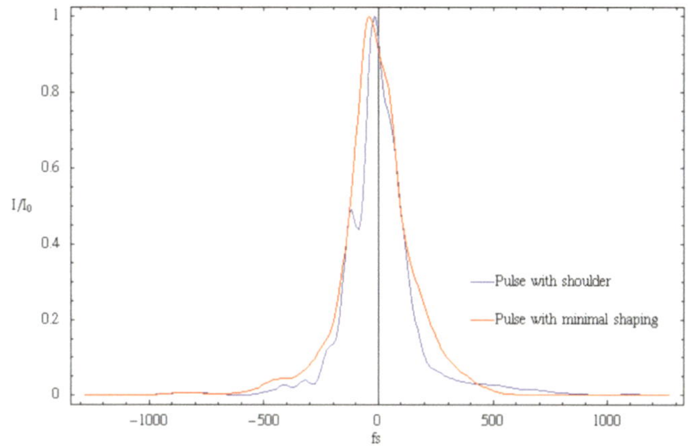

Fig. 7. Comparison between pulse shaped (blue) and plain amplifier (red)

The deposition experiments were conducted by PLD from bulk SiC target in vacuum (10^{-4} Pa), at temperatures around 750° C. The laser system was a Spectra Physics Tsunami with a BM Industries amplifier system giving 200 fs pulse duration with 600 mJ at 1 kHz and 800 nm wavelength, Fastlite Dazzler AOM system with controller and software driver running under LabView. The DAZZLER system is an acousto-optic programmable dispersive filter. It enables the separate control of both the spectral amplitude and the spectral phase. The crystal is an active optic component which, through the acousto-optic interaction, allows the spectral phase and amplitude shaping of an optical pulse. The general layout can be seen in Ref. (Verluise et al., 2000). We selected a generation regime where a pulse with a typical shape such as that shown in Fig. 7. The pulses were temporally characterized using the standard frequency-resolved optical gating (FROG) technique (Trebino and Kane, 1993; Trebino et al., 1997; DeLong et al., 1994). It enables both the phase and the amplitude of the pulse be retrieved simultaneously. More precisely, we applied the second harmonic generation (SHG) version of this technique using a thin BBO crystal as the NLO medium.

After optimization, we have chosen the following laser parameters: laser beam focused in spots of 0.07 mm^2, corresponding to a laser fluence on the target surface of 714 mJ/cm^2. For the deposition of one film we applied trains of subsequent laser pulses with a total duration of 15 min.

The SiC films obtained with unmodulated laser pulses are not fully crystallized, consisting in a nanostructured matrix incorporating well defined crystalline grains with elongated shapes (Ghica et al., 2006). A high density of {111} planar defects has been observed inside the crystalline grains, most probably formed by the dissociation of screw dislocations into partials on the {111} slip planes (Fig. 8). The dissociation of the screw dislocations and the motion of the partial dislocations on the slip planes may be triggered by the stress between adjacent growing grains or exerted by the highly energetic nanometric particles (droplets) resulting from the interaction between the target and the extremely short laser pulses.

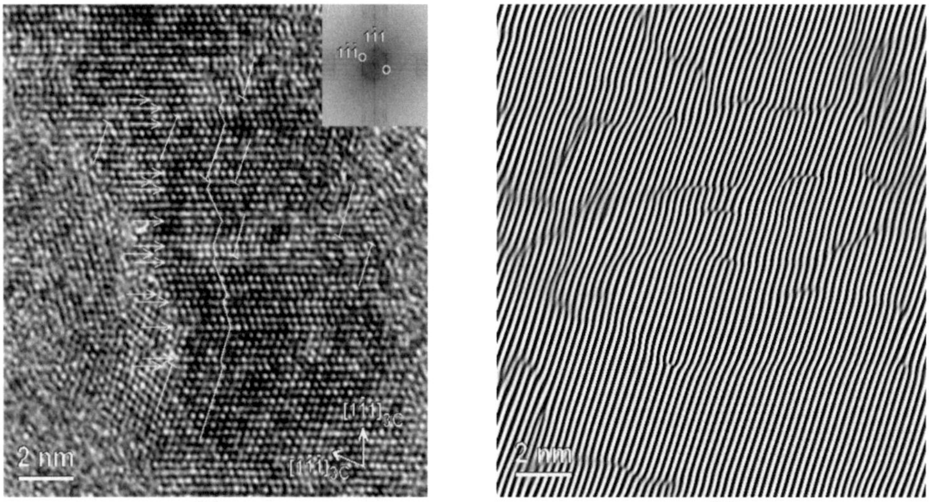

(a) (b)

Fig. 8. (a) HRTEM image along the (110)3C-SiC zone axis showing the bottom part of a SiC column; the trace of the (1-1-1) planes (the zig-zag line) and the position of the planar defects (arrows) are indicated; the Fourier transform (FFT) of the image is inserted in the upper right corner; (b) Bragg filtered image obtained by inverse FFT using the 1-1-1 and -111 pair of spots (encircled on the FFT image); the image contrast has been intentionally exaggerated in order to improve the visibility of the dislocations (Ghica et al., 2006)

In XRD patterns of the films deposited with tailored pulses, only the lines of Si (100) originating from the substrate and of β-SiC phase were visible. The formation of β-SiC was further supported by electron microscopy studies. Two important differences are to be emphasized with respect to samples deposited with unmodulated laser pulses:

i. The film surface is rather smooth, the roughness being dramatically reduced;
ii. The film is rather compact, showing no cracks, unlike the SiC films synthesized with unmodulated pulses, where a high density of fissures could be observed (Ghica et al., 2006). The cracks occurring are linked to the presence of droplets on the film surface and, further, to their high energy at the impact with the substrate. The lack of droplets or their low density leads to the growth of a compact film, free of cracks. This is precisely the case of thin structures deposited by tailored laser pulses.

For a comparison between the surface morphologies of the two types of films we give in Figs. 9a and c two characteristic SEM images. The fine structure of the surface of films obtained with unmodulated and tailored pulses is presented in the two SEM images recorded at higher magnification (Figs. 9b and d). The film synthesized with unmodulated laser pulses shows a high density of particulates (about 62 μm^{-2}), reaching up to 400 nm in size (Fig. 10a). Comparatively, a striking reduction of the droplets density can be observed for the film obtained with time tailored laser pulses, down to 8.6 μm^{-2} (Fig. 10b) The largest particulates also reach 400 nm.

We consider that this noticeable decrease of density along with the conservation of particulates dimension (in both size and distribution) is the effect of the particular pulse coupling mechanism which becomes effective in case of tailored laser pulses. The particulates generated by the first peak efficiently absorb the light in the second one, are vaporized and partially eliminated.

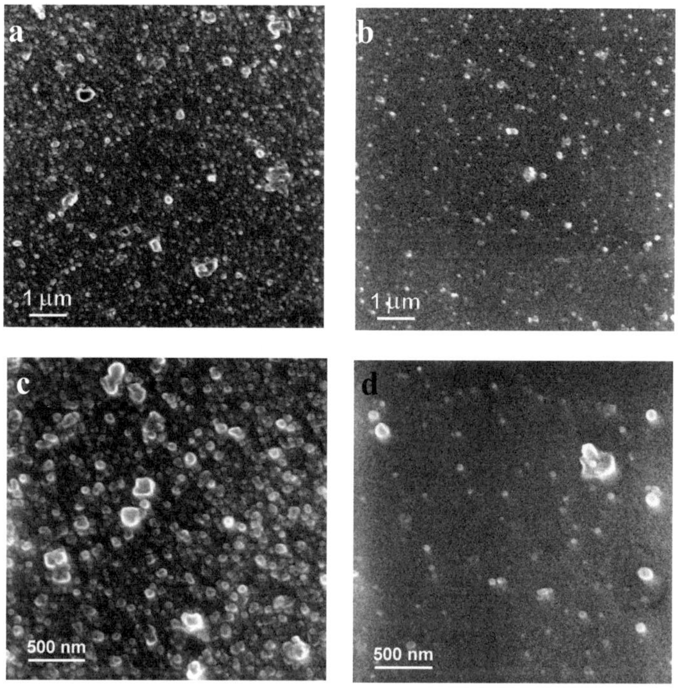

Fig. 9. SEM images showing the surface morphology of the samples obtained with unmodulated (a, c) and tailored (b, d) laser pulses (Ghica et al., 2006)

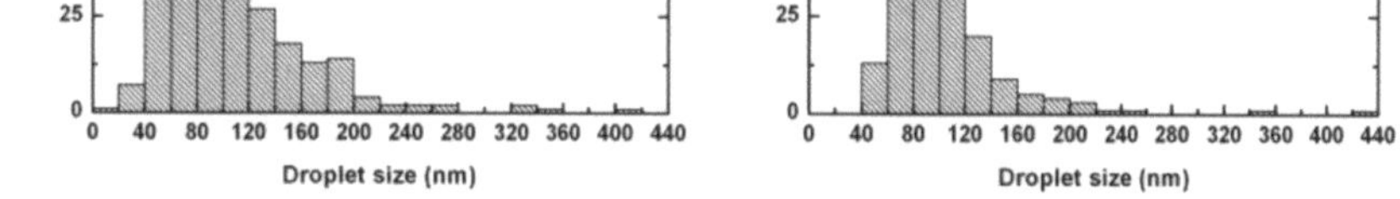

Fig. 10. Histograms of the SiC samples obtained with unmodulated (a) and tailored (b) laser pulses

5. Temporally shaped ultrashort pulse trains applied in PLD of AlN

Amplified Ti:Sapphire laser pulses at 800 nm, 1 kHz repetition rate, with durations of 200 fs were used. The repetition rate was scaled down electronically to 1 Hz. Prior to amplification, a programmable liquid crystal SLM was inserted into the Fourier plane of a 4f zero-dispersion configuration (Weiner 2000), allowing temporal pulse shaping of the incoming beam to two pulses with the temporal separation determined by phase modulation. The phase mask for the generation of the pulse shapes was determined numerically using an

iterated Fourier transform method (Schmidt et al., n.d.). These generated shapes are then amplified thus compensating for spatio-temporal and energetic fluctuations that are inherent in this system (Wefers and Nelson, 1995; Tanabe et al., 2005). We selected a generation regime where the pulse has the typical shape shown in Fig. 11. The pulses were temporally characterized a standard frequency-resolved optical gating technique (Trebino 2002). This algorithm facilitates the simultaneous retrieve of both the phase and amplitude of the pulse. We applied the second harmonic generation (SHG) version of this technique using a thin BBO single crystal as the NLO medium where the ambiguity in the temporal symmetry of the retrieved pulses was resolved separately using an etalon.
We have chosen the following laser parameters: a laser beam spot of 0.08 mm^2 and an incident laser energy on the target surface of 400 μJ. For the deposition of one film we applied trains of laser pulses with a total duration of 20 min. The deposition of AlN thin films has been carried out in vacuum (10^{-4} Pa) at 800° C substrate temperature. Three types of samples have been deposited under identical conditions, with the exception of the shape of the laser pulse: AlN-1 with unshaped laser pulses, AlN-2 and AlN-3 using the shapes 2 and 3 respectively, given in Fig. 11.

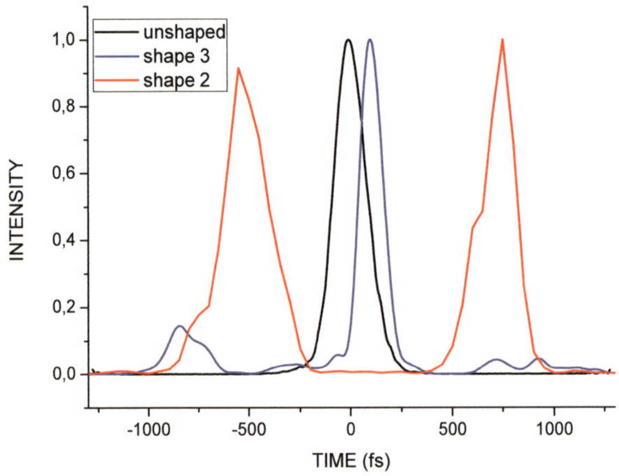

Fig. 11. Comparison between pulse shaped (blue, red) and plain amplifier (black)

For samples AlN-1 (Fig. 12a), we could identify three classes of surface particulates: particulates smaller than 100 nm, medium sized particulates up to 1 μm and large particulates, up to 2 μm. The large particulates were rather rare. The typical surfaces of the AlN-2 film (Fig. 12b) also showed large crystallites ranging up to 1.5 μm, with a rather high density. In the case of the AlN-3 samples (Fig. 12c), the particulates could be grouped in three classes on their average size: particulates around 100 nm, particulates around 500 nm and particulates larger than 1.5 μm up to 2.5 μm. The large particulates showed well defined facets.
The measured average particulates density was quite similar in the three cases, specifically $(5\pm0.8)\times10^{8}$ cm^{-2} for the AlN-1 samples, $(4.8\pm0.7)\times10^{8}$ cm^{-2} for the AlN-2 and $(5.6\pm0.8)\times10^{8}$ cm^{-2} for the AlN-3 samples, with about 15% counting error in each case. Histograms of the microparticle size distribution in the case of the 3 types of samples were presented next to the corresponding SEM image. The particulates average size resulting from the histogram analysis was 390±5 nm in case of samples AlN-1, 230±3 nm for AlN-2 and 310±4 nm in case of samples AlN-3.

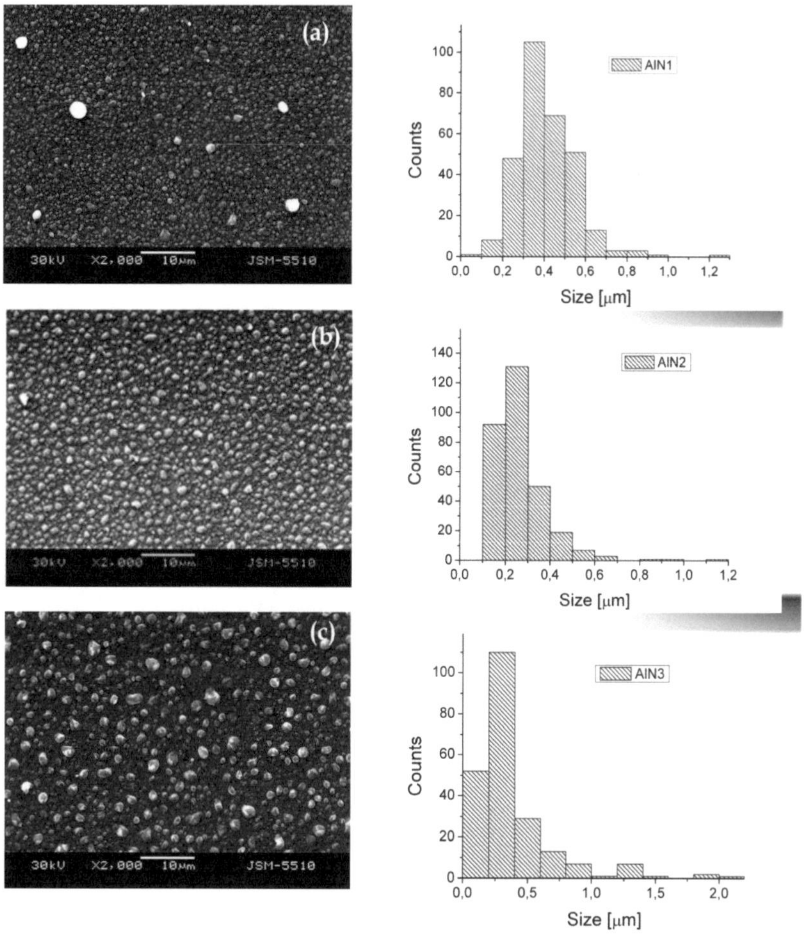

Fig. 12. SEM images showing the surface morphology of samples AlN-1 (a), AlN-2 (b) and AlN-3 (c) along with their histograms

From Fig. 11 we observed that the two pulses composing shapes 2 and 3 are separated at 1.25 ps and less than 1 ps, respectively. According to Ref. (Itina and Shcheblanov, 2010), this means that in both cases, the second pulse interacts with the plasma produced by the first one. Moreover, the 2 pulses were more or less equal as duration and intensity for shape 2, while for shape 3 the intensity of the second pulse is largely surpassing the one of the first pulse. As an effect, the second pulse is more efficient for samples AlN-2 in breaking the particulates generated under the action of the first one. This is visible in Figs. 12b and c which show larger and slightly more numerous particulates for samples AlN-3. As known, wide band-gap materials normally present a rather limited density of conduction electrons. Under intense (ultrashort) laser irradiation, the generation of a high number of conduction electrons is initiated, strongly influencing the electron interactions, and eventually determining the structural modifications of the material.

TEM investigations showed that the films are a mixture of crystalline and amorphous components. It is demonstrated that in dielectrics electron thermalization requires hundreds of fs (Bulgakova et al., 2010). The free-electron gas transfers energy to the lattice by coupling to the vibration bath, which results in heating and triggering of a whole range of phase transformation processes in the material, including melting, ablation via the different mechanisms such as phase explosion, fragmentation, and upon cooling solidification with formation of amorphous and/or polycrystalline phases.

The AlN films deposited under the action of temporally shaped or unshaped fs laser pulses consisted of a mixture of crystalline phases characterized by the prevalent presence of hexagonal AlN and existence of metallic Al traces. SEM and TEM investigations showed that when using shaped pulses the number of large crystal grains in the films was increasing. On the other hand, the average grains size decreased by about a half as an effect of shaping.

6. Temporally shaped ultrashort pulse trains applied in PLD of SiC, ZnO and Al

A reduction of number of particulates accompanied by an increase of their size was observed for SiC when applying mono-pulses of different duration or passing to a sequence of two pulses of different intensities (see Fig. 11). The SiC structures present a smoother surface as compared with the other films (Figs. 13a-c). The average dimension of particulates is 150 nm for samples obtained with unshaped pulses. When using shape 2 laser pulses, the average dimension of the particulates present on surface of SiC films is ~ 100 nm, with a higher density than the structure obtained with unshaped pulses. Large particulates of ~1 μm can be observed on the surface of the films obtained with the shape 3 pulse, with the lowest density. For this material, when using this shaped pulse (a small shoulder well separated from a higher one) we obtained better surfaces.

(a) (b) (c)

Fig. 13. SEM images showing the surface morphology of samples SiC when using unshaped (a), shape 2 (b) and shape 3 (c) laser pulses

Typical XRD pattern of the SiC films is presented in Fig. 14, wherefrom we can see only the line assigned to β-SiC phase.

On the other hand, in case of ZnO the best laser pulse which induces a dramatic decrease of particulate density was shape 2 (see Fig. 11). The deposition of ZnO thin films has been carried out in vacuum (10^{-4} Pa) at 350° C substrate temperature. We generally obtained smooth surfaces with particulates lower than 100 nm (Figs. 15 a-c). The sample obtained with unshaped pulses exhibit a reduced roughness with fine particles having dimensions around 100 nm. The shape 2 laser pulses favored the development of a surface with large porosity and particulates of ~ 100 nm diameter. The shape 3 pulse induced also a porosity of the deposited film but a decrease of the particles size to ~ 50 nm.

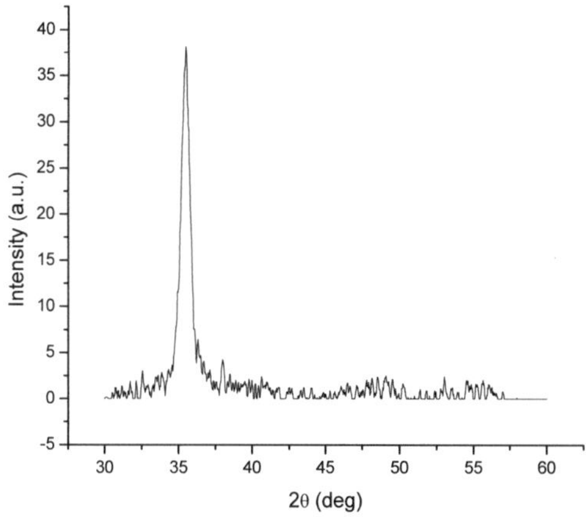

Fig. 14. Tipical XRD pattern of SiC film obtained with shape 3 laser pulses

(a) (b) (c)

Fig. 15. SEM images showing the surface morphology of samples ZnO when using unshaped (a), shape 2 (b) and shape 3 (c) laser pulses

For ZnO samples we acquired also transmission spectra. The transmission was higher than 85% for all deposited structures, irrespective the shape of the ultra-short laser pulses.

In case of metallic targets, the obtained Al films present identical morphologies, irrespective of the pulse shape. Their surfaces are rough with micro-particles having an average dimension of about 1 μm (Figs. 16 a,b). Moreover, the microparticulates are well connected to each other, suggesting that they arrive on substrate in liquid phase.

(a) (b)

Fig. 16. SEM images showing the surface morphology of samples Al when using unshaped (a) and shape 2 (b) laser pulses

7. Conclusion

We conclude that by optimization of the temporal shaping of the pulses besides the other laser parameters (wavelength, energy, beam homogeneity, fluence), one could choose an appropriate regime to eliminate excessive photo-thermal and photomechanical effects and obtain films with desired crystalline phase, number and dimension of grains/particulates, or controlled porosity.

8. Acknowledgment

Part of the experiments were carried out at the Ultraviolet Laser Facility operating at IESL-FORTH and supported by the EU through the Research Infrastructures activity of FP6 (Project: Laserlab-Europe; Contract No: RII3-CT-2003-506350). The authors are thankful to C. Ghica for the electron microscopy analyses. The financial support of the CNCSIS - UEFISCDI, project number PNII - IDEI 1289/2008 is acknowledged.

9. References

Ackermann, R., Salmon, E., Lascoux, N., Kasparian, J., Rohwetter, P., Stelmaszczyk, K., Li, S., Lindinger, A., Woste, L., Bejot, P., Bonacina, L., Wolf, J.-P. (2006) Optimal control of filamentation in air *Applied Physics Letters* volume 89, no. 17 (October 2006) pp. 171117_1-3, ISSN 1077-3118

Amoruso, S., Bruzzese, R., Vitiello, M., Nedialkov, N.N., Atanasov, P.A. (2005a) Experimental and theoretical investigations of femtosecond laser ablation of aluminum in vacuum *Journal of Applied Physics* volume 98, no. 4 (August 2005) pp 044907_1-7, ISSN 1089-7550

Amoruso, S., Ausanio, G., Bruzzese, R., Vitiello, M., Wang, X., (2005b) Femtosecond laser pulse irradiation of solid targets as a general route to nanoparticle formation in a vacuum *Physical Review B* volume 71, no. 3 (January 2005) pp 033406_1-4, ISSN 1550-235X

Amoruso, S., Bruzzese, R., Wang, X., Nedialkov, N.N., Atanasov, P.A. (2007) An analysis of the dependence on photon energy of the process of nanoparticle generation by femtosecond laser ablation in a vacuum *Nanotechnology* volume 18, no. 14 (April 2007) pp 145612, ISSN 1361-6528

Assion, A., Baumert, T., Bergt, M., Brixner, T., Kiefer, B., Seyfried, V., Strehle, M., Gerber, G., (1998) Control of chemical reactions by feedback-optimized phase-shaped femtosecond laser pulses *Science* volume 282, no. #5390 (October 1998) pp 919-922, ISSN 1095-9203

Assion, A., Wollenhaupt, M., Haag, L., Mayorov, F., Sarpe-Tudoran, C., Winter, M., Kutschera, U., Baumert, T. (2003) Femtosecond laser-induced-breakdown spectrometry for Ca^{2+} analysis of biological samples with high spatial resolution *Applied Physics B; Lasers and Optics* volume 77, no. 3 (2003) pp 391-397, ISSN 1432-0649

Banks, P.S., Dinh, L., Stuart, B.C., Feit, M.D., Komashko, A.M., Rubenchik, A.M., Perry, M.D., McLean, W. (1999) Short-pulse laser deposition of diamond-like carbon thin films *Applied Physics A: Materials Science & Processing* volume 69, supplement 1 (December 1999) pp S347–S353, ISSN 1432-0630

Barcikowski, S., Hahn, A., Kabashin, A.V., Chichkov, B.N. (2007) Properties of nanoparticles generated during femtosecond laser machining in air and water *Applied Physics A: Materials Science & Processing* volume 87, no. 1 (April 2007) pp 47-55, ISSN 1432-0630

Bartels, R., Backus, S., Zeek, E., Misoguti, L., Vdovin, G., Christov, I. P., Murnane, M. M., Kapteyn, H. C. (2000) Shaped-pulse optimization of coherent emission of high-harmonic soft X-rays *Nature* volume 406, no. 6792 (July 2000) pp 164-166, ISSN 0028-0836

Belouet, C. (1996) Thin film growth by the pulsed laser assisted deposition technique *Applied Surface Science* volume 96-98 (April 1996) pp 630-642, ISSN 0169-4332

Brown, D.M., Downey, E. Grezzo, M. Kretchmer, J. Krishnamethy, V. Hennessy, W. Michon, G. (1996) Silicon carbide MOSFET technology *Solid State Electronics* volume 39, no. 11 (November 1996) pp 1531- 1542, ISSN 0038-1101

Bulgakova, N.M., Stoian, R., Rosenfeld, A., Hertel, I.V. (2010) Continuum models of ultrashort pulsed laser ablation in *Laser-Surface Interactions for New Materials Production Tailoring Structure and Properties*, Miotello, Antonio; Ossi, Paolo M. (Eds.) pp 81-97, Springer Series in Materials Science, Vol. 130, ISBN 978-3-642-03306-3, Springer Heidelberg Dordrecht London New York

Cheung, J.T. (1994) History and Fundamentals of Pulsed Laser Deposition in *Pulsed Laser Deposition of Thin Films,* Chrisey, D. B., Hubler G. K. (Eds.) pp 1-22, John Wiley &Sons, Inc., ISBN 0471592188 9780471592181, New York

Chichkov, B., Momma, N. C., Nolte, S., von Alvensleben, F., Tunnermann, A. (1996) Femtosecond, picosecond and nanosecond laser ablation of solids *Applied Physics A: Materials Science & Processing* volume 63, no. 2 (July 1996) pp 109-115; ISSN 1432-0630

Chrisey, D. B., Hubler G. K. (Eds.) (1994) *Pulsed Laser Deposition of Thin Films* J. Wiley, ISBN 0471592188 9780471592181, New York

Claeyssens, F., Ashfold, M.N.R., Sofoulakis, E., Ristoscu, C.G., Anglos, D., Fotakis, C. (2002) Plume emissions accompanying 248 nm laser ablation of graphite in vacuum: Effects of pulse duration *Journal of Applied Physics* volume 91, no. 9 (May 2002) pp 6162-6172, ISSN 1089-7550

Colombier, J.P., Combis, P., Rosenfeld, A., Hertel, I.V., Audouard, E., Stoian, R. (2006) Optimized energy coupling at ultrafast laser-irradiated metal surfaces by tailoring intensity envelopes: Consequences for material removal from Al samples *Physical Review B* volume 74, no. 22 (December 2006) pp 224106_1-16, ISSN 1550-235X

Dachraoui, H., Husinsky, W. (2006) Thresholds of Plasma Formation in Silicon Identified by Optimizing the Ablation Laser Pulse Form *Physical Review Letters* volume 97, no. 10 (September 2006) pp 107601_1-4, ISSN 1079-7114

Dausinger, F., Lichtner, F., Lubatschowski, H. (Eds.) (2004) *Femtosecond Technology for Technical and Medical Applications,* Topics in Applied Physics, Vol. 96, Springer, ISBN 3-540-20114-9 Springer Berlin Heidelberg New York

DeLong, K.W.; Trebino, R.; Hunter, J.; White, W.E. (1994) Frequency-resolved optical gating with the use of second-harmonic generation *Journal of the Optical Society of America B* volume 11, no. 11 (November 1994) pp 2206-2215, ISSN 1520-8540

Eliezer, S., Eliaz, N., Grossman, E., Fisher, D., Gouzman, I., Henis, Z., Pecker, S., Horovitz, Y., Fraenkel, M., Maman, S., Lereah, Y. (2004) Synthesis of nanoparticles with femtosecond laser pulses *Physical Review B* volume 69, no. 14 (April 2004) pp 144119_1-6, ISSN 1550-235X

Englert, L., Rethfeld, B., Haag, L., Wollenhaupt, M., Sarpe-Tudoran, C., Baumert, T. (2007) Control of ionization processes in high band gap materials via tailored femtosecond pulses *Optics Express* volume 15, no. 26 (December 2007) pp 17855-17862, ISSN 1094-4087

Englert, L., Wollenhaupt, M., Haag, L., Sarpe-Turdoran, C., Rethfeld, B., Baumert, T. (2008) Material processing of dielectrics with temporally asymmetric shaped femtosecond laser pulses on the nanometer scale *Applied Physics A: Materials Science & Processing* volume 92, no. 4 (September 2008) pp 749-753, ISSN 1432-0630

Gamaly, E.G., Rode, A.V., Uteza, O., Kolev, A., Luther-Davies, B., Bauer, T., Koch, J., Korte, F., Chichkov, B.N. (2004) Control over a phase state of the laser plume ablated by femtosecond laser: Spatial pulse shaping *Journal of Applied Physics* volume 95, no. 5 (March 2004) pp 2250-2257, ISSN 1089-7550

Gamaly, E.G., Rode, A.V., Luther-Davies, B. (2007) Ultrafast laser ablation and film deposition in *Pulsed laser deposition of thin films: Applications-led growth of functional materials*, Edited by Robert Eason, pp 99-129, John Willey & Sons, Inc., ISBN-13: 978-0-471-44709-2, Hoboken, New Jersey

Garrelie, F., Loir, A.S., Donnet, C., Rogemond, F., Le Harzic, R., Belin, M., Audouard, E., Laporte, P. (2003) Femtosecond pulsed laser deposition of diamond-like carbon thin films for tribological applications *Surface and Coatings Technology* volume 163–164 (January 2003) pp 306-312, ISSN 0257-8972

Geretovszky, S., Kantor, Z., Szorenyi, T. (2003) Structure and composition of carbon-nitride films grown by sub-ps PLD *Applied Surface Science* volume 208–209 (March 2003) pp 547-552; ISSN 0169-4332

Ghica, C., Ristoscu, C., Socol, G., Brodoceanu, D., Nistor, L.C., Mihailescu, I.N., Klini, A., Fotakis, C. (2006) Growth and characterization of β-SiC films obtained by fs laser ablation *Applied Surface Science* volume 252, no. 13 (April 2006) pp 4672–4677, ISSN 0169-4332

Golan, B., Fradkin, Z., Kopnov, G., Oron, D., Naaman R. (2009) Controlling two-photon photoemission using polarization pulse shaping *The Journal of Chemical Physics* volume 130 no 6 (February 2009) pp 064705_1-6, ISSN 1089-7690

Grojo, D., Hermann, J., Perrone, A. (2005) Plasma analyses during femtosecond laser ablation of Ti, Zr, and Hf *Journal of Applied Physics* volume 97, no. 6 (March 2005) pp 063306_1-9, ISSN 1089-7550

Guillermin, M., Liebig, C., Garrelie, F., Stoian, R., Loir, A.-S., Audouard, E. (2009) Adaptive control of femtosecond laser ablation plasma emission *Applied Surface Science* volume 255, no. 10 (March 2009) pp 5163–5166, ISSN 0169-4332

Gunaratne, T., Kangas, M., Singh, S., Gross, A., Dantus, M. (2006) Influence of bandwidth and phase shaping on laser induced breakdown spectroscopy with ultrashort laser pulses *Chemical Physics Letters* Volume 423, no. 1-3 (May 2006) pp 197-201, ISSN 0009-2614

Gyorgy, E., Ristoscu, C., Mihailescu, I.N., Klini, A., Vainos, N., Fotakis, C., Ghica, C., Schmerber, G., Faerber, J. (2001) Role of laser pulse duration and gas pressure in deposition of AlN thin films *Journal of Applied Physics*, volume 90, no. 1 (July 2001) pp 456-461, ISSN 1089-7550

Gyorgy, E., Teodorescu, V.S., Mihailescu, I.N., Klini, A., Zorba, V., Manousaki, A., Fotakis, C. (2004) Surface Morphology Studies of Sub-Ps Pulsed-Laser-Deposited AlN Thin Film *Journal of Materials Research* volume 19, no. 3 (March 2004) pp 820-826, ISSN 2044-5326

Hergenroder, R., Miclea, M., Hommes, V., (2006) Controlling semiconductor nanoparticle size distributions with tailored ultrashort pulses *Nanotechnology* volume 17, no. 16 (August 2006) pp 4065-4071, ISSN 1361-6528

http://www.nasatech.com/Briefs/Feb04/LEW17186.html

Hu, Z., Singha, S., Liu, Y., Gordon, R.J. (2007) Mechanism for the ablation of Si(111) with pairs of ultrashort laser pulses *Applied Physics Letters* volume 90, no. 13 (March 2007) pp 131910_1-3, ISSN 1077-3118

Itina, T. E., Shcheblanov, N. (2010) Electronic excitation in femtosecond laser interactions with wide-band-gap materials *Applied Physics A: Materials Science & Processing* volume 98, no. 4 (March 2010) pp 769-775, ISSN 1432-0630

Jegenyes, N., Toth, Z., Hopp, B., Klebniczki, J., Bor, Z., Fotakis, C. (2006) Femtosecond pulsed laser deposition of diamond-like carbon film: The effect of double laser pulses *Applied Surface Science* volume 252, no. 13 (April 2006) pp 4667-4671, ISSN 0169-4332

Judson, R.S., Rabitz, H. (1992) Teaching lasers to control molecules *Physical Review Letters* volume 68, no. 10 (March 1992) pp 1500-1503, ISSN 1079-7114

Kaganov, M. I., Lifshitz, I. M., Tanatarov, L. V. (1957) Relaxation between electrons and the crystalline lattice *Soviet Physics - JETP* volume 4 (1957) pp 173-180, ISSN 0038-5646

Klini, A., Loukakos, P.A., Gray, D., Manousaki, A., Fotakis, C. (2008) Laser Induced Forward Transfer of metals by temporally shaped femtosecond laser pulses *Optics Express* Vol. 16, No. 15 (July 2008) pp 11300-11309, ISSN 1094-4087

Lozovoy, V.V., Xu, B., Coello, Y., Dantus, M. (2008) Theoretical development of a high-resolution differential-interference-contrast optic for x-ray microscopy *Optics Express* volume 16, no. 2 (January 2008) pp 592-597, ISSN 1094-4087

Luculescu, C.R., Miyake, H., Sato, S. (2002) Deposition of BN thin films onto Si(1 0 0) substrate by PLD with nanosecond and femtosecond pulses in nitrogen gas background *Applied Surface Science* volume 197-198 (September 2002) pp 499-504, ISSN 0169-4332

Mihailescu, I.N., Hermann, J. (2010) Laser Plasma Interactions *Laser Processing of Materials: Fundamentals, Applications, and Developments*, Ed. P. Schaaf, pp 51-90, Springer Series in Materials Science, ISBN 978-3-642-13280-3, Springer Heidelberg Dordrecht London New York

Miller, J.C. (Ed.) (1994) *Laser Ablation Principles and Applications*, Springer-Verlag, ISBN 3540575715 9783540575719 Berlin

Miller, J. C., Haglund, R.F. (Eds.) (1998) *Laser Ablation and Desorption*, Academic Press, ISBN 0124759750 9780124759756, San Diego

Millon, E., Albert, O., Loulergue, J. C., Etchepare, J. C., Hullin, D., Seiler, W., Perriere, J. (2000) Growth of heteroepitaxial ZnO thin films by femtosecond pulsed-laser deposition *Journal of Applied Physics* volume 88, no. 11 (December 2000) pp 6937-6939; ISSN 1089-7550

Muller, G., Krotz, G., Niemann, E. (1994) SiC for sensors and high-temperature electronics *Sensors and Actuators A: Physical* volume 43, no. 1-3 (May 1994) pp 259-268, ISSN 0924-4247

Okoshi, M., Higashikawa, K., Hanabusa, M. (2000) Pulsed laser deposition of ZnO thin films using a femtosecond laser *Applied Surface Science* volume 154–155 (February 2000) pp 424-427, ISSN 0169-4332

Palmour, J. W., Edmond, J.A., Kong, H. S., Carter Jr., C.H. (1993) 6H-silicon carbide devices and applications *Physica B: Condensed Matter* volume 185, no 1-4 (April 1993) pp 461-465, ISSN 0921-4526

Papastathopoulos, E., Strehle, M., Gerber, G. (2005) Optimal control of femtosecond multiphoton double ionization of atomic calcium *Chemical Physics Letters* volume 408, no. 1-3 (June 2005) pp 65-70, ISSN 0009-2614

Parker, D.S.N., Nunn, A.D.G., Minns, R.S., Fielding, H.H. (2009) Frequency doubling and Fourier domain shaping the output of a femtosecond optical parametric amplifier: easy access to tuneable femtosecond pulse shapes in the deep ultraviolet *Applied Physics B: Lasers and Optics* volume 94, no. 2 (February 2009) pp 181–186, ISSN 1432-0649

Perriere, J., Millon, E. Seiler, W. Boulmer-Leborgne, C. Craciun, V. Albert, O. Loulergue, J.C. Etchepare, J. (2002) Comparison between ZnO films grown by femtosecond and nanosecond laser ablation *Journal of Applied Physics* volume 91, no. 2 (January 2002) pp 690 - 696, ISSN 1089-7550

Piñon, V., Fotakis, C., Nicolas, G., Anglos, D. (2008) Double pulse laser-induced breakdown spectroscopy with femtosecond laser pulses *Spectrochimica Acta Part B: Atomic Spectroscopy* volume 63, no. 10 (October 2008) pp 1006-1010; ISSN 0584-8547

Pronko, P.P., Cutta, S.K., Squier, J., Rudd, J.V., Du, D., Mourou, G. (1995) Machining of sub-micron holes using a femtosecond laser at 800 nm *Optics Communications* volume 114, no. 1-2 (January 1995) pp 106–110, ISSN 0030-4018

Pronko, P.P., Zhang, Z., Van Rompay, P.A. (2003) Critical density effects in femtosecond ablation plasmas and consequences for high intensity pulsed laser deposition *Applied Surface Science* volume 208–209 (March 2003) pp 492-501, ISSN 0169-4332

Qian, F., Craciun, V., Singh, R.K., Dutta, S.D., Pronko, P.P. (1999) High intensity femtosecond laser deposition of diamond-like carbon thin films *Journal of Applied Physics* volume 86, no. 4 (August 1999) pp 2281-2290, ISSN 1089-7550

Ristoscu, C., Mihailescu, I.N., Velegrakis, M., Massaouti, M., Klini, A., Fotakis, C. (2003) Optical Emission Spectroscopy and Time-of-Flight investigations of plasmas generated from AlN targets in cases of Pulsed Laser Deposition with sub-ps and ns Ultra Violet laser pulses *Journal of Applied Physics* volume 93, no. 5 (March 2003) pp 2244-2250, ISSN 1089-7550

Ristoscu, C. Gyorgy, E. Mihailescu, I. N. Klini, A. Zorba, V. Fotakis, C. (2004) Effects of pulse laser duration and ambient nitrogen pressure in PLD of AlN *Applied Physics A: Materials Science & Processing* volume 79, no 4-6 (September 2004) 927-929 ISSN 1432-0630

Ristoscu, C., Socol, G., Ghica, C., Mihailescu, I.N., Gray, D., Klini, A., Manousaki, A., Anglos, D., Fotakis, C. (2006) Femtosecond pulse shaping for phase and morphology control in PLD: Synthesis of cubic SiC *Applied Surface Science* volume 252, no 13 (April 2006) pp 4857-4862, ISSN 0169-4332

Schmidt, B., Hacker, M., Stobrawa, G., Feurer, T. (n.d.) *LAB2-A virtual femtosecond laser lab* Available from http://www.lab2.de

Sheng, S., Spencer, M.G., Tang, X., Zhou, P., Wongchoitgul, W., Taylor, C., Harris, G. L. (1997) An investigation of 3C-SiC photoconductive power switching devices *Materials Science and Engineering B* volume 46, no. 1-3 (April 1997) pp 147–151, ISSN 0921-5107

Singha, S., Hu, Z., Gordon, R.J. (2008) Ablation and plasma emission produced by dual femtosecond laser pulses *Journal of Applied Physics* volume 104, no. 11 (December 2008) pp 113520_1-10, ISSN 1089-7550

Stoian, R., Boyle, M., Thoss, A., Rosenfeld, A., Korn, G., Hertel, I. V., Campbell, E. E. B. (2002) Laser ablation of dielectrics with temporally shaped femtosecond pulses *Applied Physics Letters* volume 80, no. 3 (January 2002) pp 353-355, ISSN 1077-3118

Stoian, R., Boyle, M., Thoss, A., Rosenfeld, A., Korn, G., Hertel, I. V. (2003) Dynamic temporal pulse shaping in advanced ultrafast laser material processing *Applied Physics A: Materials Science & Processing* volume 77, no. 2 (July 2003) pp 265-269, ISSN 1432-0630

Tanabe, T., Kannari, F., Korte, F., Koch, J., Chichkov, B. (2005) Influence of spatiotemporal coupling induced by an ultrashort laser pulse shaper on a focused beam profile *Applied Optics* volume 44, no. 6 (February 2005) pp 1092-1098, ISSN 2155-3165

Teghil, R., D'Alessio, L., Santagata, A., Zaccagnino, M., Ferro, D., Sordelet, D.J. (2003) Picosecond and femtosecond pulsed laser ablation and deposition of quasicrystals *Applied Surface Science* volume 210, no. 3-4 (April 2003) pp 307-317, ISSN 0169-4332

Trebino, R., Kane, D.J. (1993) Using phase retrieval to measure the intensity and phase of ultrashort pulses: frequency-resolved optical gating *Journal of the Optical Society of America A* volume 10, no. 5 (May 1993) pp 1101–1111, ISSN 1520-8532

Trebino, R., DeLong, K.W., Fittinghoff, D.N., Sweetser, J.N., Krumbügel, M.A., Richman, B.A., Kane, D.J. (1997) Measuring ultrashort laser pulses in the time-frequency domain using frequency-resolved optical gating *Review of Scientific Instruments* volume 68, no. 9 (September 1997) pp 3277-3295, ISSN 1089-7623

Trebino, R. (2002) *Frequency-Resolved Optical Gating: The Measurement of Ultrashort Laser Pulses* Kluwer Academic ISBN 1402070667 9781402070662, Boston

Verluise, F., Laude, V., Cheng, Z., Spielmann, C.H., Tournois, P. (2000) Amplitude and phase control of ultrashort pulses by use of an acousto-optic programmable dispersive filter: pulse compression and shaping *Optics Letters* volume 25, no. 8 (April 2000) pp 575-577, ISSN 1539-4794

Von Allmen, M., Blatter, A. (1995) *Laser-Beam Interactions with Materials* (2nd edition) Springer, ISBN 3540594019 9783540594017, Berlin; Heidelberg; New York; Barcelona; Budapest; Hong Kong; London; Milan; Paris; Tokyo

Wefers, M.M., Nelson, K.A. (1995) Analysis of programmable ultrashort waveform generation using liquid crystal spatial light modulators *Journal of the Optical Society of America B* volume 12, no. 7 (July 1995) pp 1343-1362, ISSN 1520-8540

Weiner, A. M. (2000) Femtosecond pulse shaping using spatial light modulators *Review of Scientific Instruments* volume 71, no. 5 (May 2000) pp 1929-1960, ISSN 1089-7623

Zhang, Z., VanRompay, P.A., Nees, J.A., Clarke, R., Pan, X., Pronko, P.P. (2000) Nitride film deposition by femtosecond and nanosecond laser ablation in low-pressure nitrogen discharge gas *Applied Surface Science* volume 154–155 (February 2000) pp 165–171, ISSN 0169-4332

Zhigilei, L. V., Garrison, B. J. (2000) Microscopic mechanisms of laser ablation of organic solids in the thermal and stress confinement irradiation regimes *Journal of Applied Physics* volume 88, no. 3 (2000) pp 1281-1298, ISSN 1089-7550

4

Laser Pulse Patterning on Phase Change Thin Films

Jingsong Wei[1] and Mufei Xiao[2]
[1]*Shanghai Institute of Optics and Fine Mechanics, Chinese Academy of Sciences*
[2]*Centro de Nanociencias y Nanotecnología, Universidad Nacional Autónoma de México*
[1]*China*
[2]*México*

1. Introduction

In the present chapter, we discuss the formation of microscopic patterns on phase change thin films with low power laser pulses. The discussions are mostly based on our recent experimental and theoretical results on the subject.

Phase change thin films are widely used as optical and electric data storage media. The recording is based on the phase change between the crystalline and amorphous states. In the writing process, a small volume in the thin film is locally and rapidly heated to above the melting point and successively quenched into the amorphous phase. In the erasing process, the material undergoes a relatively long heating to reach a temperature above the glass transition but yet below the melting point, which brings the material back to the crystalline phase.

However, during the writing process, apart from the phase changes, physical deformation of the surface occurs, which often creates bumps of various forms. In other words, low intensity laser pulses are able to microscopically form patterns on phase change films. The formed patterns modify the topographic landscape of the surface and bring about variations on the material properties of the films. The modifications can be harmful or helpful depending on what kind of applications one looks for. Therefore, in order to properly deal with the laser induced bumps, it is essential to understand the process of bump formation, and to qualitatively and quantitatively describe the created bumps as well as its relation with the laser pulse parameters, such as the beam distributions and the average intensity etc. so that one is able to closely control the formation of microscopic patterns on phase change films with low power laser pulses. Recently, we have systematically studied the formation of bumps during laser writing both experimentally and theoretically.

In the present chapter we shall round up the important results from our studies and present detailed discussions on the results. We organize the chapter as follows. In the first part, we present results of forming circular bumps as a by-production of rather conventional laser writing process for the purpose of data storage on $Ag_8In_{14}Sb_{55}Te_{23}$ chalcogenide phase change films. In this part, the detailed process of writing and erasing will be described, and the experimental and theoretical characterizations of the bumps are demonstrated. In the second part, we expand our work to intentionally form micro patterns on multilayer ZnS-SiO_2/AgO_x/ZnS-SiO_2 thin films by laser direct writing technology. We shall conclude the work in the end of the chapter.

2. Laser pulse induced bumps in chalcogenide phase change films

Chalcogenide phase change thin films are widely used as optical and electric data storage media. The recording is based on the phase change between the crystalline and amorphous states (Kolobov et al., 2004; Kalb et al., 2004; Welnic et al., 2006; Wuttig & Steimer, 2007). In the writing process, a small volume in the thin film is locally and rapidly heated to above the melting point and successively quenched into the amorphous phase. In the erasing process, the material undergoes a relatively long heating to reach a temperature above the glass transition but yet below the melting point, which brings the material back to the crystalline phase. The heat source for the phase change is usually from laser pulses in optical data storage, or electric current pulses in electric data storage. In the present work we shall selectively concentrate on the optical storage.

In the process of amorphization, i.e., the laser writing process, the material experiences a volume change due to the stronger thermal expansion in the melting state than in the crystalline state, as well as the density difference between the two states. Therefore, the amorphous recording marks are actually physically deformed as circular bumps because the amorphous recording marks inherit the volume in the melting state after a fast cooling stage. Subsequently, the bumps may cause further deformation in other thin layers stacked underneath as in the cases of optical information memory in optical storage and the electrode in electric storage. While slight deformation in the writing process is inevitable, significant bumps are harmful for the storage media as they affect dramatically the size of the marks, which eventually reduces the recording density of the media, and shorten the durability of the device. In extreme cases the bumps may grow so big that a hole is formed at the apex of the bump. Therefore, to quantitatively describe the bump formation is of great interest for storage applications.

We have established a theoretical model for the formation process, where the geometric characters of the formed bumps can be analytically and quantitatively evaluated from various parameters involved in the formation. Simulations based on the analytic solution are carried out taking $Ag_8In_{14}Sb_{55}Te_{23}$ as an example (Wei et al., 2008; Dun et al., 2010). The results are verified with experimental observations of the bumps.

2.1 Theory

Let us start by describing the amorphization process schematically in the volume-temperature diagram as shown in Fig. 1, where the principal paths for the phase changes are depicted. Initially, the chalcogenide thin film is considered in the crystalline state represented by point *a*; a laser or current pulse of nanosecond duration heats the material up to the melting state, which is represented by point *b*. Subsequently, the material is cooled quickly with a high rate exceeding $10^7\ C/s$ to the room temperature to form the final amorphous mark. During the quenching stage, the material structure does not have sufficient time to rearrange itself and remains in the equilibrium state, and thus inherits the structure and volume at the melting state. Therefore, the volume has an increase ΔV , and the mark appears as a bump. If the laser or current pulse injects energy higher than the ablated threshold corresponding to the vaporization temperature, the heating temperature reaches point *d*, and the material is then rapidly cooled to the room temperature, which is represented by point *e*; an ablated hole can be formed at the top of the bump.

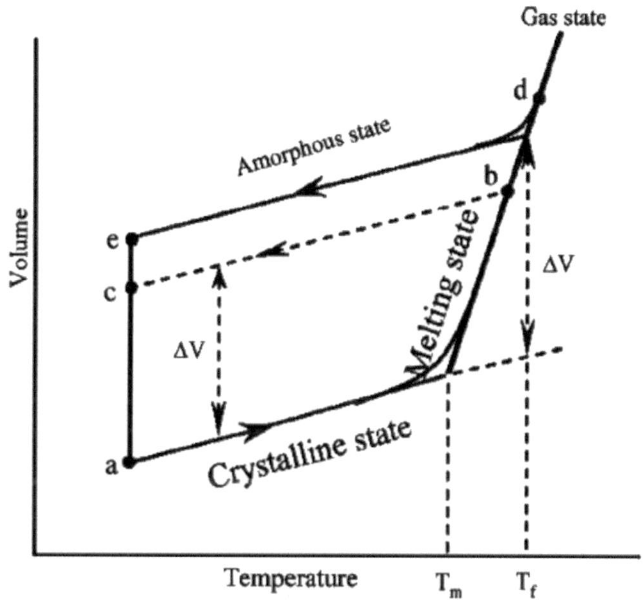

Fig. 1. Volume-temperature diagram of chalcogenide films. The film is heated by laser from point *a* to point *b* and returns to point *d*, or to point *c* and returns to point *e* after faster cooling.

The geometric characters of the bump are graphed in Fig. 2, where cross-sections of the circular bump are schematically shown respectively for the case of a bump and the case of a bump with a hole on its top. It is worth noting that, in general, the volume thermal expansion coefficient for chalcogenide thin films has two different constant values in the crystalline and melting states, respectively. In our analysis, there is assumed a Gaussian intensity profile for the incident laser pulse, and volume changes occur only in the region irradiated by the laser pulse, as shown in Fig. 2(a). If the laser pulse energy exceeds the ablated threshold, a hole is to be formed at the top of the bump, which is shown in Fig. 2(b). Mathematically, for the fast heating and amorphization process, the net volume increase can be written as $\Delta h = (\beta_m - \beta_c) \cdot V_0 \cdot (T_{surf} - T_m)$, where β_m and β_c are the volume thermal expansion coefficients in the crystalline and melting states, respectively. V_0 is the irradiated region volume. T_{surf} is the material surface temperature heated by laser pulse and T_m is the temperature corresponding to the melting point. Since the irradiated region is axially symmetric due to the Gaussian laser beam intensity profile, the bump height can be expressed as

$$\Delta h(r) = (\beta_m - \beta_c) \cdot h_0(r) \cdot (T_{surf} - T_m) \tag{1}$$

where r is the radial coordinate, and $h_0(r)$ is the height of the irradiated region.

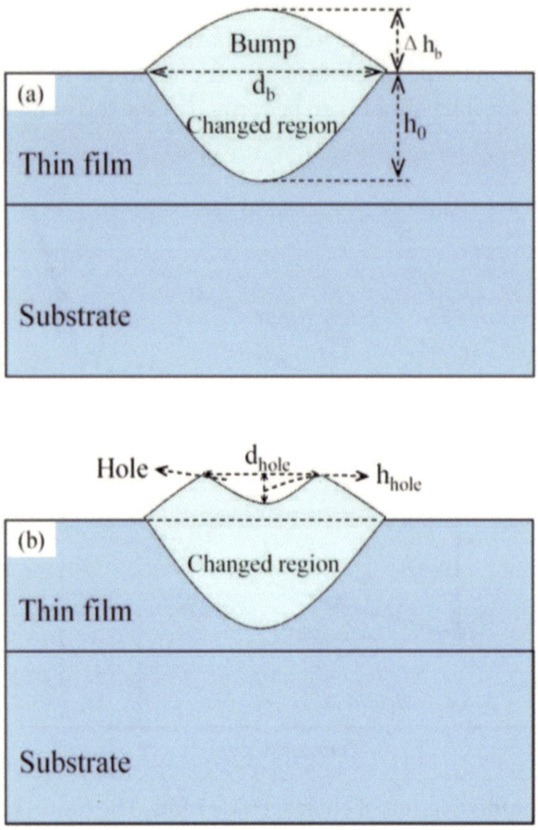

Fig. 2. Bump formation schematics: (a) bump and (b) hole on the top of bump.

Furthermore, the absorbed energy per unit volume and per unit time can be calculated by

$$g(r,z)=\alpha(1-R)\frac{2P}{\pi w^2}\exp(-\frac{2r^2}{w^2})\exp(-\alpha z) \tag{2}$$

where α is the absorption coefficient, R is the reflectivity of the material, P is the laser power, w is the laser beam radius at the $1/e^2$ of the peak intensity, and z is in the depth direction from the sample surface. In Eq. (2) the quantity $\alpha(1-R)$ is the absorbed part of the transmitted light, which decays exponentially $\exp(-\alpha z)$ along the z direction and spreads as a Gaussian function $\exp\left(-2r^2/w^2\right)$ in the r direction.

Generally for data storage, the width of the laser pulse is in the range from nanosecond to millisecond. Within this range, the temperature distribution in the irradiated region can be expressed as

$$T(r,z)=\frac{g(r,z)\tau}{\rho C_p} \tag{3}$$

where ρ is the density, C_p is the heat capacity of the material, and τ is the laser pulse width. According to (Shiu et al., 1999), the bump height $\Delta h(r)$ can be calculated, within the temperature interval $T_m < T(r,0) < T_f$, where T_f is the temperature corresponding to the vaporization point above which the material will be ablated, by

$$\Delta h(r) = \frac{\beta_m - \beta_c}{\alpha}\left[T(r,0) - T_m\right]\ln\left[\frac{T(r,0)}{T_m}\right] \tag{4}$$

and the bump diameter d_p can be calculated by setting $T(r,0) = T_m$ and $r = d_p / 2$ in Eq. (3) with

$$d_p = \sqrt{2}w\sqrt{\ln\left[F_0\frac{\alpha(1-R)}{\rho C_p}\right]\frac{1}{T_m}} \tag{5}$$

where $F_0 = 2P\tau / \pi w^2$. Similar to the derivation of bump diameter, if the laser pulse energy exceeds the ablated threshold, an ablated hole is formed when $T(r,0) > T_f$ and the hole diameter in the bump d_{hole} can be calculated as

$$d_{hole} = \sqrt{2}w\sqrt{\ln\left[F_0\frac{\alpha(1-R)}{\rho C_p}\right]\frac{1}{T_f}} \tag{6}$$

It should be noted that in our analytical model, the thermo-physical parameters of material are assumed independent from temperature.

2.2 Experimental observations

Before presenting results of simulation based on the above developed formalism, let us show some experimental observations of the bumps. The experimental results provided useful and meaningful values for choosing the parameters involved in the theoretical simulations. In the experiments, $Ag_8In_{14}Sb_{55}Te_{23}$ thin films were directly deposited on a glass substrate by dc-magnetron sputtering of an $Ag_8In_{14}Sb_{55}Te_{23}$target. The light source is a semiconductor laser of wavelength $\lambda = 650nm$, and the laser beam is modulated to yield a $50ns$ laser pulse. The laser beam is focused onto the $Ag_8In_{14}Sb_{55}Te_{23}$ thin film, and the light spot diameter is about $2\mu m$. In order to form bumps with different sizes, various laser power levels were adapted. Some of the experimental results are presented in Figs. 3–5.

Fig. 3(a) shows some bumps obtained with laser power $3.8mW$. The inset in Fig. 3(a) is an enlarged image of one bump. The bump diameter is about $0.9 - 1.0\mu m$. In order to further analyze the bump morphology, an atomic force microscope (AFM) was used to scale the bump. The results are shown in Fig. 3(b), where the top-left inset shows the same bumps as in Fig. 3(a), and the top-right inset is the cross-section profile of the bump. One notes that the bump height is about $60 - 70nm$, and the diameter is about $1\mu m$. With the increase of laser power, a round hole in the bump is formed, as shown in Fig. 4, where the laser powers are 3.85, 3.90, and 4.0 *mW*, respectively. The corresponding bumps are shown from left to right in Fig. 4.

The bumps in Fig. 5(a) were produced at laser power level 4.0 *mW*. In Fig. 5(a) the left-bottom inset is an enlarged bump image. It is found that holes are formed in the central region of the bumps. Fig. 5(b) presents the AFM analysis, where the top-right inset is the three-dimensional bump image. It can be seen that the bump diameter is about $1\mu m$, and the size of the hole is about $250 - 300nm$.

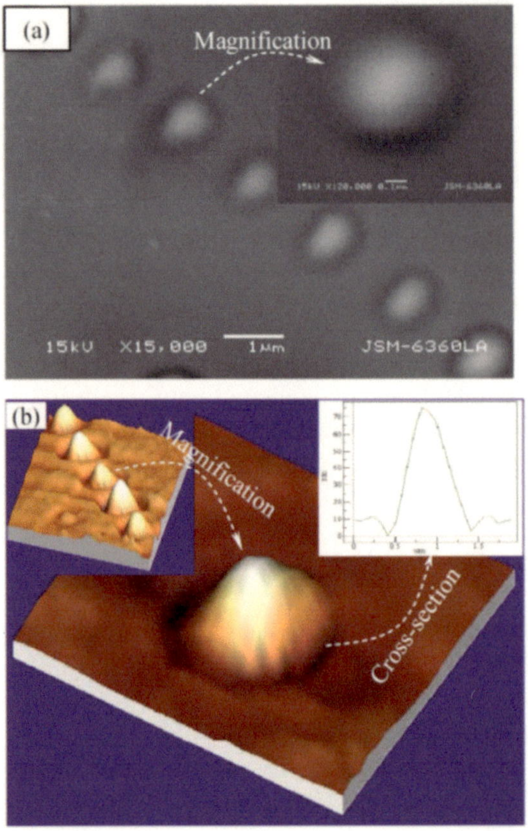

Fig. 3. Bumps formed at laser power $3.8mW$: (a) SEM analysis and (b) AFM analysis.

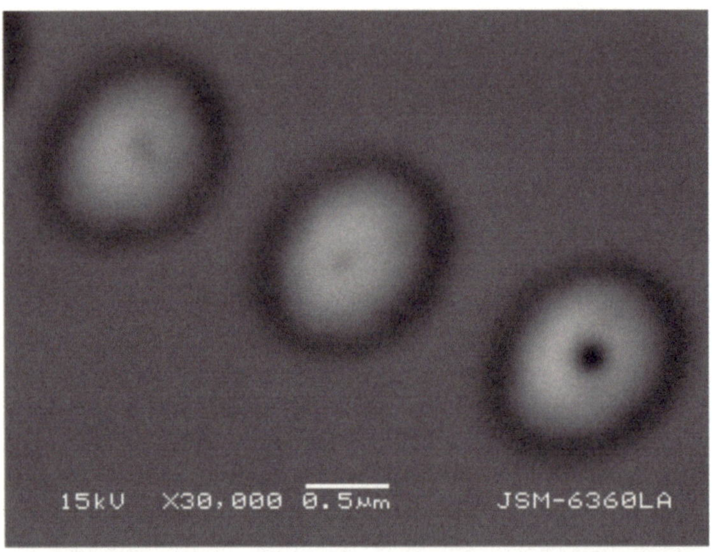

Fig. 4. SEM analysis for bumps formed at laser power of 3.85 , 3.90 and $4.0mW$.

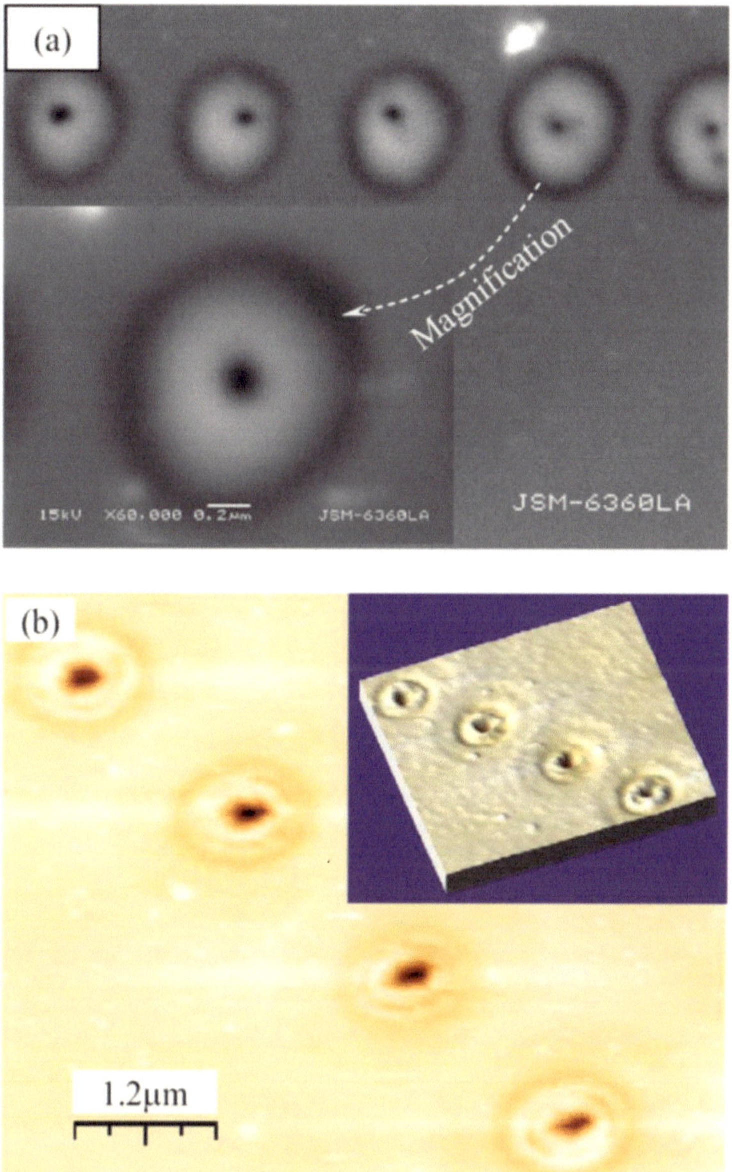

Fig. 5. Bumps formed at laser power 4.0 *mW*: (a) SEM analysis and (b) AFM analysis.

2.3 Numerical simulations

In this section, we present results of theoretical simulations based on the developed formalism. Calculations were carried out to simulate the experiments presented in the previous section, so that the numerical results can be compared with the experimental observations. Some parameters needed for the calculations were obtained from experiments. The melting and vaporization points of $Ag_8In_{14}Sb_{55}Te_{23}$ were measured by a differential scanning calorimeter (DSC), and the results are given in Fig. 6. It can be seen that the melting T_m and vaporization T_f points are 512 °C and 738 °C, respectively. It should be noted that T_f is determined by the cross point between the tangent lines of AB and CD. The capacity C_P was also measured to be about $320J/KgK$ by the DSC method. The density is obtained by $\rho = \left(\rho_{Ag} \times 8 + \rho_m \times 14 + \rho_{Sb} \times 55 + \rho_{Te} \times 23\right)/100 = 6981.2Kg/m^2$. The thermo-physical parameters used in the calculation are listed in Table I. The volume thermal expansion coefficients of $Ag_8In_{14}Sb_{55}Te_{23}$ thin film in the crystalline and melting states are difficult to measure, and we estimated that β_c and β_m were 25×10^{-6} / °C and 25×10^{-3} / °C, respectively. This is reasonable because the linear thermal expansion coefficient in liquid state is about ten times that in the solid state, therefore, the corresponding volume thermal expansion coefficient in the liquid state is about 10^3 times that in the solid state.

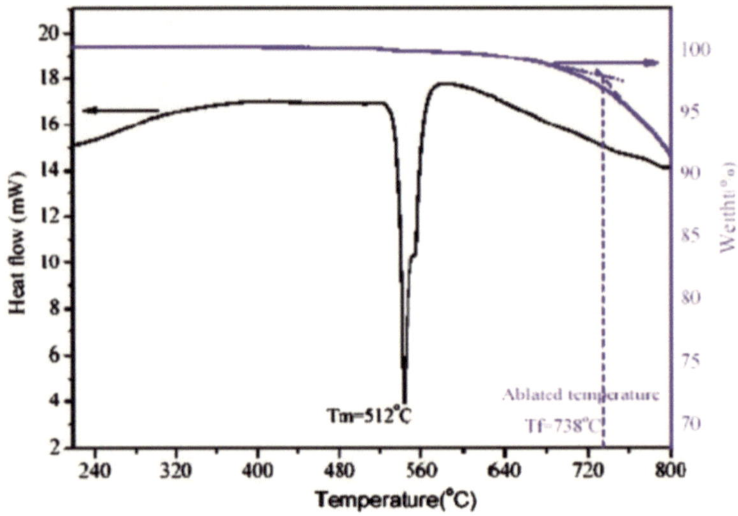

Fig. 6. DSC analysis for $Ag_8In_{14}Sb_{55}Te_{23}$ thin films.

With all the parameters assigned, simulations were carried out based on the developed formalism, and some of the simulation results in comparison with the above experimental observations are presented as follows. In Fig. 7, it shows the simulation results for laser power $P = 3.8mW$, which corresponds to the experimental situation in Fig. 3. Fig. 7(a) gives the temperature profile at different depth positions. One sees that the maximum temperature is about 720 °C at the centre of the thin film surface. At this temperature, a bump is to be formed, but the ablation is not to occur. This is shown in Fig. 7(b), where the bump height is about $70nm$. The bump diameter and area can be estimated from the top view of Fig. 7(b) to be about $847nm$ and $0.5636m^2$, respectively. These results are consistent with the experimental results in Fig. 3.

w (μm)	τ (ns)	α (m^{-1})	R	ρ (Kg m^{-3})	C_p (J Kg^{-1} K^{-1})	T_m (°C)	T_f (°C)	β_c (°C)	β_m (°C)
1.0	50	3×10^7	0.55	6981.2	320	512	738	25×10^{-3}	25×10^{-6}

Table 1. Thermo-physical and experimental parameters for the simulation.27

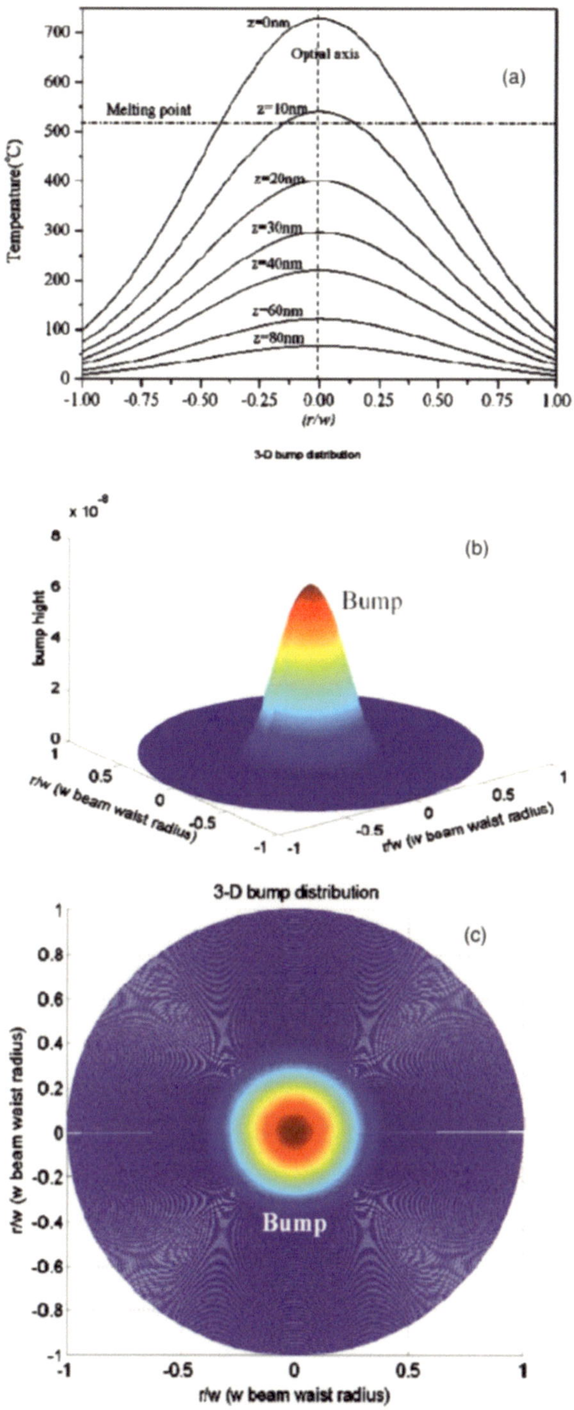

Fig. 7. Simulation results for laser power $3.8mW$: (a) temperature profile, (b) 3D image of bump, and (c) top-view of bump.

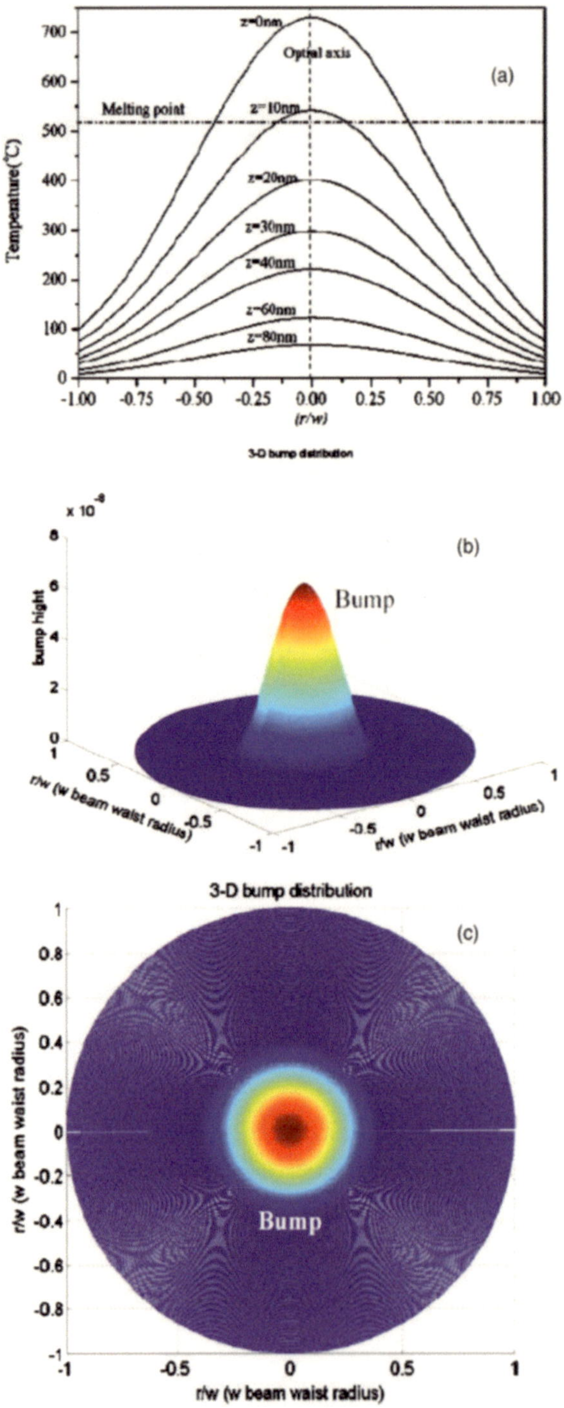

Fig. 8. Simulation results for laser power 4.0mW : (a) temperature profile and (b) hole formed at the top of bump.

With an increase of laser power, the temperature of thin films will exceed the vaporization point, and the ablation in the bump will take place. Fig. 8 shows the simulation results for laser power $P = 4.0mW$, which corresponds to the experimental situation in Fig. 5. In Fig. 8(a) the radial temperature is shown at different depth position. It can be seen that the maximum temperature at the centre of the sample surface reaches up to about 760 °C, which exceeds the vaporization point T_f (738 °C), and indicates that the ablation may occur in the centre of the spot. The resulting ablation is shown in Fig. 8(b), where a two-dimensional ablation image is given. One realizes that the bump diameter and area are about $905nm$ and $0.644m^2$, respectively. It can also be seen that an ablation hole is formed in the centre of the bump, and the diameter of the hole is about $270nm$. One compares the simulation results in Fig. 8 to the experimental results in Fig. 5 and realizes that the model simulation is consistent with the experimental observation. This confirms the correctness and usefulness of the established model and the developed formalism.

3. Patterning on multilayer thin films with laser writing

Recently, pattern structures have been used widely in many fields, such as photonic crystal and solar cell industry, owing to its advantages over the common coatings. In the last several years, pattern structures have been fabricated on silicon, quartz, and especially photo-resist by many kinds of technologies, such as ultraviolet lithography (DUV), electron beam lithography, and focused ion-beam (FIB). However, most of the technologies are not suitable to fabricate large-area structures due to the time-consuming process and high-cost equipment. One of the most attractive and competitive technologies is laser direct writing technology, in which the structures are usually written on photo-resist. But photo-resist is often followed by developing and etching procedures after writing by laser beam, which definitely increases the time-consuming and cost and restricts the application of the structure.

AgOx material has been applied to photoluminescence (PL) emission field, nonlinear optics, and superconductive magnetic levitation due to its better performance. One of the most important applications is optical storage mask layer in super-resolution near-field structure (Super-RENS), and (Tominaga et al., 1999; Liu et al., 2001) have applied this structure to optical storage field using different recording layers, respectively. In this special structure, AgOx thin film layer is usually sandwiched by two protective layers (ZnS–SiO_2), i.e., (ZnS–SiO_2)/AgOx/(ZnS–SiO_2). In the present work, we used this structure to fabricate pattern by laser direct writing. Compared with photo-resist, the materials do not need developing and etching process, and the laser power is required to be in a very low range, so it is suitable to fabricate a large-area pattern structure in very short time and very low cost, which largely decrease the time-consuming and industrial cost.

3.1 Principle

It is well known that in an open system the AgO_x material is chemically known to decompose into Ag particles and O_2 at about 160°C. When the film structure is thermally heated beyond this temperature, AgOx layers will decompose to Ag and O_2 according to AgOx $\rightarrow$Ag+x_2O_2. The decomposition reaction has been verified by many methods. When the laser beam irradiates on the AgOx film, a small volume of thin film is locally and rapidly heated to above the decomposing temperature, and then the reaction happens. The oxygen released by the decomposition is stayed in the enclosed system, so it will apply pressure to surface. Generally speaking, the AgOx thin film with the thickness of about 10 nm was used in optical storage field as the mask film. While when the thickness is increased to more than 100 nm, it may produce a big oxygen bubble by the pressure and heat following the AgOx decomposition induced by the focused laser beam, and a huge volume expansion is formed at last, just as shown in Fig. 9(a). Fig. 9(b) shows the interior situation when AgOx decomposes into silver and oxygen. The O_2 and Ag particles are rough and tumble and filled the whole room. After the AgOx cooling down to the room temperature, the expanded volume will be left as bump. If we precisely control the laser parameters, the regular and uniform bump array pattern structure can be obtained.

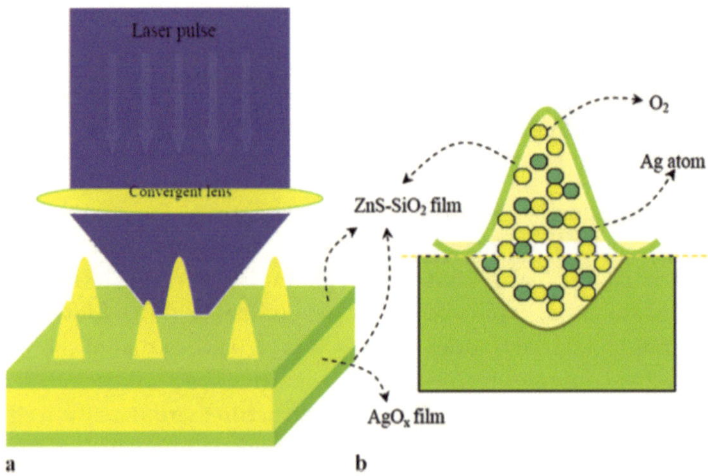

Fig. 9. Schematic of laser direct writing multilayered AgOx thin film. (a) Laser irradiated the thin film and the bubble formed. (b) Decomposition of AgOx and the formation of ZnS-SiO_2 bubble.

In fact, the layer structure is not a completely enclosed system; inter-diffusion between the as in the bump and the air outside occurs, which causes the pressure inside and outside the bump to reach up to balance. However, if the laser energy is very high and exceeds the ablation threshold of AgOx, the bumps may grow so big that a hole will form at the apex.

3.2 Experiments

According to the principle, the samples with a multilayer thin film structure "ZnS-SiO_2(10 nm)/AgOx(100 nm)/ZnS-SiO2(10 nm)" were prepared on glass substrates by radio frequency (RF) reactive magnetron sputtering. A pure Ag target with a diameter of 60 cm was bombarded by a gas mixture of Ar/O_2 plasma. In order to make more Ag particles react with O_2, we finally chose the ratios of O_2/(O_2+Ar) at 0.9 and the sputtering power 50W. Then the AgOx film with a thickness of 100 nm was prepared, and the structural phase of the as-deposited AgOx film was identified by X-ray Diffraction (XRD). The ZnS-SiO_2 films were prepared by RF magnetron sputtering. The sputtering power was 100 W, and the thickness was 10 nm, correspondingly. After the samples were prepared, the laser direct writing was carried out with a laser wavelength of 488 nm. And the pattern structures were observed by atomic force microscope (AFM).

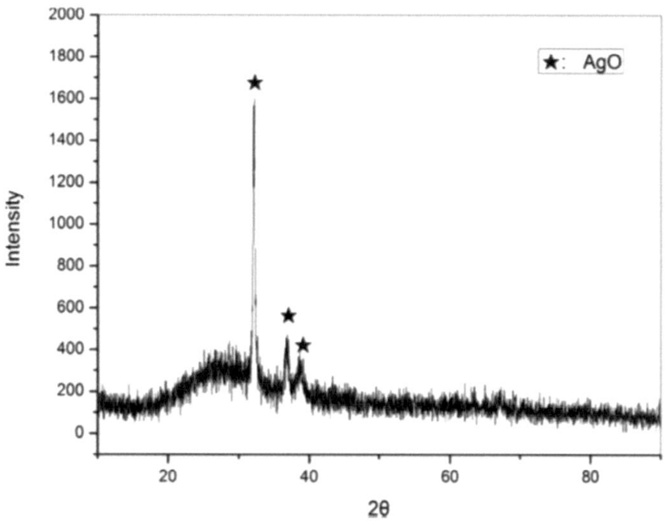

Fig. 10. XRD patterns of the as-deposited AgOx films prepared by RF reactive magnetron sputtering.

3.3 Results

Fig. 10 shows the XRD pattern of the as-deposited AgOx films. It was found that the main constituent is AgO, which agrees with the results of (Liu et al., 2001). Fig. 11 shows the pattern structures fabricated on multilayered "ZnS-SiO_2/AgOx/ZnS-SiO_2" sample, where the range of laser power is from 3.0 mW to 5.0 mW. From the Fig. 11 we can see that the pattern structure appears to be taper shape and very regular and uniform. The boundary between the area with and without laser irradiation is well defined as shown in Fig. 11(c), and the patterns are very steep with smooth wall. Fig. 11(a) and 11(c) show the three dimensional (3D) photos written by higher and lower laser powers, respectively.

Fig. 11(b) and 11(d) are the lateral photos of Figs. 11(a) and 11(c), respectively. One can find that the different pattern height can be realized by tuning the laser power. The larger the laser power, the higher the pattern. When the laser power is 5.0 mW, the height reaches the largest value. As the laser power decreases, the height gradually decreases to the lowest value at the laser power of 3.0 mW, where the pattern almost is undistinguishable. Fig. 12 shows the dependences of pattern structure height, diameter, and aspect ratio (aspect ratio = height/diameter) on laser power. We can find that both height and diameter increase with the laser power, as shown in Figs. 12(a) and 12(b). The range of the height is from 6 nm to 183 nm, and the diameter is from 482 nm to 912 nm, correspondingly. The aspect ratio is an important factor in pattern structure application. Generally speaking, the higher aspect ratio will possess a better performance. In this work, we find that the aspect ratios rapidly increase from the minimum of 0.012 at laser power of 3.0 mW to the maximum of 0.201 at laser power of 5.0 mW, which indicates that the better aspect ratio can be obtained in higher laser powers.

In order to obtain more details about the pattern structure, we amplify a small area from Fig. 11(a), and the result is shown in Fig. 13(a). It can be seen that the pattern structures appear taper shape and are very regular and uniform. The boundary between the area with and without laser irradiation is well defined, and the patterns are very uniform and smooth.

We also chose six pattern units (marked by line in Fig. 13(b)) to measure the height and diameter of the structures, and the result is shown in Fig. 13(c). One notes that the height of pattern is about 150 nm, and the diameter is around 650 nm, accordingly.

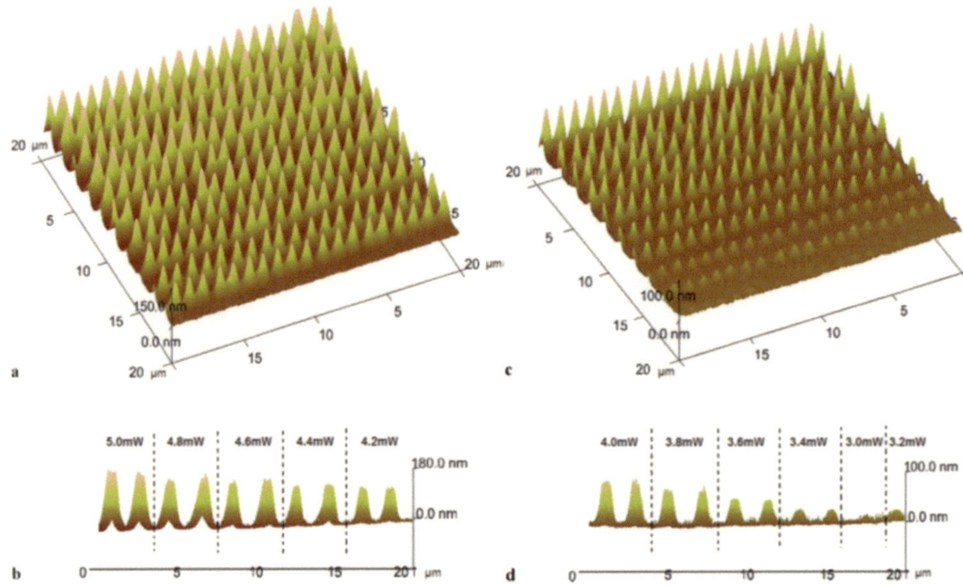

Fig. 11. Pattern structures written on multilayered ZnS-SiO2/AgOx/ZnS-SiO2 films by green laser (λ = 488 nm) in different laser processing parameters. (a) The AFM 3D photos of pattern structures written by higher laser power. (b) The lateral photos of the pattern structures in (a). (c) The AFM 3D photos of pattern structures written by lower laser power. (d) The lateral photos of the pattern structures in (c).

In order to test the stability of the pattern structure, we heated the sample shown in Fig. 11(a) in the furnace and kept the temperature at 100°C for 1 h. The result measured by AFM is shown in Fig. 14(a), and Fig. 14(b) is the lateral photo of Fig. 14(a). As shown in the photos, the taper shape does not change even if the temperature is kept at 100°C , and the regular and uniform pattern structures are almost the same as Fig. 11(a). Besides of that, one can see both the diameters and the heights gradually decreased with the laser power decreasing. The range of the height is from 100 nm to 180 nm, and the diameter is from 500 nm to 900 nm, correspondingly, which is very close to the values in Fig. 11(a), so we can conclude that the pattern structure was stable even if the temperature is higher than room temperature. We think that there are two main reasons. One is that AgOx material in this micro-zone is all decomposed under the laser irradiation with only Ag particles left. The other is that the layer structure is not a completely enclosed system, inter-diffusion between the gas in the bump and the air outside causes the pressure inside and outside the bump to reach a balance, and also the temperature is gradually increased. That is to say, in this gradual heating process, the gas in and out of the structure has enough time to inter-diffuse to reach a balance, so the structures can keep the same as before. The further explanation will be studied in our next work.

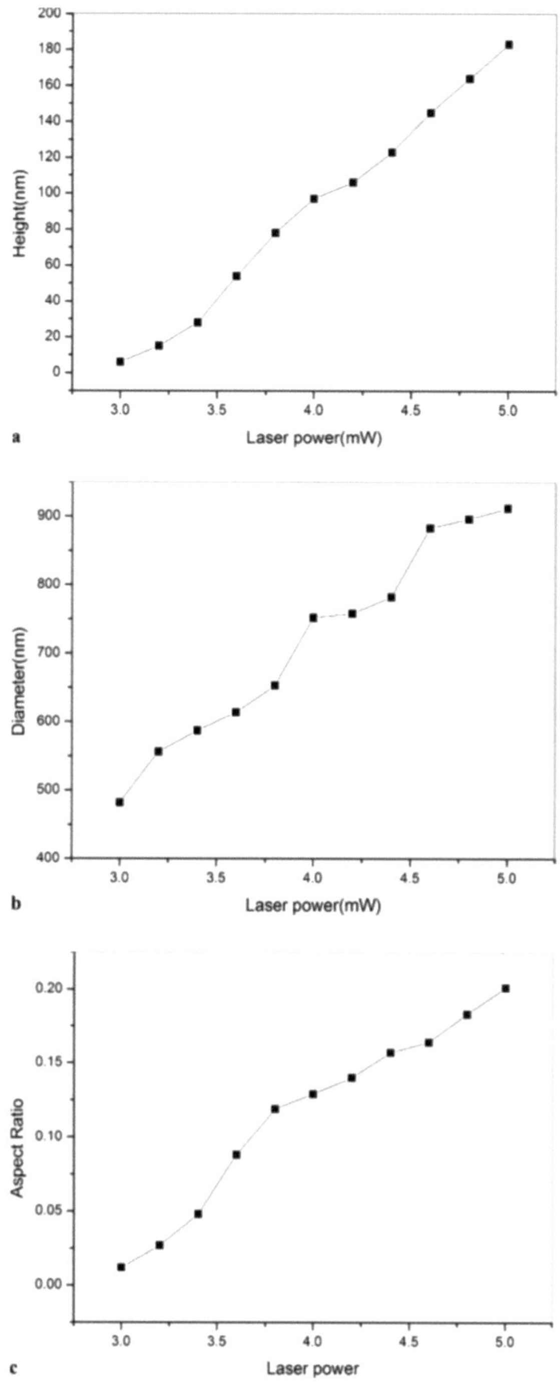

Fig. 12. Dependence of the pattern structures parameters on the laser power. (a) Dependence of height on laser power. (b) Dependence of diameter on laser power. (c) Dependence of aspect ratio on laser power

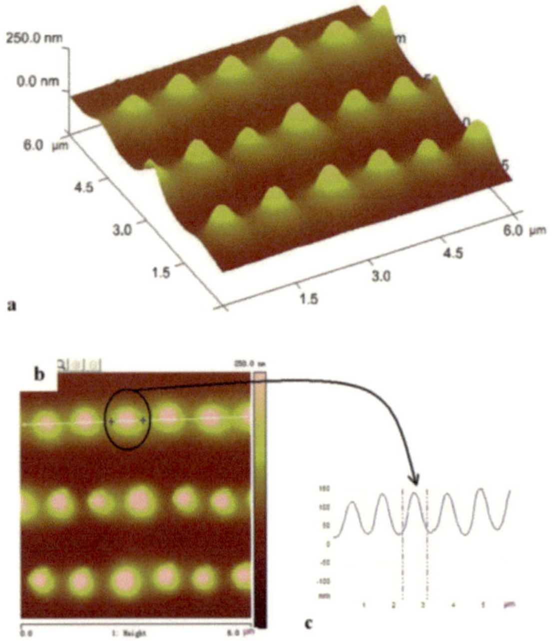

Fig. 13. Amplification photos of pattern structures chosen from Fig. 3(a). (a) The AFM 3D photos of pattern structure. (b) The chosen area of AFM analysis (marked by the line). c) AFM analysis of the pattern structure in (b).

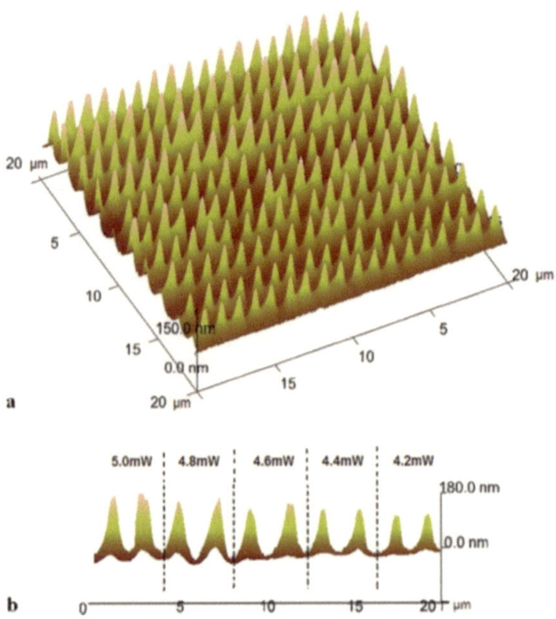

Fig. 14. AFM 3D photos of pattern structures in Fig. 3(a) kept the temperature at 100°C for 1 h. (a) The AFM 3D photos of pattern structures. (b) The lateral photos of the pattern structures in (a).

4. Conclusion

A theoretical model has been established for the bump formation in the optical writing process. Based on the developed formalism, geometric characters of the formed bumps can be analytically and quantitatively evaluated from various parameters involved in the formation. Simulations based on the analytic solution have been carried out taking $Ag_8In_{14}Sb_{55}Te_{23}$ as an example. The results have been verified with experimental observations of the bumps. It has been verified that the results from the simulations are consistent with the experimental observations. Micro/nanometric pattern structures have been fabricated on "ZnS-SiO2/AgOx/ZnS-SiO2" multilayer thin film sample by laser direct writing method. The pattern structures with different shapes and sizes could be directly written by very low laser power without developing and etching procedures, which could largely decrease the time-consuming and cost.

5. Acknowledgment

The work is partially supported by National Natural Science Foundation of China (Grant Nos. 50772120, 60507009, 60490290, and 60977004). This work is supported by the Natural Science Foundation of China (Grant Nos. 50772120 and), Shanghai rising star tracking program (10QH1402700), and the Basic Research Program of China (Grant No. 2007CB935400), and UNAM-DGAPA Mexico Grant No. IN120406-3. Support from supercomputer DGSCA-UNAM is gratefully acknowledged.

6. References

Wuttig R. and Steimer C. (2007). Phase change materials: From material science to novel storage devices. *Applied Physics A*, Vol.87, No.3, (June 2007), pp. 411-417, ISSN 1432-0630

Kolobov A. V., Fons P., Frenkel A. I., Ankudinov A. L., Tominaga J., and Uruga T. (2004). Understanding the phase-change mechanism of rewritable optical media. *Nature Materials*, Vol.3, No.10, (October 2004), pp. 703-708, ISSN 1476-1122

Welnic W., Parnungkas A., Detemple R., Steimer C., Blugel S., and Wuttig M., (2006). Unravelling the interplay of local structure and physical properties in phase-change materials. *Nature Materials*, Vol.5, No.1, (January 2006), pp. 56-62, ISSN 1476-1122

Kalb J., Spaepen F., and Wuttig M. (2004). Atomic force microscopy measurements of crystal nucleation and growth rates in thin films of amorphous Te alloys. *Applied Physics Letters*, Vol.84, No.25, (June 2004), pp. 56-62, ISSN 0003-6951

Wei J., Jiao X., Gan F. and Xiao M. (2008). Laser pulse induced bumps in chalcogenide phase change films. *Journal of Applied Physics*, Vol.103, No.12, (June 2008), pp. 124516-5, ISSN 0021-8979

Dun A. Wei J. And Gan F. (2010). Pattern structures fabricated on ZnS-SiO2/AgOx/ZnS-SiO2 thin film structure by laser direct writing technology. *Applied Physics A*, Vol.100, No.2, (August 2010), pp. 401-407, ISSN 1432-0630

Shiu T., Grigoropoulos C. P., Cahill D. G., and Greif R. (1999). Mechanism of bump formation on glass substrates during laser texturing. *Journal of Applied Physics*, Vol.86, No.3, (August 1999), pp. 1311-6, ISSN 0021-8979

Tominaga J., Haratani S., Uchiyama K., and Takayama S. (1992). New Recordable Compact Disc with Inorganic Material, AgOx. *Japaness Journal of Applied Physics*, Vol.31, No.9A, (September 1992), pp. 2757-2759, ISSN 0021-4922

Liu W.C., Wen C.Y., Chen K.H., Lin W.C., and Tsai D.P. (2001). Near-field images of the AgOx-type super-resolution near-field structure. *Applied Physics Letters*, Vol.78, No.6, (February 2001), pp. 685-687, ISSN 0003-6951

5

Laser Patterning Utilizing Masked Buffer Layer

Ori Stein and Micha Asscher
Institute of Chemistry and the Farkas Center for light induced processes, The Hebrew University of Jerusalem
Israel

1. Introduction

Laser-matter interaction has been the focus of intense research over the past three decades with diverse applications in the semiconductor industry (photolithography), sensing and analytical chemistry in general. Pulsed laser ablation of adsorbates under well controlled ultra high vacuum (UHV) conditions has enabled detection in the gas phase of large (mostly biologically important) molecules via mass spectrometry, but also to study the remaining species on the surface. In this chapter we will focus our report on these remaining atoms and molecules following selective laser ablation of weakly bound buffer layers as a novel tool for patterning of adsorbates on solid surfaces.

1.1 Patterning of adsorbates for diffusion measurements

Laser Induced Thermal Desorption (LITD) of adsorbates has developed as an important technique for surface diffusion measurements. In the hole-refilling method, a hole was burnt within an adsorbate covered surface. Subsequent time delayed laser pulse was employed to measure the refilling rate due to surface diffusion process (Brand et al., 1988, Brown et al., 1995). Accurate analysis of data acquired that way is not straight forward since the diffusion measured this way is two dimensional (and not necessarily isotropic). The actual hole size burnt into the surface is typically in the order of ~100μm, limiting the diffusion measurement to relatively fast occurring processes with low energy barrier compared to the activation energy for desorption.

A different method, utilizing two interfering laser beams to an adsorbate covered surface, has resulted in a sinusoidal spatial temperature profile and selective desorption of the adsorbates, thus creating a density modulation grating on the surface. In this way the typical measured diffusion length can decrease down to sub-micrometer scale.

The grating formed on the surface obeys Bragg law:

$$w = \frac{\lambda}{2\sin(\theta)} \tag{1}$$

w - grating period

λ - desorbing laser wave length

θ - angle between one of the incident laser beams and the surface normal.

Such grating formation can be explored optically by recording a diffraction pattern from it. The decay of the measured 1st order diffraction due to smearing of the grating formation is indicative of one dimensional diffusion process- at the direction normal to the grating stripes. In this way, anisotropic diffusion can be measured simply by changing the direction of the substrate with respect to the grating symmetry. Second Harmonic Generation (SHG) diffraction from one monolayer (1ML) of CO on Ni(111), (Zhu et al. 1988, 1989) and on Ni(110), (Xiao et al. 1991). Coverage dependent diffusion coefficient models were found necessary to understand the experimental data (see e.g. Rosenzwig et al, 1993, Verhoef and Asscher, 1997, Danziger et al. 2004). An alternative way, utilizing optical linear diffraction method combined with polarization modulation techniques (Zhu et al. 1991, Xiao et al, 1992, Wong et al. 1995, Fei and Zhu, 2006) has yielded a more sensitive and accurate calculation of the anisotropic diffusion coefficient of CO on Ni(110), (Xiao et al, 1993).

Selective patterning of H on top of Si(111) surface (Williams et al. 1997) was demonstrated via pre-patterning a thin layer of Xe adsorbed on the Si surface that has reduced the sticking coefficient of H on Si by more than an order of magnitude. This way the authors were able to pattern chemisorbed H while avoiding high power laser pulses impinging on the surface thus preventing possible laser induced surface damage.

We have recently introduced a procedure that adopts the concept of laser-induced ejection of a weakly bound, volatile layer, applied for generation of size-controlled arrays of metallic clusters and sub- micron wide metallic wires. This buffer layer assisted laser patterning (BLALP) procedure utilizes a weakly bound layer of frozen inert gas atoms (e.g., Xe) or volatile molecules (e.g., CO_2 and H_2O) that are subsequently exposed to metal atoms evaporated from a hot source. It results in the condensation of a thin metal layer (high evaporation flux) or small clusters (low flux) on the top surface of the buffer layer. The multi-layered system is then irradiated by a short single laser pulse (nsec duration) splits and recombines on the surface in order to form the interference pattern. It results in selective ablation of stripes of the volatile buffer layer along with the metallic adlayer deposited on it. This step is followed by a slow thermal annealing to evaporate the remaining atoms of the buffer layer with simultaneous soft landing of metallic stripes on the substrate. In other words, this procedure combines the method for generating grating-like surface patterns by laser interference (Zhu et al. 1988, Williams et al. 1997) with a buffer-assisted scheme for the growth of metallic clusters (Weaver and Waddill, 1991, Antonov et al., 2004).

Employing a single, low power laser pulse, the BLALP technique has been utilized to form parallel stripes of potassium (Kerner and Asscher, 2004a, Kerner et al., 2006), as well as continuous gold wires (Kerner and Asscher, 2004b) strongly bound to a ruthenium single crystal substrate.

An extensive study of surface diffusion of gold nanoclusters on top of Ru(100) and p(1x2)-O/Ru(100) was preformed utilizing the BLALP technique (Kerner et al., 2005). The authors discuss the smearing out of gold clusters density grating deposited on the substrate due to one dimensional diffusion process. Figure 1 describes the smearing out of a density grating created after evaporating 1nm of gold onto 60ML of Xe adsorbed on Ru(100) surface.

Heating a similar grating structure in air to 600K for 2h has resulted in no noticeable effect on the metal clusters forming the grating. It is believed that heavy oxidation of the Ru substrate under these conditions acts as an anchor and inhibited the cluster diffusion. Smearing out of the density grating had little or no effect on the size distribution of the gold clusters, suggesting no significant sintering and coalescence of the clusters under these conditions.

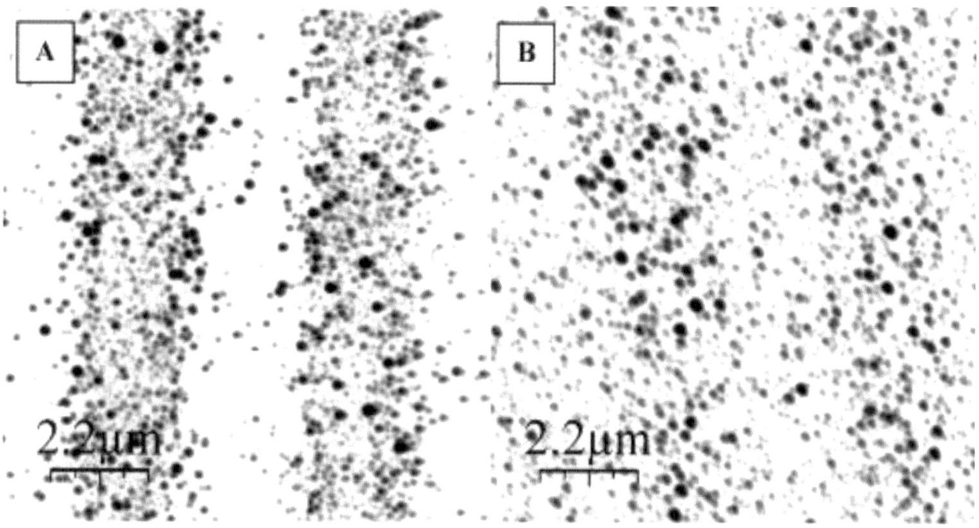

Fig. 1. AFM images of a high density gold cluster coverage grating created via BLALP scheme, evaporating 1nm of gold on top of 60ML of Xe. All images were taken at ambient environment. A) After annealing in vacuum to 300K and kept at room temperature. B) After annealing to 450K at 3K/s and quenched back to room temperature. Images are courtesy of Kerner et al., 2005.

Monitoring clusters' diffusion in-situ is possible by simultaneously recording the first order linear diffraction signal decay resulting from shining low power (5mW) He-Ne cw laser on such grating while heating the substrate. The 1st order diffraction decay can be correlated to the diffusion coefficient of the clusters on the substrate (Zhu et al., 1991, Zhu, 1992). Due to the large temperature range in which diffusion takes place in this system (~250K), performing isothermal measurements is impractical. Introducing a novel, non-isothermal diffusion method has enabled Kerner et al. to circumvent the complexity of isothermal diffusion measurements in this system and has provided the authors a method to measure the diffusion of a range of cluster sizes and density distributions on top of Ru(100) and on top of p(1x2)-O/Ru(100). On both surfaces, it was found that the diffusion coefficient is density (coverage) independent. The activation energy for diffusion was sensitive to the cluster size on the bare Ru(100) surface but only weakly dependent on cluster size on the p(1x2)-O/Ru(100) surface. This arises from the weak interaction of the gold clusters with the oxidized surface and in particular the incommensurability of the clusters with the under laying oxidized substrate.

1.2 Pulsed laser driven lithography and patterning

Direct laser interference lithography/patterning involving selective removal of material from the surface of a solid sample employing two or more interfering laser beams has been used in a large variety of applications. These techniques were utilized for polymers patterning, micromachining, semiconductor processing, oxide structure formation and for nano-materials control over magnetic properties(Kelly et al., 1998, Ihlemqnn & Rubahn, 2000, Shishido et al., 2001, Chakraborty et al., 2007, Lasagni et al., 2007, 2008, Leiderer et al., 2009, Plech et al., 2009).

A modified version of the BLALP technique that involves laser patterning of the clean volatile buffer layer prior to the deposition of the metal layer has also been introduced to generate smooth metallic stripes on metallic (Kerner et al., 2004c, 2006) as well as oxide (SiO_2/Si(100)) substrates. The unique advantage of BLALP is the low laser power needed for patterning, which prevents any damage to the substrate.

The importance of laser-driven ejection of a layer of weakly bound material from light absorbing substrates has motivated a number of experimental (Kudryashov & Allen 2003, 2006, Lang & Leiderer, 2006, Frank et al.,2010) and computational studies (Dou et al., 2001a, 2001b, Dou et al., 2003, Smith et al., 2003, Gu & Urbassek, 2005, 2007, Samokhin, 2006) targeted at revealing the fundamental mechanisms responsible for the layer ejection. The physical picture emerging from these investigations suggests that fast vaporization (explosive boiling) and expansion of the superheated part of the layer adjacent to the hot substrate provides the driving force for the ejection of the remaining part of the layer.

In this paper, we report the results of utilizing a single pulse laser patterning, all-in vacuum procedure that can produce practically any sub- micron resolution pattern using an optical system consisting of a masked imaging system.

2. Experimental

The experimental setup has been described elsewhere in detail (Kerner et al., 2005a, 2005b). Briefly, a standard UHV chamber at a base pressure of $5x10^{-10}$mbar, equipped with Ne^+ sputter gun for sample cleaning and a quadrupole mass spectrometer (QMS, VG SX-200) for exposure and coverage determination and calibration, are used in the experiments. In addition, separate Au, Ag and Ti deposition sources are used, with in-situ quartz microbalance detector for flux calibration measurements. A native oxide SiO_2/Si(100) sample is attached via copper rods to a closed cycle helium cryostat (APD) that cools the sample down to 25 K with heating capability up to 800 K (Stein & Asscher, 2006). 700eV Ne^+ ion sputtering for sample cleaning was carried out prior to each patterning experiment.

In order to perform laser assisted ablation and patterning measurements, a p-polarized Nd:YAG pulsed laser working at the second harmonic wavelength was used (Surlight, Continuum λ = 532 nm, 5 ns pulse duration). The laser power absorbed by the silicon substrate was kept lower than 80 MW/cm^2 (160mJ/pulse) to avoid surface damage (Koehler et al., 1988). During the experiments we assumed complete thermalization between the SiO_2/Si layers with no influence of the thin oxide layer (~2.5 nm thick) on the heat flow towards the adsorbates. Details of Xe template formation via laser induced thermal desorption (LITD) and its characterization are given elsewhere (Kerner & Asscher 2004a, 2004b, Kerner et al., 2005, 2006). After patterning the physisorbed Xe, 12±1 nm thick film of metal, typically Au or Ag, is deposited on the entire sample. Subsequently, a second uniform laser pulse strikes the surface, ablating the stripes of Xe buffer layer remaining on the substrate together with the deposited metal film/clusters on top and leaving behind the strongly bound metal stripes that are in direct contact with the SiO_2 surface. A 2±1 nm thick layer of Ti deposited over the SiO_2 surface prior to the buffer layer adsorption and metal grating formation, ensures good adhesion of the noble metals to the silicon oxide substrate and avoid de-wetting (Bauer et al., 1980, George et al., 1990, Camacho-López et al., 2008). The Ti adhesion layer does not affect the optical properties of the substrate (Bentini et al., 1981).

Patterning through a mask is introduced here for the first time, utilizing a single uniform laser pulse. The mask is a stainless steel foil 13μm thick that contains the laser engraved word "HUJI" (Hebrew University Jerusalem Israel). An imaging lens was used in order to transfer the object engraved on the mask onto the sample plane while reducing its size according to the lens formula:

$$\frac{1}{u}+\frac{1}{v}=\frac{1}{f} \qquad (2)$$

u- mask-lens distance (120 cm).
v- lens- sample distance (24 cm).
f- focal length of the lens (20 cm) .

Ex-situ characterization of the resulting patterns was performed by HR-SEM (Sirion, FEI), AFM in tapping mode (Nanoscope Dimension 3100, Veeco) and an optical microscope (Olympus BX5).

3. Results and discussion

3.1 Metallic line patterning via laser interference

Metallic lines were patterned directly on the SiO_2/Si sample using Lift-off (Kerner et al., 2004c, 2006) and BLALP schemes. Using CO_2 as the buffer material it was possible to perform a BLALP patterning process under less stringent cooling requirements than those previously used with Xe as the buffer material (Rasmussen et al. 1992, Funk et al., 2006). Figure 2 demonstrates the results of patterning 12 nm thick layer of Au using 10ML of CO_2 as the buffer material.

Although metal stripes obtained this way demonstrate good continuity, their texture is corrugated since these stripes are composed of metal clusters soft-landed on the substrate after annealing the sample to room temperature, according to buffer layer assisted growth (BLAG) procedure (Weaver & Waddill 1991). Using this scheme, metal clusters are evenly distributed in the areas between the metal stripes. Molecular dynamics (MD) simulations describing the laser ablation of the buffer material from a silicon surface have indicated that under the experimental conditions adopted in the current study, evaporative buffer material removal scheme is dominant (Stein et al., 2011). This evaporative mode of ablation, unlike the abrupt or explosive ablation that dominates at higher laser power, does not necessarily removes all the metal layer or clusters that reside on top. In this case, therefore it is likely that some of the metal evaporated on top of the buffer could not be removed by the laser pulse, and was finally deposited on the surface as clusters.

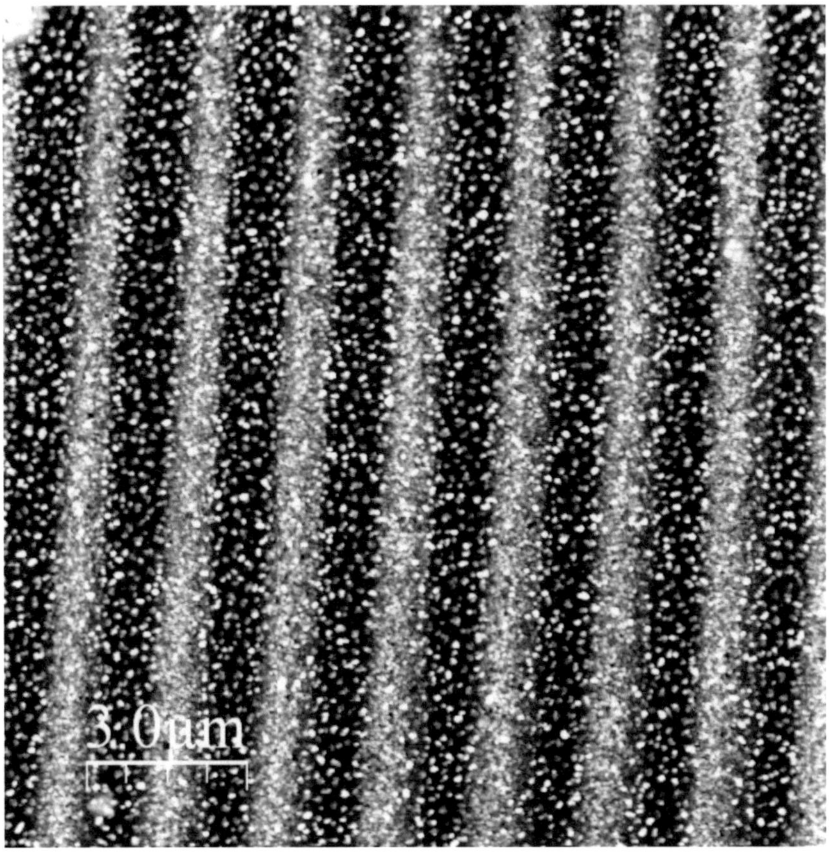

Fig. 2. AFM image of BLALP patterning of 12 nm thick layer of Au deposited on top of 10ML CO_2 buffer material on a SiO_2/Si(100) sample at 25 K. Laser power was 14 MW/cm^2.

Figure 3 illustrates the two different Xe removal mechanisms. Figures 3A and 3B demonstrate intense evaporation and explosive desorption of Xe from Si(100) surface, respectively (Stein et al., 2011).

Figure 4 demonstrates lift-off patterning: after patterning 80ML of Xe using laser power of 12MW/cm^2, 18 nm of Au were deposited on the sample. A second, uniform pulse at a power of 9MW/cm^2 was subsequently applied in order to remove the remaining Xe and metal on top.

The fragmented and discontinuous nature of the metal stripes resulting from this patterning procedure on SiO_2/Si samples is apparent. This shape is due to the poor adhesion (and de-wetting) of Au on SiO_2 (Bentini et al., 1981, George et al., 1990, Lani et al., 2006). Overcoming this problem requires evaporation of 2±1 nm Ti on top of the entire SiO_2 surface as an adhesion and wetting layer (Bentini et al., 1981). Figure 5 displays the effect of Ti evaporation on the integrity and smoothness of the metal stripes patterned via the lift off procedure.

The images in figure 5 reveal a clear power effect which is a characteristic feature of the lift-off patterning scheme. Raising the laser power leads to widening of the ablated buffer troughs as the sinusoidal temperature profile increases. Into these wider troughs metal is evaporated, eventually (after the second pulse) forming smooth and continuous wires, ideally across the entire laser beam size. Increasing the pulse power by 40% has led to wider stripes from 700 nm to 1300 nm, see Fig. 5A and 5B.

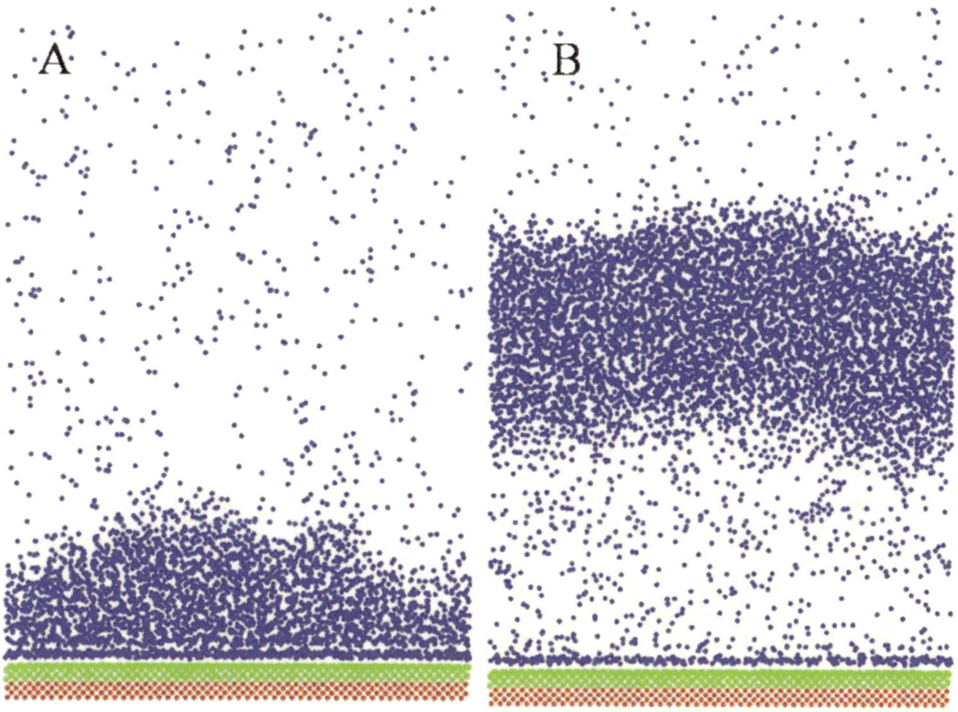

Fig. 3. Snapshots from MD simulations performed on 7744 Xe atoms adsorbed on top of Si(100) surface. A and B represent evaporative and explosive desorption while irradiating the surface by 12 and 16MW/cm^2 pulse power, respectively. Snapshots were taken at 9.4 ns (A) and 6.6 ns (B) from the onset of the laser pulse.

Electrical resistance measurements were performed on these metallic wires. On a patterned sample a set of 100X100μm metallic pods with ohmic contact to the patterned wires were prepared by e-bean lithography in order to ex-situ measure the resistivity of the silver metal wires. The resistivity measurements were calibrated against a similar measurement performed using Au wires of identical dimensions, produced via e-beam lithography.Measurements have revealed that the resistivity of the laser patterned wires were about 40% (on average, calculated from four different measurements performed at different locations on the sample) higher compared to the e-beam prepared Au, 197 and 140Ω for the laser-patterned Ag and the e-beam Au over a line distance of 24.2μm, respectively. Annealing the patterned sample at 600K for two hours in ambient conditions has led to higher resistivity by 60%, as a result of oxidation and aggregation of the Ag wires, increasing from 197 to 318Ω. In contrast, the annealed Au wires have shown a 75% drop in resistivity, from 140 to 79Ω, as expected since no oxidation takes place in the case of gold. Figure 6 demonstrates the aggregation occurs within the Ag stripes to form spherical clusters caused by annealing the sample to 600K for two hours in ambient conditions.

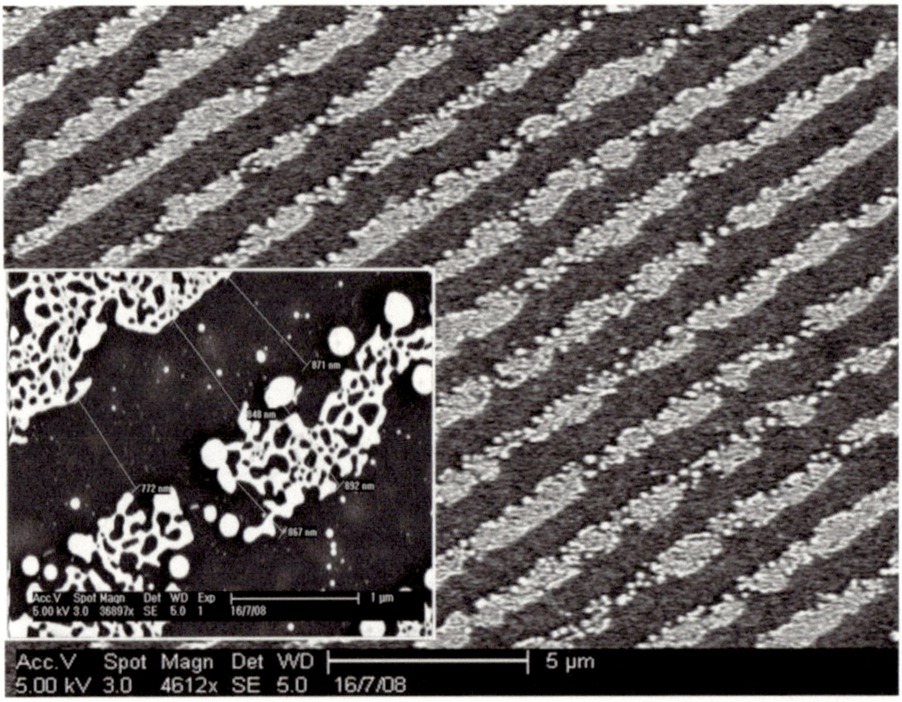

Fig. 4. SEM image of lift- off patterning of 18 nm Au on top of SiO_2/Si surface. The power of the first and second laser pulses was 12 and 9MW/cm^2, respectively. Inset depicts the corrugated (and fragmented) texture of the resulting metal wires.

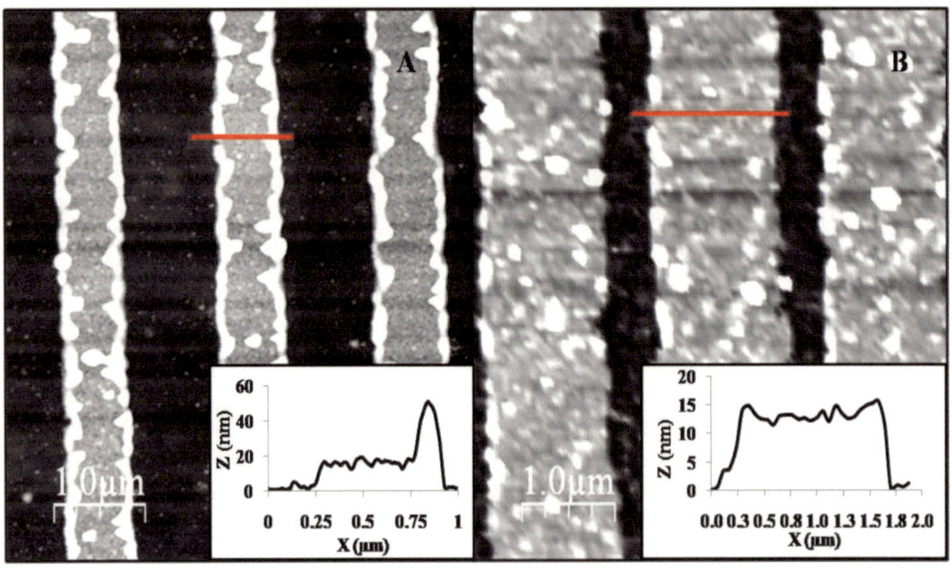

Fig. 5. AFM images of lift-off patterning procedure including line scan along the red line. A) 15 nm of Ag on top of a coverage grating formed via a 50ML Xe on $Ti/SiO_2/Si$ surface. First and second pulse power were both of 10 MW/cm^2. B) 12 nm of Ag on top of grating produced with 70ML Xe on top of $Ti/SiO_2/Si$ surface. Both the first and second pulses were at 14 MW/cm^2.

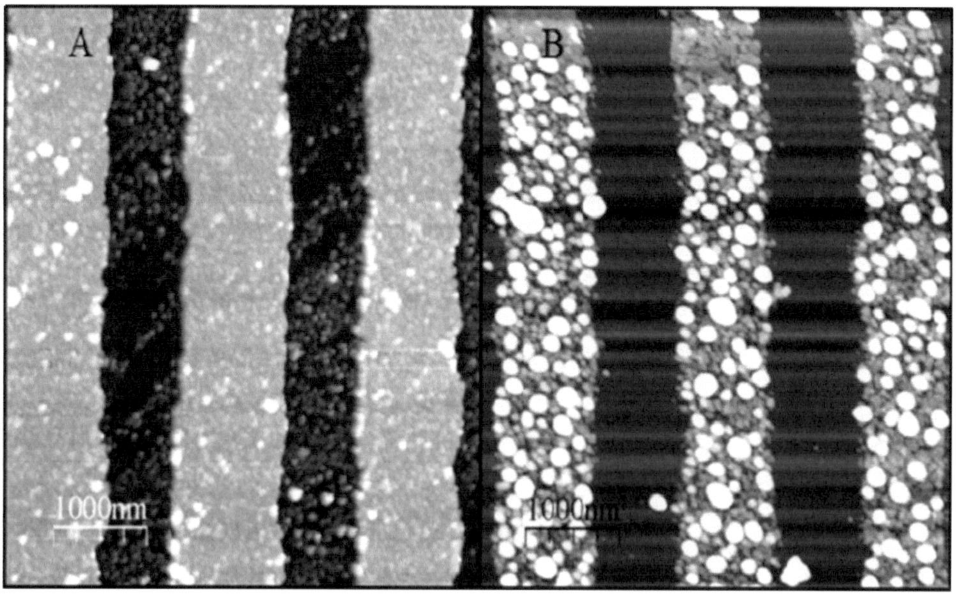

Fig. 6. Annealing effect on a lift-off patterned sample consisting of 20 nm of Ag evaporated on 70ML of Xe grating on $Ti/SiO_2/Si$ surface. A and B: AFM images in ambient conditions before and after annealing to 600K, respectively.

3.2 Laser patterned mask imaging

General application of the buffer layer assisted laser patterning scheme requires the ability to perform any desired shape and structure. This can be achieved by striking the buffer covered substrate with a laser beam that has been partially blocked by a patterned mask. In order to demonstrate the ability to pattern via a mask, a stainless steel foil, 12.7μm thick that contains the laser engraved word "HUJI" as our mask, the size of the word-object was 4X1.3mm. After passing through the mask, the laser pulse traveled through a lens in order to reduce-image the HUJI word on the sample's plane. Five times reduction required a 20 cm focal length lens at a distance of 120 and 24 cm from the mask and sample, respectively. Using this imaging lens required a dramatic reduction of laser power in order to avoid surface damage. Figure 7 demonstrates the lift-off lithography of the word "HUJI" on top of $Ti/SiO_2/Si$ surface.

A 60 ML Xe deposited on $Ti/SiO_2/Si(100)$ sample was prepared to demonstrate the mask-laser patterning. A single pulse, 0.8 MW/cm^2 (2.5 mJ/pulse) penetrating through the mask and the lens system was employed as described above. Prior to the second, uniform laser pulse striking the entire sample without the mask and the lens, 12±2 nm Au was evaporated on top of the HUJI patterned Xe buffer layer covered substrate. The sample was subsequently heated to room temperature and removed from the vacuum chamber for characterization using AFM and optical microscopy.

Employing a weaker first laser pulse at a power of 0.5 MW/cm^2 (1.5 mJ/pulse) prior to the evaporation of gold, has resulted in a narrower line width of the final pattern although with somewhat poorer quality (not shown), as was previously demonstrated in the case of two interfering beams forming parallel metallic stripes (see Fig. 5 above). This behavior, of

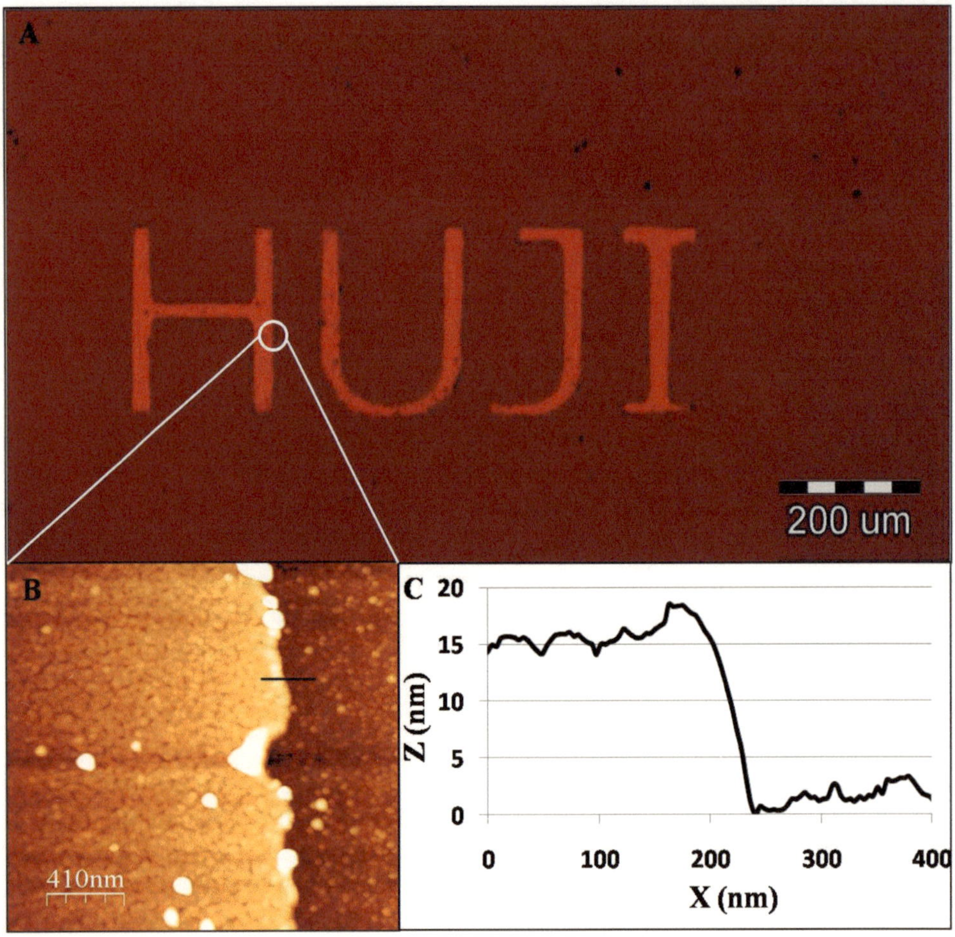

Fig. 7. A) Optical microscope image of the word "HUJI" following lift-off lithography written by 12±2 nm thick Au on $Ti/SiO_2/Si$ surface. First pulse 0.8 MW/cm^2, Xe buffer thickness was 60ML. B) AFM image demonstrating the edge of the patterned letter "H". C) A height profile taken along the line in image B.

narrowing the line width of a given feature while lowering the pulse power is a result of the laser pulse Gaussian spatial profile. As the first pulse power goes up, a wider part of the pulse reaches ablation threshold of the buffer material (Xe in this case), allowing more buffer material to be removed from the surface. Line narrowing through pulse power lowering is one of the characteristics of the lift-off patterning scheme. This power knob is a unique, very practical and easy to use for various applications, allowing patterning far from substrate laser induced damage threshold. One should bear in mind that the total pattern quality is highly dependent on the first pulse power uniformity as random variations in the laser pulse profile will be manifested in overall lower quality pattern, especially while employing near ablation threshold power.

By tuning the first pulse power up from 0.5 MW/cm^2 (1.5 mJ/pulse) to 0.8 MW/cm^2 (Fig. 7A) we were able to significantly improve the image quality while introducing a minor increase in the size of the object, all without changing the optical imaging parameters.

Image 7B represents a characteristic edge image of the patterned object, utilizing a tapping mode AFM. One can clearly notice the corrugated texture of the evaporated gold film on top of the $Ti/SiO_2/Si$ surface, featuring the 3D growth of multilayer Au on top of metal surfaces. Looking at the line profile presented in fig. 7C, the sharp drop representing the edge of the letter "H", as shown in the image. The sharp drop from the top of the gold film to the bottom of the Ti surface occurs in a lateral distance of ~50 nm, ten times smaller than the 532 nm wavelength used in this experiment, evidence to the abrupt, temperature exponential dependent ablation of the Xe buffer. Even in our simple, basic optical design consisting of a mask and lens, we were able to arrive at the sharply resolved lines shown in Fig. 7A. Simple reduction in the ablating laser wavelength and by meticulously measure the relevant distances (objective- lens, lens- surface) one can further enhance this process' resolution.

This simple, all-in-vacuum fast and clean patterning procedure does require highly accurate and robust, through vacuum imaging technique in order to avoid standard diffraction based distortions of the desired features to be patterned.

4. Conclusions

The role of weakly bound atomic and molecular buffer layers in forming periodic coverage density has been discussed as a versatile tool to study in-vacuum metallic nano-particles growth and their surface diffusivity, an important aspect of catalysis. In addition, we have demonstrated the application of the buffer layer method to pattern a $Ti/SiO_2/Si$ surface using pulsed laser lithography through a simple optical system consisting of a mask and an imaging lens. Feature (the letters HUJI) size reduction of 1:5 has been demonstrated with AFM imaged sharp edges that are three orders of magnitude narrower than a letter size. Focusing on weakly bound buffer materials for the patterning method has enabled us to use low power laser, significantly below surface damage threshold. Employing the buffer assisted laser patterning method there is virtually no limit to the pattern that can be transferred to practically any (light absorbing) substrate.

5. Acknowledgments

Partial support for this research by the US-Israel Binational Science Foundation and the Israel Science Foundation is acknowledged. The authors thank Uriel Levi for insightful discussions and help regarding in-vacuum imaging.

6. References

Antonov, V. N., J. S. Palmer, P. S. Waggoner, A. S. Bhatti, and J. H. Weaver. (2004). Nanoparticle diffusion on desorbing solids: The role of elementary excitations in buffer-layer-assisted growth. *Physical Review B* 70 (4) ISSN: 1098-0121.

Bauer, R. S., R. Z. Bachrach, and L. J. Brillson. (1980). Au and Al interface reactions with SiO_2. *Applied Physics Letters* 37 (11):1006-1008 ISSN: 0003-6951.

Bentini, G. G., C. Cohen, A. Desalvo, and A. V. Drigo. (1981). Laser annealing of damaged silicon covered with a metal-film - test for epitaxial-growth from the melt. *Physical Review Letters* 46 (2):156-159 ISSN: 0031-9007.

Brand, J. L., A. A. Deckert, and S. M. George. (1988). Surface-diffusion of hydrogen on sulfur-covered Ru(001) surfaces studied using laser-induced thermal-desorption. *Surface Science* 194 (3):457-474 ISSN: 0039-6028.

Brown, D. E., D. S. Sholl, R. T. Skodje, and S. M. George. (1995). Surface-diffusion of H and Co on Cu/Ru(001) - evidence for long-range trapping by copper islands. *Chemical Physics* 201 (1):273-286 ISSN: 0301-0104.

Camacho-Lopez, S., R. Evans, L. Escobar-Alarcon, and M. A. Camacho-Lopez. (2008). Polarization-dependent single-beam laser-induced grating-like effects on titanium films. *Applied Surface Science* 255 (5):3028-3032 ISSN: 0169-4332.

Chakraborty, S., H. Sakata, E. Yokoyama, M. Wakaki, and D. Chakravorty. (2007). Laser-induced forward transfer technique for maskless patterning of amorphous V2O5 thin film. *Applied Surface Science* 254:638-643 ISSN: 0169-4332.

Danziger, I. M., B. Hallac, and M. Asscher. (2004). Surface diffusion of potassium coadsorbed with CO on Ru(001): A coverage grating-optical second harmonic diffraction study. *Journal of Physical Chemistry B* 108 (46):17851-17856 ISSN: 1520-6106.

Dou, Y. S., N. Winograd, B. J. Garrison, and L. V. Zhigilei. (2003). Substrate-assisted laser-initiated ejection of proteins embedded in water films. *Journal of Physical Chemistry B* 107 (10):2362-2365 ISSN: 1089-5639.

Dou, Y. S., L. V. Zhigilei, Z. Postawa, N. Winograd, and B. J. Garrison. (2001). Thickness effects of water overlayer on its explosive evaporation at heated metal surfaces. *Nuclear Instruments & Methods in Physics Research Section B-Beam Interactions with Materials and Atoms* 180:105-111 ISSN: 1520-6106.

Dou, Y. S., L. V. Zhigilei, N. Winograd, and B. J. Garrison. (2001). Explosive boiling of water films adjacent to heated surfaces: A microscopic description. *Journal of Physical Chemistry A* 105 (12):2748-2755 ISSN: 0295-5075.

Fei, Y. Y., and X. D. Zhu. (2006). Surface diffusion of hydrogen atoms on Cu(111) studied by optical diffraction from hydrogen density patterns formed on removable templates of xenon monolayers. *Europhysics Letters* 76 (5):877-883 ISSN: 0295-5075.

Frank, P., J. Graf, F. Lang, J. Boneberg, and P. Leiderer. (2010). Laser-induced film ejection at interfaces: Comparison of the dynamics of liquid and solid films. *Applied Physics a-Materials Science & Processing* 101 (1):7-11DOI 10.1007/s00339-010-5770-5 ISSN: 0947-8396.

Funk, S., B. Hokkanen, J. Wang, U. Burghaus, G. Bozzolo, and J. E. Garces. (2006). Adsorption dynamics of CO2 on Cu(110): A molecular beam study. *Surface Science* 600 (3):583-590 ISSN: 0039-6028.

George, M. A., Q. C. Bao, I. W. Sorensen, W. S. Glaunsinger, and T. Thundat. (1990). Thermally induced changes in the resistance, microstructure, and adhesion of thin gold-films on Si SiO_2 substrates. *Journal of Vacuum Science & Technology a-Vacuum Surfaces and Films* 8 (3):1491-1497 ISSN: 0734-2101.

Gu, X., and H. M. Urbassek. (2005). Atomic dynamics of explosive boiling of liquid-argon films. *Applied Physics B-Lasers and Optics* 81 (5):675-679 ISSN: 0946-2171.

Gu, X., and H. M. Urbassek. (2007). Laser-induced desorption of overlayer films off a heated metal substrate. *Applied Surface Science* 253 (9):4142-4149 ISSN: 0169-4332.

Ihlemann, J., and K. Rubahn. (2000). Excimer laser micro machining: fabrication and applications of dielectric masks. *Applied Surface Science* 154:587-592 ISSN: 0169-4332.

Kelly, M. K., J. Rogg, C. E. Nebel, M. Stutzmann, and S. Katai. 1998. High-resolution thermal processing of semiconductors using pulsed-laser interference patterning. *Physica Status Solidi a-Applied Research* 166 (2):651-657 ISSN: 0031-8965.

Kerner, G., and M. Asscher. (2004). Laser patterning of metallic films via buffer layer. *Surface Science* 557 (1-3):5-12 ISSN: 0039-6028.

Kerner, G., and M. Asscher. (2004). Buffer layer assisted laser patterning of metals on surfaces. *Nano Letters* 4 (8):1433-1437 ISSN: 1530-6984.

Kerner, G., O. Stein, M. Asscher. (2004). "Physisorbed buffer layer as a template for pulsed laser patterning of metallic thin films: an alternative approach for photolithography", *Proc. SPIE*, vol. 5531, pp. 128-136 DOI: 10.1117/12.557621.

Kerner, G., Y. Horowitz, and M. Asscher. (2005). Diffusion of buffer layer assisted grown gold nanoclusters on Ru(100) and p(1 x 2)-O/Ru(100) surfaces. *Journal of Physical Chemistry B* 109 (10):4545-4553 ISSN: 1520-6106.

Kerner, G., O. Stein, and M. Asscher. (2006). Patterning thin metallic film via laser structured weakly bound template. *Surface Science* 600 (10):2091-2095 ISSN: 0039-6028.

Kerner, G., O. Stein, Y. Lilach, and M. Asscher. (2005). Sublimative desorption of xenon from Ru(100). *Physical Review B* 71 (20) ISSN: 1098-0121.

Koehler, B. G., C. H. Mak, D. A. Arthur, P. A. Coon, and S. M. George. (1988). Desorption-kinetics of hydrogen and deuterium from Si(111) 7x7 studied using laser-induced thermal-desorption. *Journal of Chemical Physics* 89 (3):1709-1718 ISSN: 0021-9606.

Kudryashov, S. I., and S. D. Allen. (2003). Optical transmission measurements of explosive boiling and liftoff of a layer of micron-scale water droplets from a KrF laser-heated Si substrate. *Journal of Applied Physics* 93 (7):4306-4308 ISSN: 0021-8979.

Kudryashov, S. I., K. Lyon, and S. D. Allen. (2006). Near-field thermal radiative transfer and thermoacoustic effects from vapor plumes produced by pulsed CO2 laser ablation of bulk water. *Journal of Applied Physics* 100 (12) ISSN: 0021-8979.

Lang, F., and P. Leiderer. (2006). Liquid-vapour phase transitions at interfaces: sub-nanosecond investigations by monitoring the ejection of thin liquid films. *New Journal of Physics* 8, 14 ISSN: 1367-2630.

Lani, S., A. Bosseboeuf, B. Belier, C. Clerc, C. Gousset, and J. Aubert. (2006). Gold metallizations for eutectic bonding of silicon wafers. *Microsystem Technologies-Micro-and Nanosystems-Information Storage and Processing Systems* 12 (10-11):1021-1025 ISSN: 0946-7076.

Leufke, P. M., S. Riedel, M. S. Lee, J. Li, H. Rohrmann, T. Eimuller, P. Leiderer, J. Boneberg, G. Schatz, and M. Albrecht. (2009). Two different coercivity lattices in Co/Pd multilayers generated by single-pulse direct laser interference lithography. *Journal of Applied Physics* 105 (11) ISSN: 0021-8979.

Plech, A., P. Leiderer, and J. Boneberg. (2009). Femtosecond laser near field ablation. *Laser & Photonics Reviews* 3 (5):435-451 ISSN: 1863-8880.

Rasmussen, P. B., P. A. Taylor, and I. Chorkendorff. (1992). The interaction of carbon-dioxide with Cu(100). *Surface Science* 270:352-359 ISSN: 0039-6028.

Rosenzweig, Z., I. Farbman, and M. Asscher. (1993). Diffusion of ammonia on Re(001) - a monolayer grating optical 2nd-harmonic diffraction study. *Journal of Chemical Physics* 98 (10):8277-8283 ISSN: 0021-9606.

Samokhin, A. A. (2006). Estimate of pressure produced during explosive boiling of a liquid film on a substrate heated by laser pulses. *Quantum Electronics* 36 (7):684-686 ISSN: 1063-7818.

Shishido, A., I. B. Diviliansky, I. C. Khoo, T. S. Mayer, S. Nishimura, G. L. Egan, and T. E. Mallouk. (2001). Direct fabrication of two-dimensional titania arrays using interference photolithography. *Applied Physics Letters* 79 (20):3332-3334 ISSN: 0003-6951.

Smith, K. M., M. Y. Hussaini, L. D. Gelb, and S. D. Allen. (2003). Modeling laser-assisted particle removal using molecular dynamics. *Applied Physics a-Materials Science & Processing* 77 (7):877-882 ISSN: 0947-8396.

Stein, O., and M. Asscher. (2008). Adsorption of H2O, CO2 and Xe on soft surfaces. *Journal of Physical Chemistry B* 112 (13):3955-3962 ISSN: 1520-6106.

Stein O., Z. Lin, L.V. Zhigilei, M, Asscher, (2011)"Selective ablation of Xe from Silicon surfaces: MD simulations and experimental laser patterning", *Journal of Physical Chemistry A., Accepted* .

Verhoef, R. W., and M. Asscher. (1997). Effect of lateral interactions on coverage grating formation on surfaces. *Surface Science* 376 (1-3):389-394 ISSN: 0039-6028.

Weaver, J. H., and G. D. Waddill. (1991). Cluster assembly of interfaces - nanoscale engineering. *Science* 251 (5000):1444-1451 ISSN: 0036-8075.

Williams, P. A., G. A. Reider, L. P. Li, U. Hofer, T. Suzuki, and T. F. Heinz. 1997. Physisorbed template for spatial patterning of adsorbates. *Physical Review Letters* 79 (18):3459-3462 ISSN: 0031-9007.

Wong, A., A. Lee, and X. D. Zhu. (1995). Coverage dependence of quantum tunneling diffusion of hydrogen and deuterium on Ni(111). *Physical Review B* 51 (7):4418-4425 ISSN: 0163-1829.

Xiao, X. D., Y. L. Xie, and Y. R. Shen. (1992). Surface-diffusion probed by linear optical diffraction. *Surface Science* 271 (1-2):295-298 ISSN: 0039-6028

Xiao, X. D., Y. L. Xie, and Y. R. Shen. (1993). Coverage dependence of anisotropic surface-diffusion - CO/Ni(110). *Physical Review B* 48 (23):17452-17462 ISSN: 1098-0121.

Xiao, X. D., X. D. Zhu, W. Daum, and Y. R. Shen. (1991). Anisotropic surface-diffusion of CO on Ni(110). *Physical Review Letters* 66 (18):2352-2355 ISSN: 0031-9007.

Zhu X. D. (1992). Optical Diffractions as Probes to Surface Diffusion of Adsorbates. *Modern Physics Letters B*, 6 (20) 1217-1235.

Zhu X. D., A. Lee & A. Wang, (1991). Detection of Monolayer Gratings of Adsorbates by Linear Optical Diffraction. *Applied Physics A.* 52. 317-322, PACS: 78.65-s, 07.60. Hv, 68.35. Fx

Zhu, X. D., T. Rasing, and Y. R. Shen. (1988). Surface-diffusion of CO on Ni(111) studied by diffraction of optical second-harmonic generation off a monolayer grating. *Physical Review Letters* 61 (25):2883-2885 ISSN: 0031-9007.

Zhu, X. D., and Y. R. Shen. (1989). Generation and detection of a monolayer grating by laser desorption and 2nd-harmonic generation - CO on Ni(111). *Optics Letters* 14 (10):503-505 ISSN: 0146-9592.

Part 2

Laser-Matter Interaction

6

Interaction Between Pulsed Laser and Materials

Jinghua Han[1] and Yaguo Li[2,3]
[1]College of Electronics & Information Engineering, Sichuan University, Chengdu,
[2]Fine Optical Engineering Research Center, Chengdu,
[3]Department of Machine Intelligence & Systems Engineering, Akita Prefectural University, Yurihonjo,
[1,2]China
[3]Japan

1. Introduction

The research on laser-matter interaction can bridge the gap between practical problems and applications of lasers, which offers an important way to study material properties and to understand intrinsic microstructure of materials. The laser irradiation-induced effects on materials refer to numerous aspects, including optical, electromagnetic, thermodynamic, biological changes in material properties. The laser-matter interaction is an interdisciplinary and complicated subject [1]. When the material is irradiated with lasers, the laser energy will be firstly transformed into electronic excitation energy and then transferred to lattices of materials through collisions between electrons and lattices. The deposition of laser energy will produce a series of effects, such as temperature rise, gasification and ionization. The physical processes of interactions between lasers and matters can be grouped into linear and nonlinear responses of materials to laser pulses, namely thermal effects, nonlinear interactions, laser plasma effects and so forth [2,3]. This chapter aims at analyzing the above-mentioned major effects due to laser irradiation.

2. Thermodynamics

Laser ablation entails complex thermal processes influenced by different laser parameters, inclusive of laser pulse energy, laser wavelength, power density, pulse duration, etc (Fig. 1). According to the response of material to incident laser, the responses can be categorized into two groups: thermal and mechanical effects. Thermal effects refer to melting, vaporization (sublimation), boiling, and phase explosion while mechanical response involves deformation and resultant stress in materials. Different thermal processes will induce different mechanical responses, which will be detailed in the following.

2.1 Thermal effects

Materials subjected to laser irradiation will absorb the incident laser energy, raising the temperature and causing material expansion and thermal stress in materials. When the stress exceeds a certain value, the material may fracture and/or deform plastically. Material expansion will induce various changes in refractive index, heat capacity, etc.

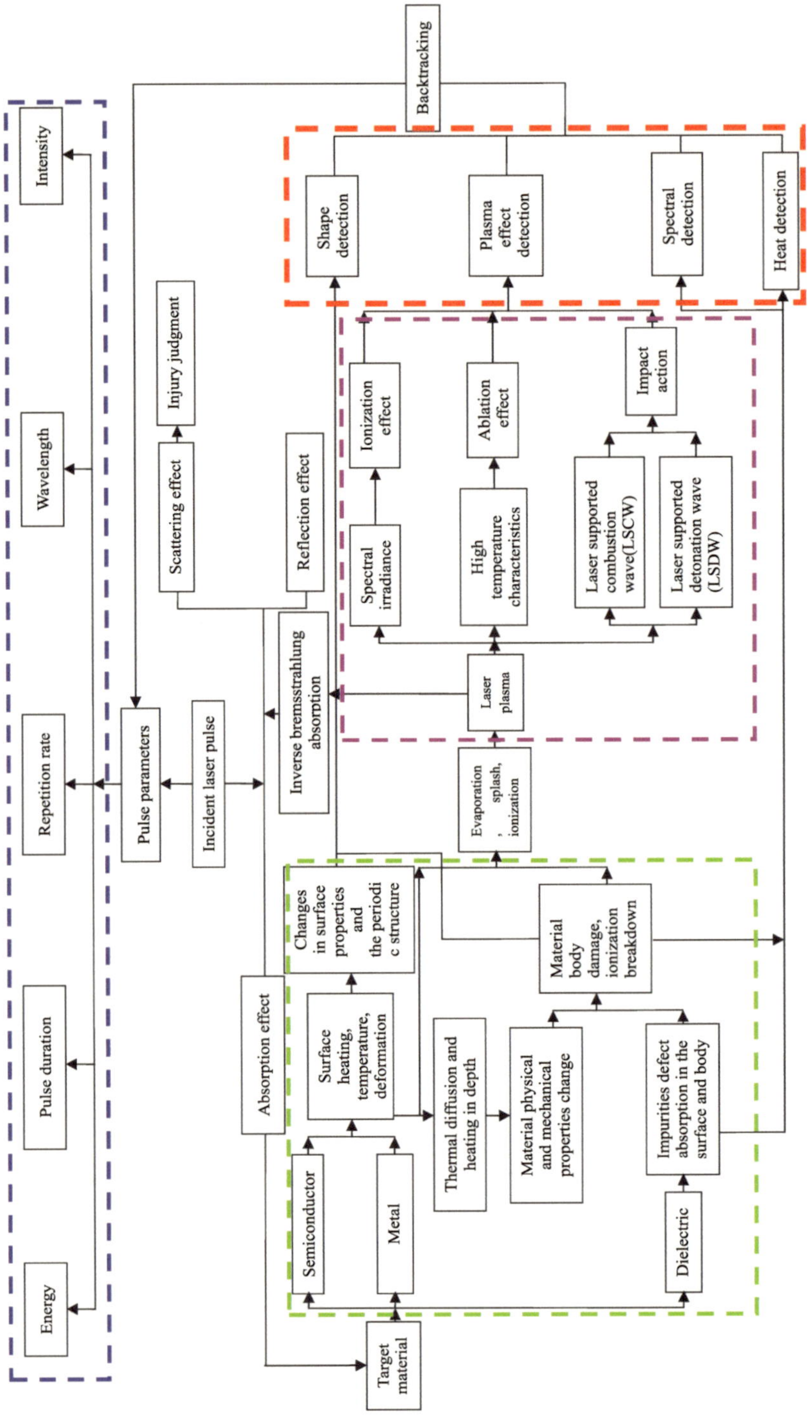

Fig. 1. Laser-matter interactions involve numerous complicated processes, inclusive of physical, mechanical, thermal, optical effects, etc. A full understanding of laser-matter interactions continues to be elusive.

The deposition of the laser pulse energy can heat the materials and raise the temperature of materials. Given that laser beam is perpendicular to the surface of materials (flat surface), the temperature with respect to time t and depth x will be:

$$\Delta T(x,t) = 2(1-R)\alpha I_0 \frac{t}{\pi k \rho C} ierfc \frac{x}{2\sqrt{kt/\rho C}} \tag{2.1}$$

where, t is the laser pulse irradiation time, R is the reflectivity, α is the absorptivity, I_0 is the spatial distribution of laser intensity, k is thermal conductivity, ρ is the density of irradiated materials. When $x > 4\sqrt{kt/\rho C}$, the surface temperature will be simplified as:

$$\Delta T(t) = \frac{2\alpha I_0 \sqrt{t}}{\sqrt{\pi k \rho C}} \tag{2.2}$$

The temperature rise may alter physical and optical properties of materials. The influence of temperature rise will be discussed in more detail.

A Analysis of damage threshold

If the laser energy level at which the irradiated materials start to melt is referred to as the damage threshold (LIDT) of the materials, it is clear that the LIDT is directly proportional to $\sqrt{\tau}$ as shown in Eqt. (2.2). A number of experiments evidence that for laser pulses that τ >10ps, the proportional relationship is applicable to vast majority of semiconductor materials, metals, and dielectric thin films coated on optical components, etc. However, the damage threshold increases with decreasing pulse duration for the laser pulses <10ps. The variation is due to different damage mechanisms of materials when subjected to ultra-short laser pulses[4], since the heat diffusion does not accord with the Fourier's heat conduction law.

B Thermal distortion and stress in solid-state lasers

Materials can absorb the energy of the incident laser, a part of which will be converted into heat. Non-uniform temperature distribution will appear because of the uneven heat

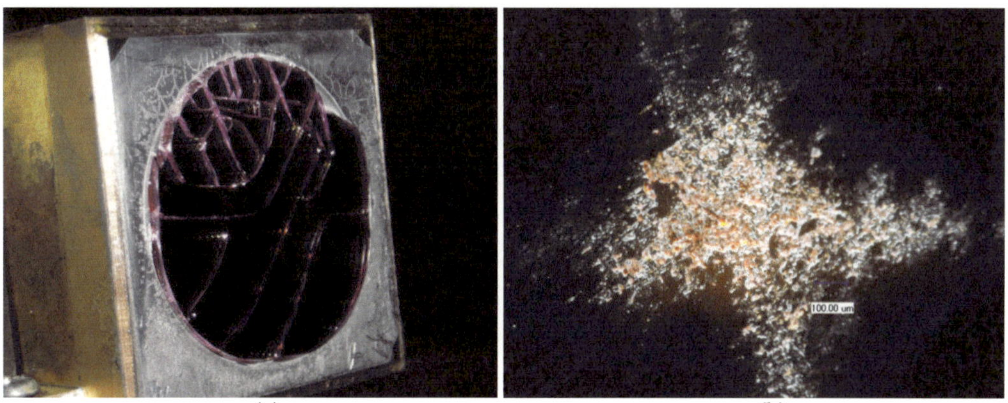

(a) (b)

Fig. 2.1. Thermal fractures of Nd:YAG and melting of SiO_2 thin film coated on Nd:YAG in a high-energy laser (Courtesy of Dr. Huomu Yang) diffusion. Consequently, expansion and contraction will lead to laser-induced thermal stress. The stress can limit the average workable power of solid-state lasers (Fig. 2.1). Thermo-aberration can seriously affect the uniformity of the output laser field and therefore induce the phase distortion (Fig. 2.2).

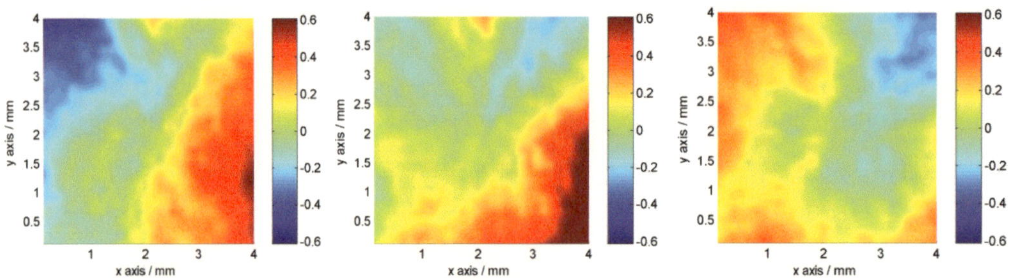

(a)original (b) irradiated for 15min (c) distortion

Fig. 2.2. The distorted wavefront in laser heated K9 glass (The wavelength was 635nm and Shack Hartmann sensor was used to record the wavefront distortion. Courtesy of Dr.Yongzhao Du)

C Frequency doubling

The deposition of laser pulse energy can result in thermal depolarization in optical crystals for doubling/tripling frequency and also degrade the efficiency of frequency doubling. Self-thermal-effect resulting from pump loss will influence the harmonic conversion of the incident laser.
During the process of harmonic conversion, crystals inevitably absorb the energy of fundamental frequency light and frequency-doubled/tripled light. Part of the absorbed energy will convert into heat leading to uniform temperature rise in crystals, which will give rise to a refractive index ellipsoid and disturb phase matching. Furthermore, harmonic conversion efficiency will drop and the quality of output beam will deteriorate [5].

2.2 Melting and solidification

With the increase of laser pulse energy, materials will absorb more laser energy and the deposited energy will cause the material to melt in the case that materials temperature

Fig. 2.3. Morphologies of melting damage on the end surface of end-pumped fiber laser. The material is continuously heated with repetitive pumped laser pulses and finally damaged due to non-uniform thermal stress. (Courtesy of Dr. Xu Han)

exceeds the melting point (Fig. 2.3). Melting followed by solidification will change the atomic structure of materials and can realize the mutual transformation between crystalline and amorphous state.

2.3 Ionization and gasification

Laser-induced gasification can be divided into surface gasification and bulk gasification. As the temperature continues to increase to the vaporization point, part of the absorbed laser energy is converted into the latent heat of evaporation, the kinetic energy of gasification and the quality of spray steam. With increasing the laser intensity, the melted materials will be gasified and/or ionized. The gasification is discussed based mainly on liquid-gas equilibrium. Gaseous particles with the Maxwell distribution will splash out from the molten layer. The gasified particles are ejected several microns away from the surface. The space full of particles is the so-called Knudsen layer.

The ionization will greatly enhance the absorption and deposition of the laser energy. After ionization is completed, the inverse bremsstrahlung absorption dominates the absorption of plasma. Re-crystallization of the ionized materials may cause changes in material structure. The damage of SiO_2 thin film coated on $LiNbO_3$ crystal is taken as an example (Fig. 2.4):

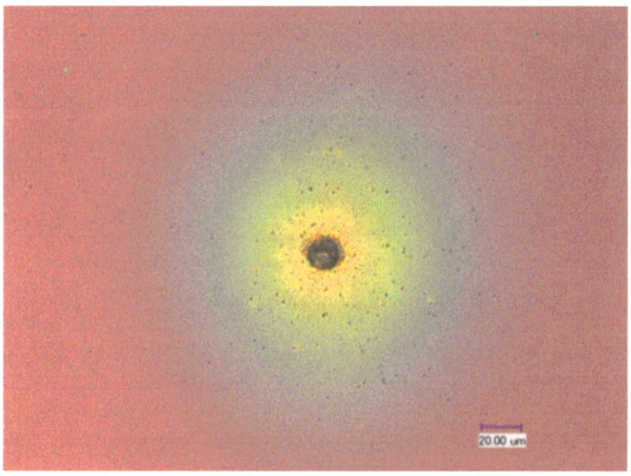

(a) The whole damage morphology

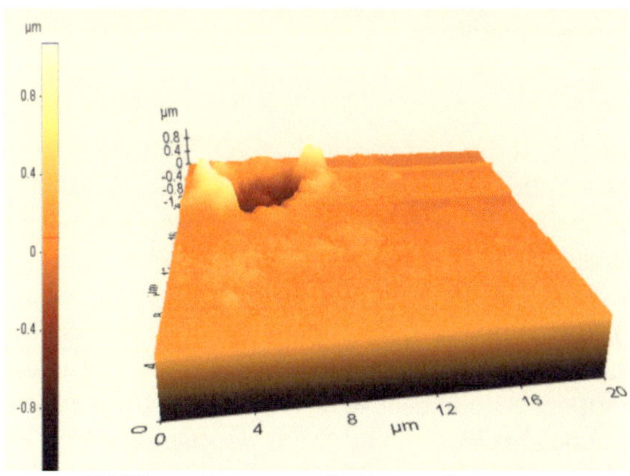

(b) The micro-morphology of a crater

Fig. 2.4. Damage morphologies of laser induced SiO_2 thin film. (Courtesy of Ms. Jin Luo)

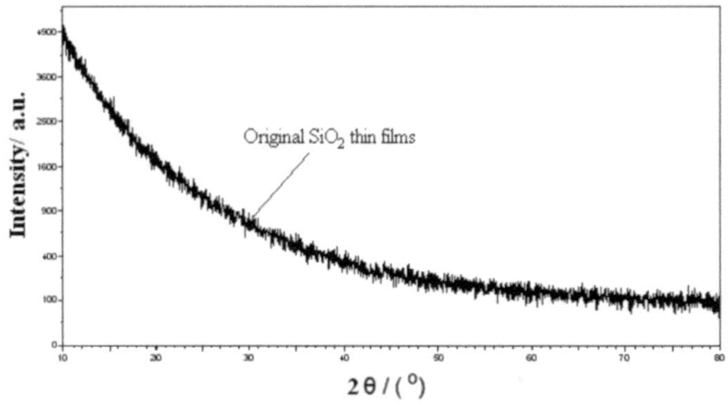

(a) Original SiO_2 thin film

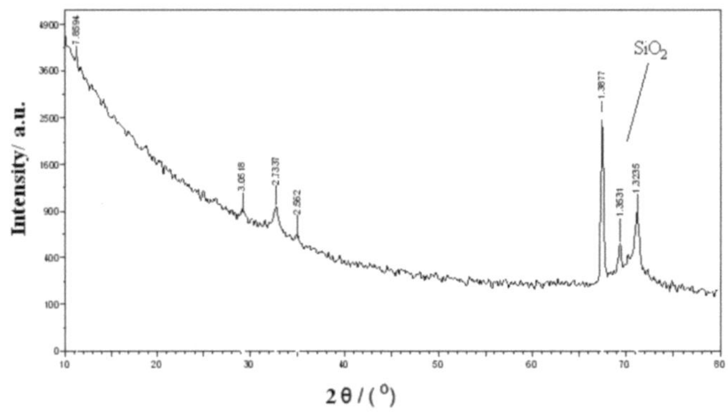

(b) Damaged SiO_2 thin film

Fig. 2.5. The XRD spectra of SiO_2 thin films on lithium niobate crystal (Courtesy of Dr. Ruihua Niu)

Figure 2.5 (a) shows that the film without being damaged is amorphous in that no diffraction peaks appear in the XRD spectrum whilst several apparent peaks are apparent in Fig. 2.5 (b), indicating the appearance of crystalline silica. It can be concluded that ionization can cause material to be re-crystallized.

2.4 Phase explosion

Phase explosion is another important thermal effect. The occurrence of phase explosion follows the stages: the formation of super-heated liquid owing to laser energy deposition; then the generation and growth of nucleation in super-heated liquid and explosion of nucleation.

The physical process is depicted in Fig. 2.6. Upon the irradiation of laser, the temperature of materials will rise and the deposited energy diffuses into the bulk of materials to a certain depth (Figure 2.6(a)); the temperature of melted materials sharply increase to over the boiling point due to the heavy deposition of laser energy; nevertheless, the boiling does not start and the liquid is super-heated because of the absence of nucleation (Figure 2.6(b)); the disturbance will bring about nucleation and the super-heated liquid thickens as the size and the number of bubbles grow (Figure 2.6(c)); the startling boiling will arise once the size of bubbles is sufficiently large and afterwards the super-heated liquid and particles will be ejected. This way, the phase explosion takes place.

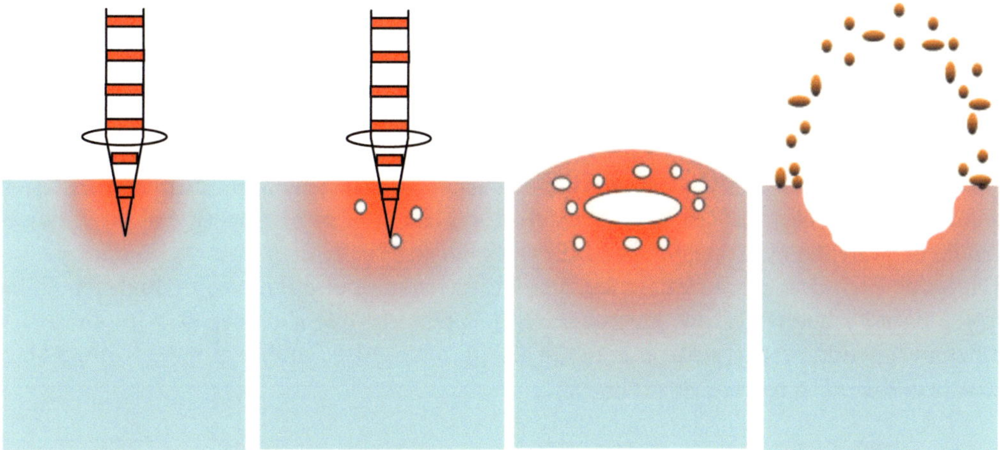

(a) (b) (c) (d)

Fig. 2.6. The generation of phase explosion

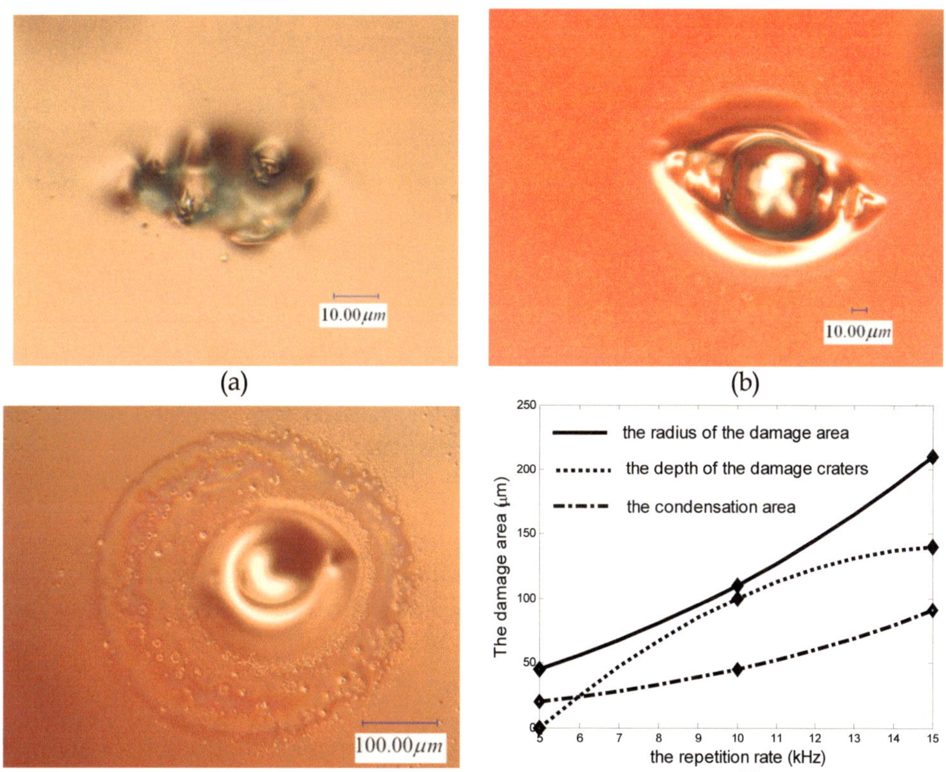

Fig. 2.7. The damage morphology induced by different repetition rate laser pulses. (a) The damage morphology induced by pulses with repetition rate of 5 kHz. (b) The damage morphology induced by pulses with repetition rate of 10 kHz. (c) The damage morphology induced by pulses with repetition rate of 15 kHz. (d) The dependence of the depth, size and of the damaged craters on the repetition rate.

In order to generate phase explosion, three requirements must be met: 1. the fast creation of super-heated liquid, the temperature of which should at least be (0.8-0.9) Tcr (Tcr is the critical temperature) [6]; 2. the thickness of super-heated liquid is large enough to accommodate the nuclear, usually on the order of tens of microns; 3. sufficient time t_c during which the size of nucleation reaches the critical size r_c, generally several hundreds of picoseconds. All the three factors are indispensable [7]. The generation of phase explosion requires specific laser pulses and material properties. The power density of laser pulses should be more than the threshold of materials (~10^{10}W/cm^2).

The phase explosion can be generated not only by single pulse but also by high-repetition rate pulses [8]. Shown below are the morphologies of craters damaged with pulses of different repetition rates (pulse energy Q= 42.7μJ, total pulse number N = 3.6 × 10^6) 5 kHz, 10 kHz and 15 kHz, respectively (Fig. 2.7).

(a) Phase explosion damage (b) The center of the depression pit (c) Molten zone and the microparticles

Fig. 2.8. Damage morphology induced by phase explosion (15kHz)

Fig.2.8(a) through 2.8(c) present the damage morphologies of materials exposed to high-repetition pulsed laser. There exists successively micro-size particles populated region and melting region from the center of the crater. Numerous micro-granules can be seen in the melting region. The set of pictures imply that the material was damaged due to phase explosion induced by the high-repetition-rate pulsed laser.

3. Effects of nonlinear interaction

Irradiated by high-intensity laser, the material exhibits a variety of nonlinear effects, such as self-focusing, multi-photon ionization, avalanche ionization, etc. The following analyzes the processes of small-scale self-focusing and nonlinear ionization.

3.1 Nonlinear ionization

When the laser beam of low energy is incident onto transparent material, linear absorption happens alone. The electrons in valence band will absorb incident laser and transit from bound states to free states when materials are irradiated with high energy lasers, which is referred to as nonlinear ionization containing two different modes: photo-ionization and avalanche ionization.

The band gap in dielectrics is wide and a single photon is not able to induce ionization and the material cannot directly absorb incident laser of low intensity. Photo-ionization consists of multi-photon ionization (MPI) and tunnel ionization: if the electric field is strong enough to make the electrons overcome potential barrier and ionize, the ionization is called tunnel ionization; multi-photon ionization is the process that the electron absorbs more photons at a time to gain enough energy beyond potential trap and to be ionized.

The Keldysh parameter can be used to classify multi-photon ionization and tunnel ionization, depending on the frequency and intensity of the incident laser and material band-gap[9].

$$\gamma = \frac{\omega}{e}\left[\frac{mcn\varepsilon_0 E_g}{I}\right]^{1/2} \tag{3.1}$$

where, ω is laser frequency, I is the laser intensity at focal point, m is the reduced mass, e is electron charge, c is the speed of light, n is the refractive index, ε_0 is material dielectric constant, E_g is material energy gap.

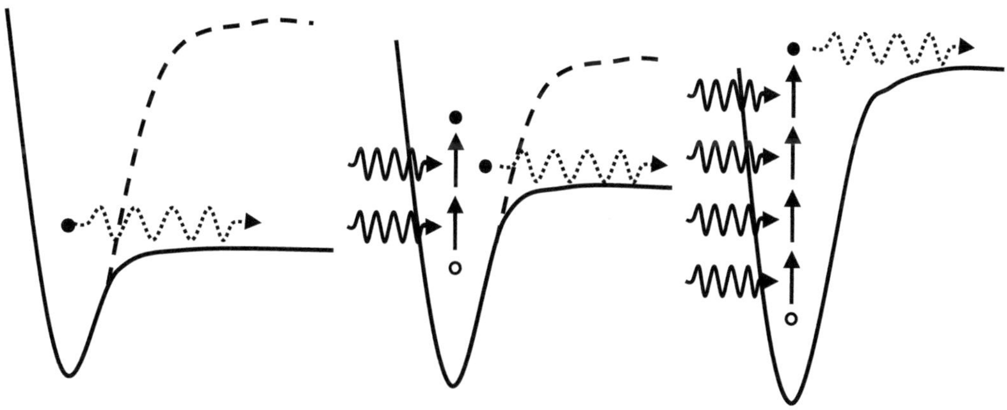

γ<1.5 tunneling γ=1.5 intermediate γ>1.5 MPI

Fig. 3.1. Schematic of photo-ionization for different Keldysh parameters.

As $\gamma > 1.5$ the primary effect is multi-photon ionization; while $\gamma < 1.5$ the main effect is tunnel ionization (Fig.3.1). Both effects should be considered for the transitional state. It also can be seen that when the material is exposed to low frequency and high power laser, tunnel ionization plays the leading role in nonlinear photo-ionization; otherwise, multi-photon ionization is the primary effect.

Conduction band electrons (seed electrons) in material can absorb subsequent photons to raise its energy. When the energy of conduction band electrons rise to a certain degree, the energized electrons can excite electrons in valance band to conduction band through collisions with other valance band electrons and produce a pair of conduction band electrons with lower kinetic energy. The number of conduction band electrons increases exponentially. The above process is the avalanche ionization (Fig. 3.2).

Nonlinear ionization can cause the increase in the density of free electrons which they strongly absorb laser energy, and in turn the density of free electrons increases sharply, which eventually induces the laser plasma and results in breakdown damage.

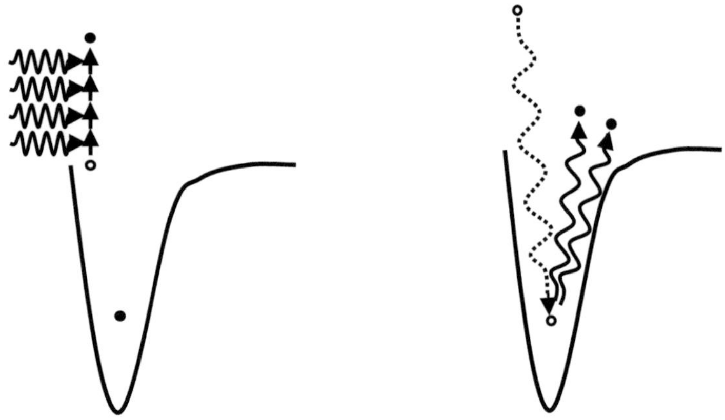

Fig. 3.2. Schematic diagram of the avalanche ionization

3.2 Self-focusing

The refractive index varies accordingly with the increase of the laser density, which can be written as $n = n_0 + n_2 I$, where $I = \frac{1}{2}\varepsilon_0 c n_0 |E|^2$ and $n_2 = \frac{3\chi(3)}{4\varepsilon_0 c n_0^2}$ .The parameter n_2 is related to laser self-focusing and self-phase modulation. When $n_2 > 0$, the medium can be considered a positive lens and self-focusing occurs when the beam travels through the medium; otherwise the defocusing happens. In light of the difference in pulse duration and nonlinear polarization time, self-focusing can be grouped into steady-state self-focusing (continuous wave of invariable amplitude), quasi-steady self-focusing (both field and power are functions of the delayed time), and transient self-focusing (when pulse duration is shorter than or similar to medium response time, the medium response time must be taken into account). In addition, the small scale self-focusing caused by the incident beam with uneven distribution of intensity or irregular modulations can result in beam splitting, medium filamentous damage, and spectrum detuning, etc.

Kerr lens effect is continuously pronounced with increasing the pulse power of laser and self-focusing becomes conspicuous until the laser power approaches the critical power at which a balance is struck between the wave-front bending caused by the diffraction and self-focusing lens. In this way, the light beam will transmit in the form of filament (Fig. 3.3). When self-focusing occurs, the nonlinear ionization can produce laser plasma and lead to filamentous destruction (Fig. 3.4). In addition, the self-defocusing of laser plasma is an obstacle to further self-focusing.

The mechanism of small scale self-focusing has been studied since early 1970s. The classical theory is B-T theory [10]. The B integration characterizes the size of self-focusing damage, which is named after Breakup-integral $B = \left(2\pi/\lambda_0\right)\int \gamma I_0(z)dz$. B integral is a criterion for determining the extent of small scale self-focusing and the causes of additional phase as well as the sources of phase modulation and spectral broadening.

B-T theory remains the basic theory for nonlinear optical transmission. The world's largest high-power solid laser–'National Ignition Facility' (NIF) is designed based on B-T theory [11].

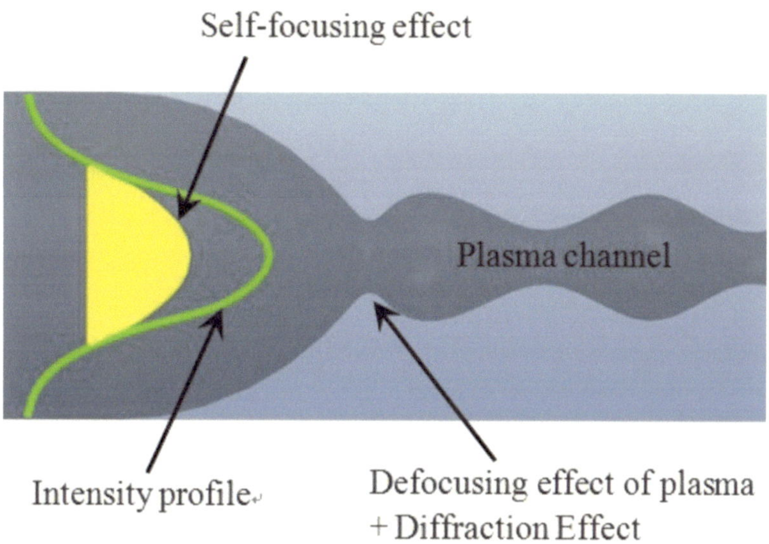

Fig. 3.3. The illustration of self-focusing filaments

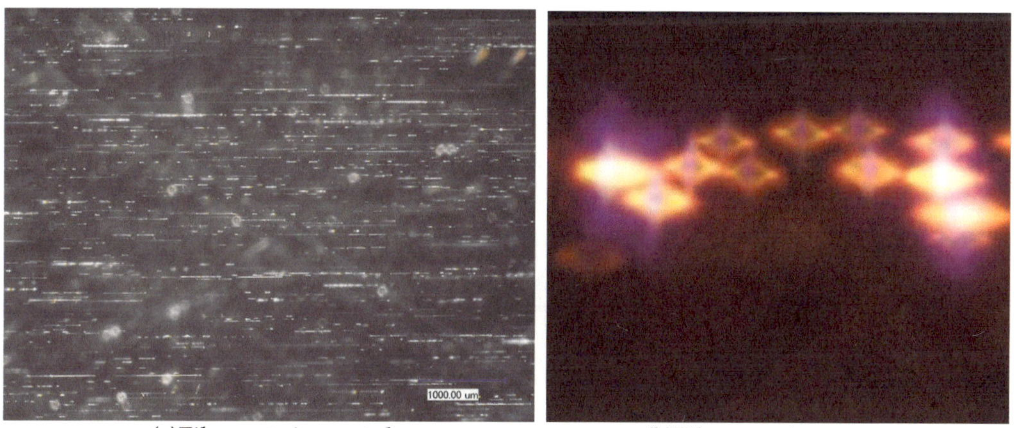

(a)Filaments in crystals (b)Filaments in water

Fig. 3.4. Small-scale self-focusing. (Courtesy of Dr. Ruihua Niu and Dr. Binhou Li)

3.3 Extrinsic damage

Dielectrics have wide band-gap and low absorptive capacity and possess high intrinsic damage threshold. However, the damage factually occurs at the laser intensity several orders of magnitude lower than the intrinsic threshold of materials, which is due mostly to the extrinsic damage. In other words, the impurities of the narrow band gap material can severely lower the damage threshold of dielectrics. When impurities of narrow band gap exist in dielectrics, the impurities can absorb laser strongly and sharply increase energy deposition locally. The rapid deposition of laser energy can result in melting, gasification ionization of dielectrics and laser plasma and therefore local damage (Fig. 3.5).

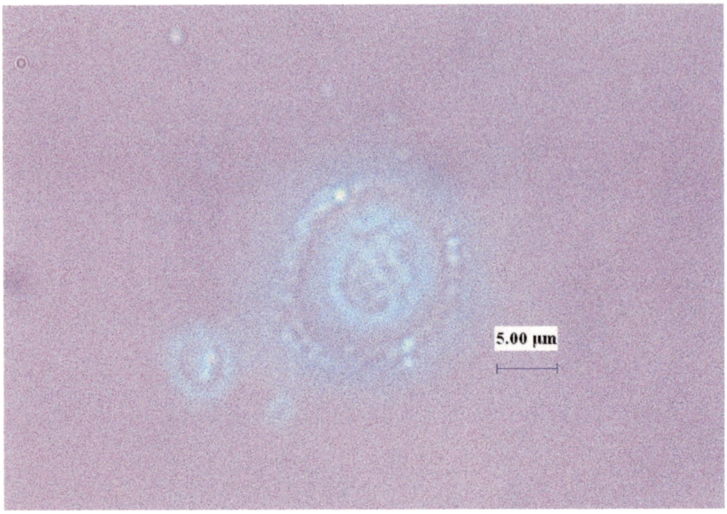

Fig. 3.5. The ripples of SiO_2 antireflection coating due to laser damage

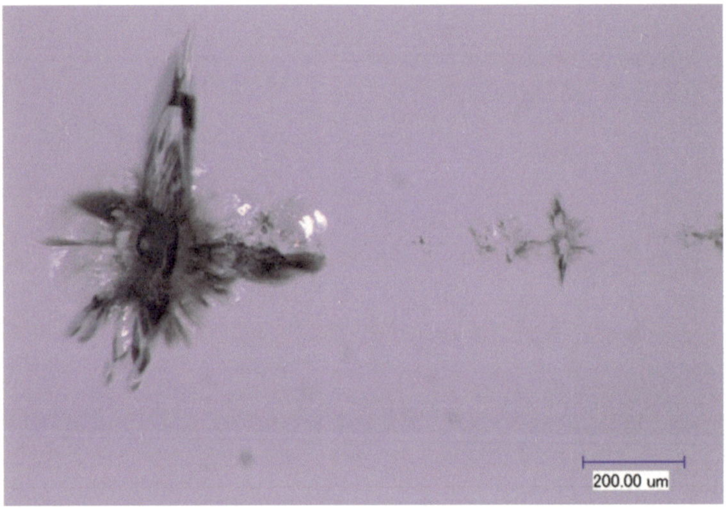

Fig. 3.6. The laser damage in the bulk of K9 glass (1064nm, 13.6ns) (Courtesy of Dr. Guorui Zhou and Dr. Shutong Wang)

The self-focusing filamentous damage in K9 glass is characterized by the connection of filamentous destruction and burst damage caused by particles that strongly absorb the laser energy (Fig. 3.6) [12]. In high-power laser systems, the elimination of platinum inclusions in Nd:Glass is of great importance so as to improve the damage threshold of Nd:glass [13].

4. Laser induced plasma shock wave

4.1 Shock wave formation

As the laser plasma with high temperature and pressure expands outward, shock waves will be formed. In fluid dynamics, the shock wave generated by the blast in early 1930s has been studied in detail and the point explosion model was proposed. Based on the model of point explosion, zel'dovich and Raiser systematically studied the laser plasma expansion and developed the Sedov-Taylor instantaneous point explosion [14].

Taking into account the lasting time of real explosion, the process of shock wave is considered to consist of two stages.

1. When $t \le \tau_0$, shock wave starts due to the ablation of laser to target materials. High-energy pulsed laser ablates and sputters the target materials to form plasma; the plasma expands immediately and rapidly and forms shock wave. In the meantime, shock wave continues to absorb the laser energy, which keeps expediting the shock waves. When $t = \tau_0$, the speed of shock wave is maximized at the end of laser action. The first stage of the formation of shock wave ends.

$$R(t) = A_0\left(\alpha E(t)\right)^{1/2+v} t^{2/2+v} \qquad t \le \tau_0 \tag{4.1}$$

where $A_0 = A(\rho_1)^{-1/2+v}$, α is the ratio of the energy transferred to shock waves to the total energy of a laser pulse.

2. When $t > \tau_0$, the stage is characterized with the expansion and propagation of shock waves in the air. It is experimentally proved that shock wave eventually decays into acoustic pulses to propagate in sonic speed in the air.

$$R(t) = A_1(t-\tau_0)\left(1-\alpha_0 \exp\left[-k\left(\frac{\tau}{t-\tau_0}\right)^n\right]\right) + R_0 \qquad t > \tau_0 \tag{4.2}$$

where $A_1 = \dfrac{2A_0}{2+v}\tau_0^{-v/2+v}\left(\alpha E(t)\right)^{1/2+v}$, k, α_0, n are undetermined coefficients.

No more energy is replenished during the propagation of shock wave because the laser irradiation discontinues. Thereby, although the radius of shock wave front keeps increasing, the velocity of shock wave propagation slows down and the intensity of shock wave is dwindling.

4.2 Laser supported absorption wave (LSAW)

Ionized vapor can absorb partial or entire energy of the incident laser to form plasma on the surface, so the temperature and intensity in gasification surface is the highest. A part of plasma expands outwards and the rest is constrained within the light path, which is different from normal gases in motion and referred to as laser supported absorption wave (LSAW) [1].

The plasma is ionized in part by low energy pulses and absorbs partial energy of pulses and the plasma shock wave propagates at the subsonic speed due to thermal conductivity, which is laser supported combustion wave (LSCW). If the laser pulse intensity is increased, the pressure, temperature and velocity of absorption wave increase correspondingly and the absorptivity of wave will be further enhanced and thus the wave consumes most of the laser energy. Then the plasma will contract and plasma propagates at supersonic speed, which is the laser supported detonation wave (LSDW).

The LSDW damaged the target materials seriously due to its exceedingly high temperature and pressure. Assuming that expansion process is isentropic and one-dimensional, the velocity and pressure of plasma wave with respect to target materials can be formulated as [1]:

$$p_2 = p_1[(\gamma+1)/2\gamma]^{2\gamma/(\gamma-1)} \tag{4.3}$$

$$p_1 = \rho_0 u^2 / (\gamma+1) \tag{4.4}$$

$$u = [2(\gamma^2-1)I / \rho_0]^{1/3} \tag{4.5}$$

where p_1, p_2 and u are the pressure of LSDW on the target surface, wave rear pressure of LSDW wave and the expansion rate of LSDW wave front, respectively, ρ_0 is the density of ambient atmosphere , γ is adiabatic coefficient of the plasma, I is the power density of laser on target surface. The temperature also depends on the power density of laser pulses [15]. $T = \alpha_T I^{1/(\beta+4)}$,where α_T and β are constants, I is the power density of incident laser in W/cm^2 .

4.3 Effect of laser plasma spectral irradiance

Laser plasma with a very wide spectrum from ultraviolet to infrared is composed of background continuous spectrum and linear spectrum due to elements including in the plasma. The continuous spectrum of radiation is mainly attributed to bremsstrahlung and recombination radiation process [16]. The bremsstrahlung radiation is the process that the electromagnetic wave is emitted due to the transition of free-state to free-state contributed to the collision between free electrons and ions. The temperature of free electron of plasma descends quickly during the process. Recombination radiation process is the process during which free electrons are captured to be bound-state by ions in electron-ion collision; meanwhile excessive energy is emitted in the form of electromagnetic radiation. The continuous spectrum with shorter wavelength and more significant intensity results from the bremsstrahlung, while the longer can be ascribed to recombination radiation. The laser plasma also has irradiant ionization effects. According to Keldysh ionization theory, the shorter wavelength, the higher the photon energy and the more likely the materials will be ionized; the short wavelength of laser plasma is much shorter than incident laser, so the ionization effect of short wavelength laser is more apparent.

The laser plasma effects act on the materials synergically. As an example, the breakdown of polycrystalline silicon by 1064nm focused laser beam is demonstrated in Fig 4.1. The laser used was 18ns in pulse duration (FWHM) and 500mJ of pulse energy. The focused spot is ~150 microns in diameters.

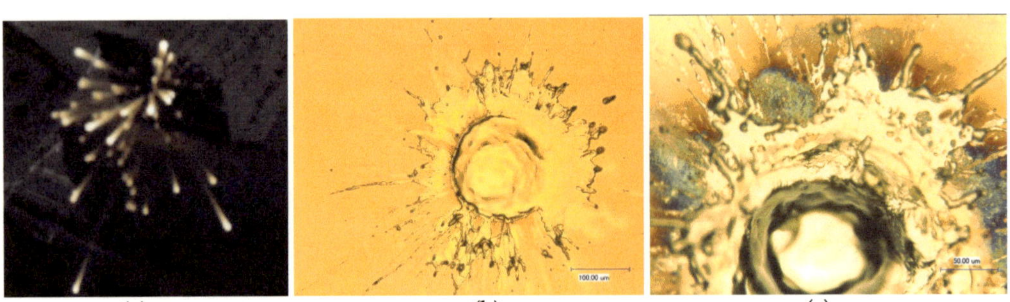

(a) (b) (c)

Fig. 4.1. The laser induced damage in polycrystalline silicon. (a) the sputtering of laser-induced plasma. (b) the damaged area is comprised of a crater and splashed materials. (c) the damage morphology created with 7Hz repetition rate laser for 1s. (Courtesy of Dr. Yanyan Liu)

The process of laser damage can be described by the action of LSDW，ionization，etc. The results are shown as follows (Fig. 4.2).

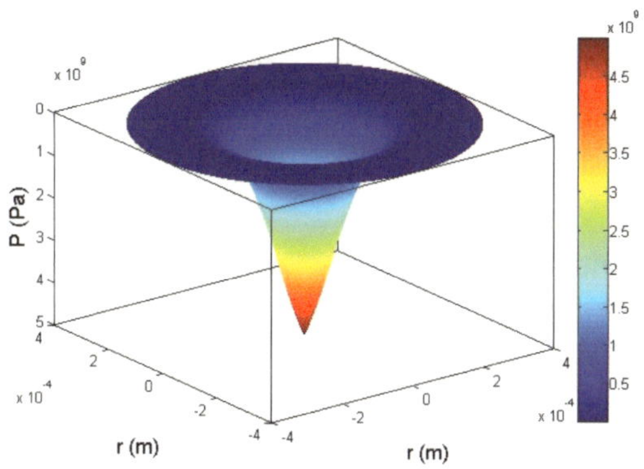

(a) The pressure distribution of LSDW

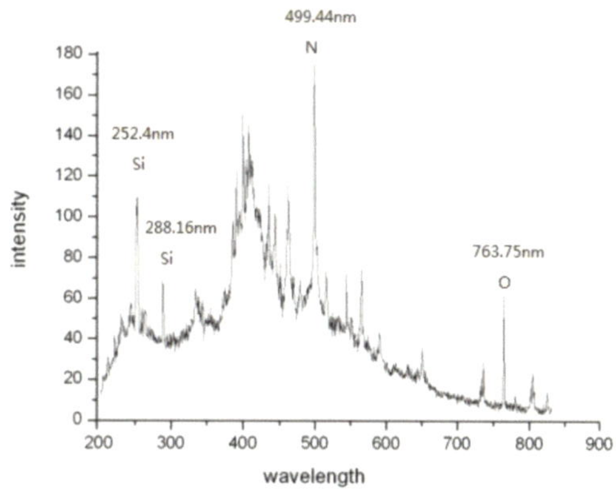

(b) Plasma spectra.

Fig. 4.2. LSDW and laser-induced plasma spectra. (Courtesy of Dr. Lingdong Bao)

The thermal and ionization cause materials to be fully melted and ionized and then the mixture is pushed out by LSDW. The laser plasma enhances material ionization. Thus the comprehensive effects form the damage pit surrounded by cooled material.

5. The femtosecond laser effects

Ultrashort pulse laser distinguished from nanosecond pulse laser lies in its peculiar laser-matter mechanism, supercontiuum generation and color-center accumulation etc. Then we will start with the ultra-short laser-matter mechanism. The ulatrafast laser refers to femtosecond pulsed laser in this chapter unless otherwise specified.

5.1 Mechanism of interaction between femtosecond laser and matters

The pulse duration of femto-second pulse is of the order of 10^{-15}s much shorter than conventional pulsed lasers. The electrons in materials absorb the energy of incident laser and their kinetic energy will increase. The activated electrons transfer energy to lattice by means of electron-lattice collision. This way, the temperature of materials rises. It needs several femtoseconds for electrons to absorb the laser energy (interactions between the photons and electrons) and the time of electron-lattice collision is of the order of picoseconds while lattices of materials melt within several nanoseconds.

There are plenty of free electrons in metals and semiconductors and the laser energy can be deposited by absorbing the energy of photons directly. The laser action process can be formulated with equation [17].

$$\begin{cases} C_e \dfrac{\partial T_e}{\partial t} = K\nabla^2 T_e - g(T_e - T_i) + A(r,t) \\ C_i \dfrac{\partial T_i}{\partial t} = g(T_e - T_i) \end{cases} \quad (5.1)$$

where, C_e、C_I are the heat capacity of electrons and lattices, K the conductivity of materials, A the thermal source relating to laser pulse, g the coefficient of electron-phonon coupling.

The first equation represents the deposition of laser energy to electrons and the other is the coupling of electron energy to lattices.
As to dielectrics with very low free electron density, dielectrics must be ionized by high intensity laser. The ionization effect of femto-second laser on dielectrics can be divided into multiphoton ionization and avalanche ionization. Then ionized electrons can absorb the incident laser energy dramatically; however there is not adequate time for the transferring of energy to lattices and thus the heat diffusion is impossible. By contrast, the energy in nanosecond laser pulses has enough time to be transferred from electrons to lattices, leading to thermodynamic damage.
Speaking generally, the craters created with femtosecond laser are smooth on the edge as compared to nanosecond pulsed lasers (Fig. 5.1). The femtosecond laser damage is more deterministic than ns lasers because of no obvious thermal effects and can be employed to accurately inscribe microstructures[18].

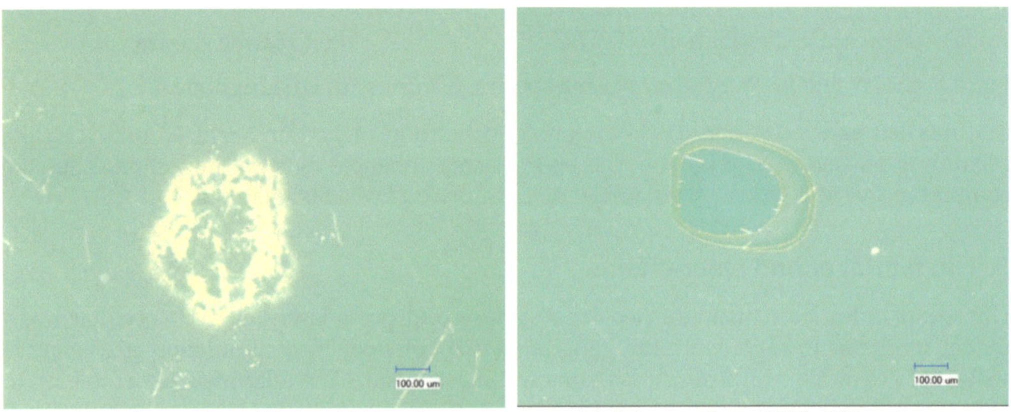

(a) ns laser pluse(13.6ns,1064nm) (b) fs laser pluse(135fs,800nm)

Fig. 5.1. The damage morphologies of SiO_2 thin film induced by nanosecond and fs laser pulse.

5.2 Supercontiuum generation

Supercontinuum generation (SCG) is another important property of fs laser, which belongs to nonlinear optical phenomenon. SCG involves a broad spectrally continuous output when narrow-band incident pulses undergo extreme nonlinear spectral broadening. The SCG has many novel applications in telecommunication, high precision frequency metrology, carrier phase stabilization, medical imaging and pulse compression for its spatial coherence, high brightness, and broad bandwidth [19,20].
The broadband pulse propagation in waveguide can be described by the nonlinear envelope equation. The supercontinuum generation mechanisms includes self-phase modulation (SPM), induced-phase modulation (IPM), crossed-phase modulation (XPM), soliton effects, Raman shift and coupling with dispersive waves, modulation instability, four wave mixing (FWM), the main effects leading to the generation of a broad spectrum starting from a narrow laser line.
The propagation of laser pulse in nonlinear materials can be expressed with nonlinear Schrödinger equation[21]:

$$\frac{\partial E}{\partial z}=\frac{ic}{2n_0\omega_0}(1+\frac{i}{\omega_0}\frac{\partial}{\partial t})^{-1}\nabla_{\perp}^2 E-\frac{i}{2}\beta_2\frac{\partial^2 E}{\partial t^2}+\frac{\beta_3}{6}\frac{\partial^3 E}{\partial t^3}-\frac{i}{24}\beta_4\frac{\partial^4 E}{\partial t^4}$$
$$+i\gamma\left[|E|^2 E+\frac{i}{\omega_0}\frac{\partial}{\partial t}(|E|^2 E)-T_R E\frac{\partial|E|^2}{\partial t}\right]-\frac{\alpha}{2}E \quad (5.2)$$

where, E is the amplitude, c the speed of light, ω_0 the pulse center frequency, n_0 the linear refractive index, β_2 the group velocity dispersion, β_3 third-order dispersion constant, β_4 fourth-order dispersion constant, α the loss coefficient. The first term describes the diffraction and the second, third and fourth terms are material dispersion terms.
There are a variety of nonlinear media that can be used to generate wide supercontinuum, including solids, organic and inorganic liquids and gas [22]. PCF (Photonic crystal fiber) has become an important media to generate wide supercontinuum because of high nonlinearity and low peak power pump threshold (Fig. 5.2).

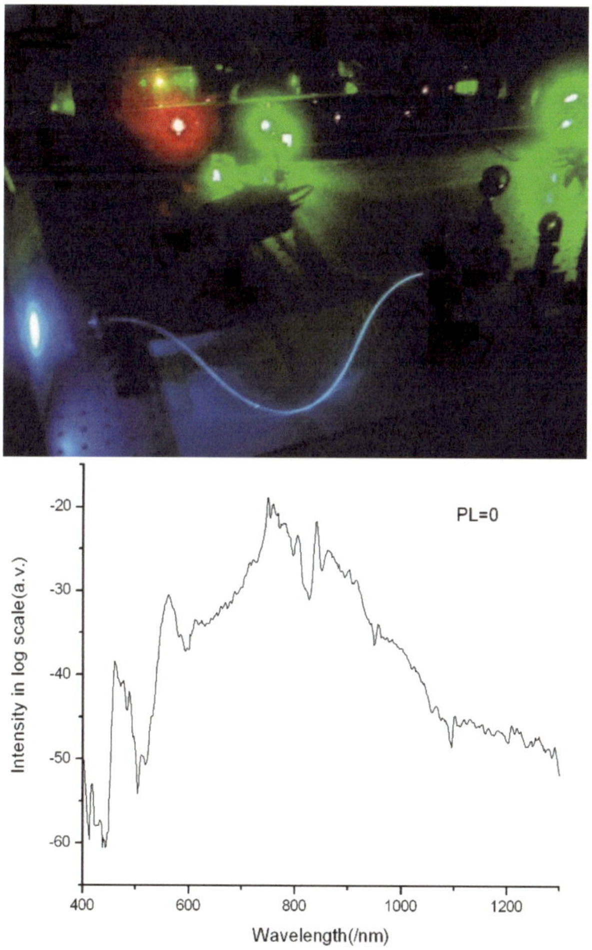

Fig. 5.2. Photonic crystal fiber and supercontinuum spectrum. (Courtesy of Dr. Ping Ying and Dr. Zairu Ma)

5.3 Color centers

Laser induced color center is a major type of material fatigue. It can be reduced by fs laser with high intensity or ns laser pulse with shorter wavelength. The color centers in materials will form if the electron-hole pairs ionized by laser are captured by defects or impurities.

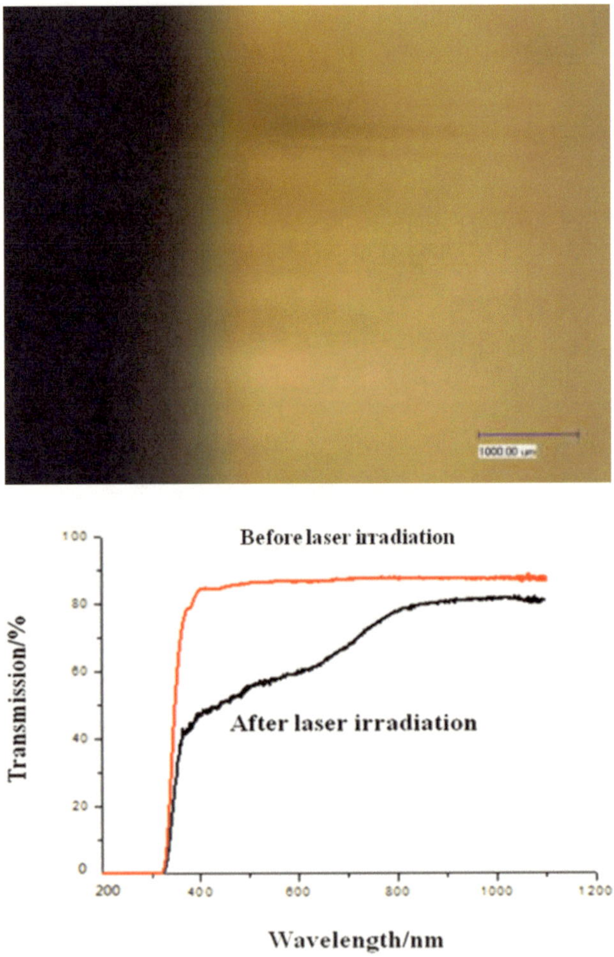

Fig. 5.3. Femtosecond laser induced darkening and transmission spectrum in K9 glass

The figures 5.3 & 5.4 show the grey tracking and corresponding absorption spectra in K9 glass and KTP crystals induced by high-repetition laser pulses with wavelength of 335nm, pulse duration of 13.6ns.

The absorption spectra exhibit several absorption peaks. The absorption at 475 nm is due to charge transferring transitions in Ti^{3+}–Ti^{4+} pairs[1], and the other two are due to the Jahn-Teller effect[23]. For UV ns pulse, free electrons generated by single-photon ionization are the main cause for color centers; in contrast, for fs laser, it is due to multiphoton ionization and avalanche ionization.

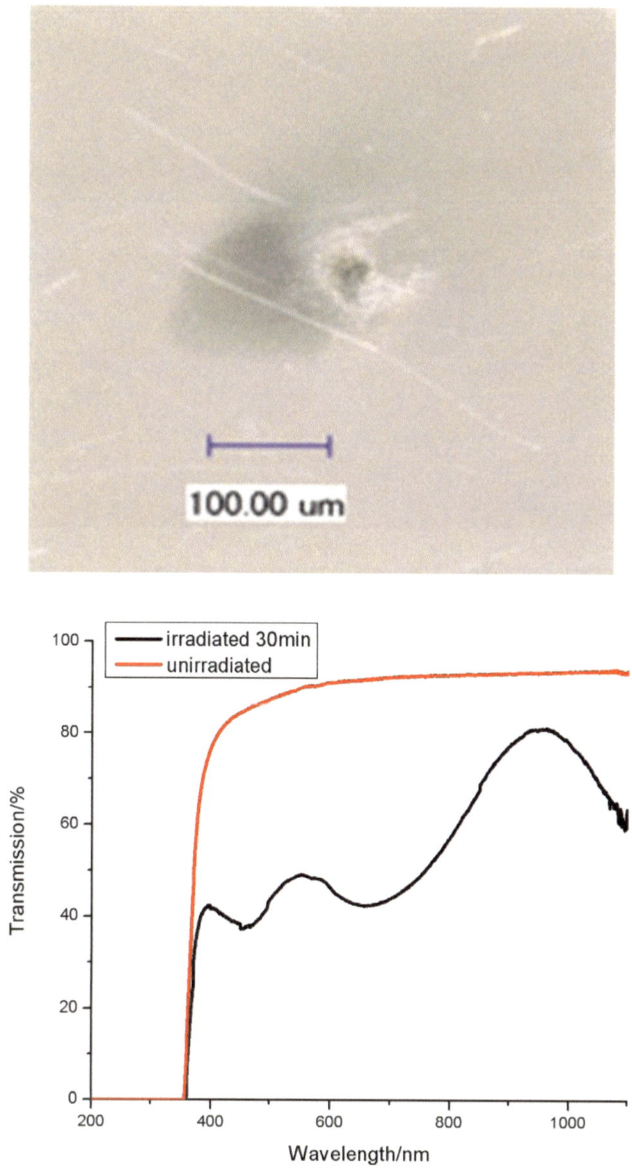

Fig. 5.4. Gray tracking in KTP crystal by 355nm laser radiation for 30min. and the transmission spectra (Courtesy of Dr. Qiuhui Zhang and Dr. Xiang Gao)

5.4 Microstructure induced by femtosecond

Irradiated by ultra-short pulse, semiconductor or metal surface can form a variety of micro and nano-scale structures. The micro-structures are influenced greatly by the laser pulse parameters, such as laser pulse number, pulse wavelength, pulse energy as well as the different external conditions (vacuum, liquid, air)[25,26].We conducted preliminary experiments on the surface periodic structure in silicon induced by repetitive femtosecond pulses, which are in the air or water, and the results obtained are as follows (Fig. 5.5):

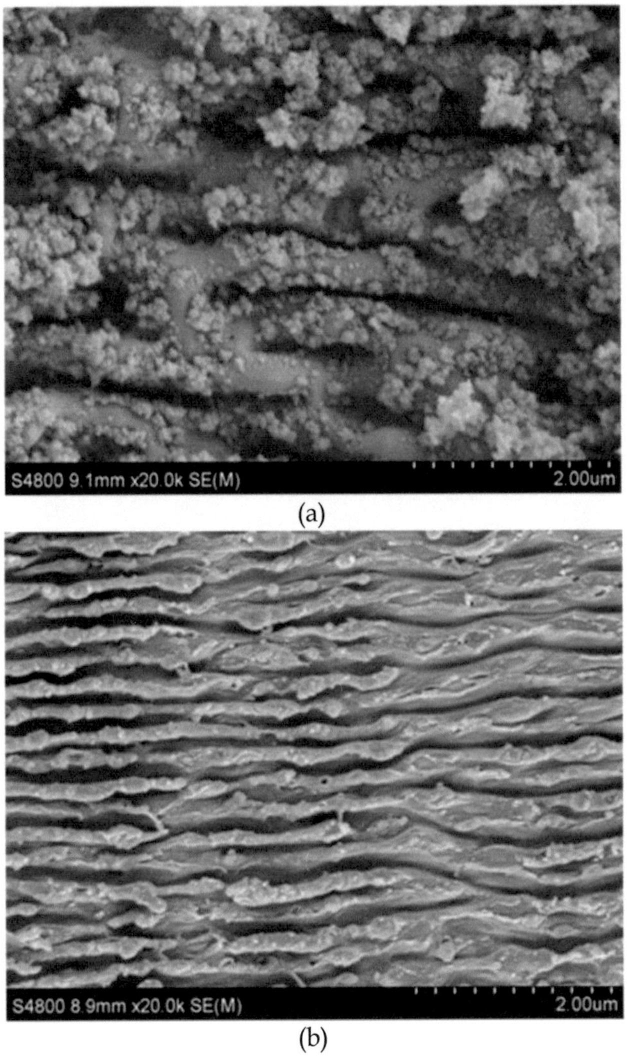

Fig. 5.5. Periodic structure on the Si surface produced in the air (a) and water (b), respectively with femtosecond pulse (repetitive rate 10Hz, single pulse energy 1mJ,pulse duration 35 fs, 500 pulses).

The study manifests that micro and nano-scale structure are due to the interference of the incident light with the scattered light and/or plasma wave, while nano-particles result from the melting and phase explosion from the shock wave.

6. Acknowledgements

This work was financed by the National Natural Science Foundation of China (Grant No. 60890203 and 10676023) and the Young Faculty Research Fund of Sichuan University (Grant No. 2009SCU11008).

7. References

[1] M. von Allmen and A. Blatter, *Laser-Beam Interactions with Materials: Physical Principles and Applications*, 2nd Edition, Springer-Verlag, Berlin & Heidelberg, Germany, 1995.

[2] R. M. Wood, *Laser-induced damage of optical materials*, IOP Publishing Ltd., London, UK, 2003.

[3] N. Bloembergen, "Laser-induced electric breakdown in solids," IEEE J. Quantum. Electron. 10, 375-386, 1974.

[4] B. C. Stuart, M. D. Feit, S. Herman, A. M. Rubenchik, B. W. Shore, and M. D. Perry, "Nanosecond-to-femtosecond laser-induced breakdown in dielectrics," Phys. Rev. B 53, 1749-1761, 1996.

[5] H. Wang and A. M. Weiner, "Efficiency of short-pulse type-I second-harmonic generation with simultaneous spatial walk-off, temporal walk-off, and pump depletion," IEEE J. Quant. Electron. 39, 1600-1618, 2003.

[6] D. Bleiner and A. Bogaerts, "Multiplicity and contiguity of ablation mechanisms in laser-assisted analytical micro-sampling," Spectrochimica Acta Part B 61, 421-432, 2006.

[7] J. H. Yoo, S. H. Jeong, R. Greif, and R. E. Russo, "Explosive change in crater properties during high power nanosecond laser ablation of silicon," J. Appl. Phys. 88, 1638-1649, 2000.

[8] J. Han, Y. Li, Q. Zhang, Y. Fu, W. Fan, G. Feng, L. Yang, X. Xie, Q. Zhu, and S. Zhou, "Phase explosion induced by high-repetition rate pulsed laser," Appl. Surf. Sci. 256, 6649-6654, 2010.

[9] L. V. Keldysh, "Ionization in the field of a strong electromagnetic wave," Sov. Phys. JETP 20, 1307-1314, 1965.

[10] V. I. Bespalov and V. I. Tanlanov, "Filamentary structure of light beams in nonlinear media," JETP Lett. 3, 307-310, 1966.

[11] "National Ignition Facility Functional Requirements and Primary Criteria," Lawrence Livermore National Laboratory Report NIF-LLNL-93-058, USA, 1994.

[12] J. O. Porteus and S. C. Seitel, "Absolute onset of optical surface damage using distributed defect ensembles," Appl. Opt. 23, 3796-3805, 1984.

[13] J. H. Campbell and T. I. Suratwala, "Nd-doped phosphate glasses for high-energy/high-peak-power lasers," J. Non-Crystal. Solids 263-264, 318-341, 2000.

[14] Ya. B. Zel'dovich and Yu. P. Raizer, *Physics of Shock Waves and High-Temperature Hydrodynamic Phenomena*, 1966 (in Russian) [edited by W. D. Hayes and R. F. Probstein, Dover Publications, Inc., N.Y., USA, 2002.]

[15] T. X. Phuoc, "An experimental and numerical study of laser-induced spark in air," Opt. Lasers Eng. 43, 113-129, 2005.

[16] G. Bekefi, *Radiation processes in plasmas*, John Wiley & Sons, Inc., N.J., 1966.

[17] C. K. Sun, F. Vallee, L. Acioli, E. P. Ippen, and J. G. Fujimoto, "Femtosecond investigation of electron thermalization in gold," Phys. Rev. B 48, 12365-12368, 1993.

[18] B. N. Chichkov, C. Momma, S. Nolte, F. Von Alvensleben, and A. Tunnermann, "Femtosecond, picosecond and nanosecond laser ablation of solids," Appl. Phys. A 63, 109-115, 1996.

[19] D. J. Jones, S. A. Diddams, J. K. Ranka, A. Stentz, R. S. Windeler, J. L. Hall, and S. T. Cundiff, "Carrier envelope phase control of femtosecond mode-locked lasers and direct optical frequency synthesis," Science 288, 635-639, 2000.

[20] I. Hartl, X. Li, C. Chudoba, R. Ghanta, T. Ko, J. G. Fujimoto, J. K. Ranka, R. S. Windeler, and A. J. Stentz, "Ultrahigh-resolution optical coherence tomography using continuum generation in an air-silica microstructure optical fiber," Opt. Lett. 26, 608-610, 2001.

[21] V. P. Kandidov, O. G. Kosareva, I. S. Golubtsov, W. Liu, A. Becker, N. Akozbek, C. M. Bowden, and S. L. Chin, "Self-transformation of a powerful femtosecond laser pulse into a white-light laser pulse in bulk optical media (or supercontinuum generation)," Appl. Phys. B 77, 149-165, 2003.

[22] J. M. Dudley, G. Genty, and S. Coen, "Supercontinuum generation in photonic crystal fiber," Rev. Mod. Phys. 78, 1135–1184, 2006.

[23] M. N. Satyanarayan, H. L. Bhat, M. R. Srinivasan, P. Ayyub, and M. S. Multani, "Evidence for the presence of remnant strain in grey-tracked $KTiOPO_4$," Appl. Phys. Lett. 67, 2810-2812, 1995.

[24] G. M. Loiacono, D. N. Loiacono, T. McGee, and M. Babb, "Laser damage formation in $KTiOPO_4$ and $KTiQAsO_4$ crystals: Grey tracks," J. Appl. Phys. 72, 2705-2712, 1992.

[25] M. Huang, F. Zhao, Y. Cheng, N. Xu, and Z. Xu, "Origin of Laser-induced Near-Subwavelength Ripples: Interference between Surface Plasmons and Incident Laser," ACS Nano 3, 4062-4070, 2009.

[26] P. Lorazo, L. J. Lewis, and M. Meunier, "Thermodynamic Pathways to Melting, Ablation and Solidification in Absorbing Solids Under Pulsed Laser Irradiation," Phys. Rev. B. 73, 134108, 2006.

7

Pulse Laser Ablation by Reflection of Laser Pulse at Interface of Transparent Materials

Kunihito Nagayama, Yuji Utsunomiya,
Takashi Kajiwara and Takashi Nishiyama
Kyushu University
Japan

1. Introduction

This chapter is devoted to a series of studies on a peculiar kind of pulse laser ablation which is found to take place at the interface of two transparent materials. Transparent media treated here include air (gas), liquid and solid medium.

This research started from the discovery of appreciable reduction of laser ablation threshold for roughened solid surface by our group. (Nagayama et al, 2005) As will be explained later in detail, precise study on this phenomenon revealed that

i. peculiar effects are observed for the interface of transparent and roughened solid surface,
ii. the phenomena have no appreciable dependence on the surface roughness,
iii. the phenomena depend strongly on the direction of the incident laser beam, i.e., whether the incident beam comes from the material of higher refractive index to that of lower one.
iv. this effect is not a microscopic one, but a macroscopic one,
v. since the phenomenon can be qualitatively described by optics equations.

Appreciable enhancement of laser absorption or the reduction of ablation threshold occurs only when the laser beam is incident from the material of higher refractive index to that of lower index. In other words, the phenomena are represented by the difference in refractive index of adjacent media in both sides of the material interface, and not by the mechanical properties of media.

This chapter consists of three topics, which are (i) the enhancement of laser beam absorption by roughened surface, and its evidence through high speed imaging of the ground glass surface ablation, and its unique application to the initiation of high explosive powder (Nagayama et al, 2005, 2007, 2007a), (ii) after a simple theoretical discussions on physics involved in the phenomena, pressure wave generation from material interface of glass or PMMA surface in air or in water, and its high speed imaging, (Nakahara et al, 2008) are described and finally, (iii) liquid jet formation from liquid-air surface will be described. (Utsunomiya et al, 2009, Kajiwara et al, 2009, Utsunomiya et al, 2010) Key issue of the physics involved in all the phenomena treated in this chapter is the difference in refractive index of a transparent material from that of the adjacent medium.

2. Reduction of pulse laser ablation threshold fluence for roughened surface of a transparent material

2.1 Evidences of enhanced absorption of laser energy at roughened surface

We have found appreciable reduction of pulse laser ablation threshold of transparent materials, when the laser beam is focused through the material to the interface of the material and air, if the output surface is a roughened surface. (Nagayama et al, 2005) Peculialities of ablation of non-smooth surfaces, however, have been reported by several authors. (Ben-Yakar et al, 2003, Kane and Halfpenny, 2000, Petr-Chylek et al, 1986) This effect is striking in that this happens only when the laser beam is focused on the surface of the transparent material not from the air but from inside the material. In other words, the effect is laser direction dependent.

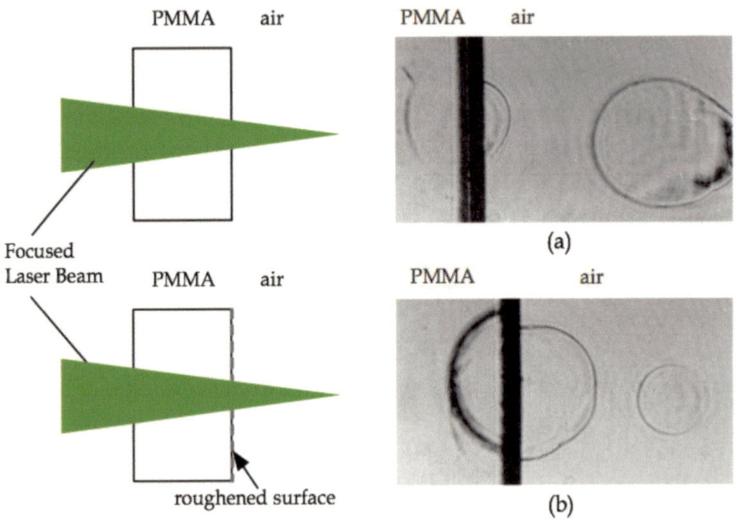

Fig. 2.1 High speed imaging of laser focus into air through PMMA plate with smooth (a) and roughened surface (b). Surface of PMMA was roughened by using #800 paper. Delay time of photos are 0.53 and 0.54 μs, respectively with a focused laser pulse of 180 mJ.

Figure 2.1 shows typical demonstration showing the difference in specific features of laser energy absorption at smooth and roughened surface of a transparent plate through which pulsed laser beam is focused into air ahead of the plate. In order to detect small difference in the change in density, pulse laser shadowgraphy with very small aperture has been adopted. Pulse laser used in these experiments is an Nd:YAG laser of fundamental frequency, and the energy per pulse is around 180 mJ. Pulse duration of the pulse is 4 ns. One may see an air shock wave front produced from the focused point in air. One may also see other waves emanated from the air-PMMA plate interface. Stronger stress wave front in PMMA can be seen in case of roughened PMMA surface. Very weak stress wave in PMMA and in air can still be seen even in case of smooth surface of PMMA. Procedures of observing these phenomena will be discussed in a later section.
Similar experiments are performed for the combination of a silica glass plate in contact with distilled water layer. Figure 2.2 shows the typical results. In this case, pulse laser is focused

into water. For glass plate with smooth surface, one may see weak pressure wave around the beam waist, and also a weak wave from the interface. On the contrary, much larger disturbance is produced in case of roughened glass surface.

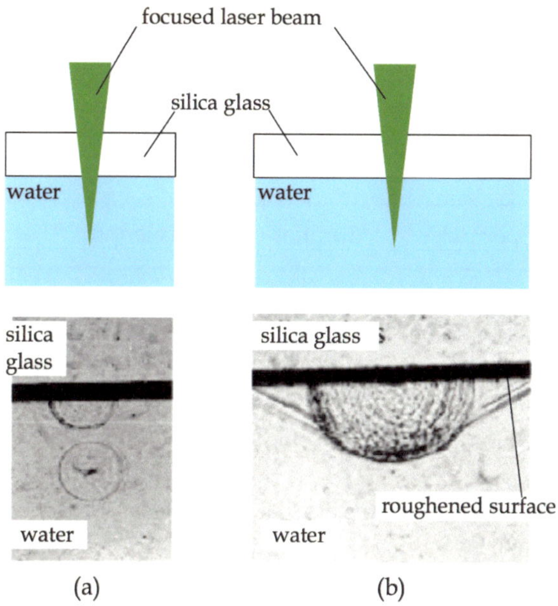

Fig. 2.2 High speed imaging of laser focus into water through silica glass plate with smooth (a) and ground surface (b). Silica glass plate was roughened by #1000 powders. Delay time of photos are 0.56 and 1.0 μs, respectively with a focused laser pulse of 100 mJ.

Two examples shown in Fig. 2.1 and 2.2 suggest the following; (i) roughening treatment of transparent material enhances laser energy absorption resulting in the reduction of ablation threshold, (ii) similar phenomena will take place for two transparent media combination including air-PMMA and water-glass. It is stressed here that direction of laser beam to material interface is from the material of higher refractive index to that of lower one. These results are quite different from the common understanding of pulse laser deposition of thin films. (Chrisey and Hubler, 1994)

In the phenomenon of pulse laser ablation, surface energy density or laser fluence is of essential importance. In this sense, determination of laser fluence must be made very carefully. In case of laser beam with Gaussian profile, it is possible for the fluence at beam waist to calculate theoretically from the value of focal length of focusing lens. In the present study, we have measured energy of incident laser pulse and the cross section of the ablated area in the following way. For the measurement of the ablated surface, we have used the damaged surface area of aluminum plate just in place of the ablation target plane, i.e., at the position of the material interface. Estimated fluence obtained in this manner is normally slightly smaller than those estimated by the theory of Gaussian beam. We used the method throughout this series of studies, since it can be a simple and basically reliable measure of the real fluence value in any situation of the experiments, including in air or in water or in any fluid, applicable for any beam profiles adopted in this study. As is noted later, the present effect has strong dependence on the laser beam pattern.

In this study, three Nd:YAG laser systems with ns duration and of fundamental frequency has been used. One of them (laser #3) was used as a light source for pulse laser shadowgraphy with an SHG crystal. They are summarized in Table 1. First two lasers in Table 1 were used solely for ablation energy source with different beam pattern. Laser #1 can be used altering its beam pattern of near Gaussian and tophat shape by the use of different output mirror. Where tophat pattern means almost uniform beam intensity distribution. Torus shape in Laser #2 indicates intensity distribution with low intensity region in the middle of the beam pattern. They can be synchronized in the precision of less than 20 ns to alter the delay time of taking pictures of the events.

laser #	max energy per pulse	duration	wavelength	beam profile
1	800 mJ	10 ns	1064 nm	Gaussian or Tophat
2	200 mJ	4 ns	1064 nm	Torus
3	200 mJ	7 ns	532 nm (SHG)	Gaussian

Table 1. Nd:YAG laser systems used in this study

2.2 Fragment cloud generation in ground glass ablation

Especially in the case of ground glass, we have observed not only ablation threshold reduction, but also burst of glass fragments from the focused region of the beam. (Nagayama et al, 2007) Figure 2.3 shows high speed imaging of the cloud of fragments burst from the ground surface of glass plate in air. We have used an intensified CCD camera capable of acquiring two frames per one shot. (Hamamatsu C-7972-11) Each frame in Fig. 2.3 corresponds to separate experiments. One notices the reproducibility of the phenomena. One may see a shock front in air followed by burst of material cloud. Appreciable amount of glass fragments are seen to move at high velocity. In most of pulse laser ablation process, material ejection is somewhat a common phenomena normally a final stage of ablation. In case of ground glass ablation, however, amount of material ejection is much larger. This is quite peculiar only in the case of ground glass.

Extensive studies of the ground glass ablation have revealed that (i) the effect has little dependence on the surface roughness, (ii) reduction of ablation threshold fluence is more than ten-fold, (iii) ejected glass particles have the velocity of around 1.5 km/s at least in the initial stage. High-speed glass particle cloud have the ability of initiating high explosive charge, PETN powder. (Nagayama et al, 2007a) We have used special ground glass plates commercially available for optics to diffuse light, called diffusion plates. They are specified by roughness numbers from #240 to #1500.

Figure 2.4 shows high speed imaging of the cloud of fragments burst from the ground surface of glass plate in vacuum. In the figure, it is apparent that air shock precursor does not exist, but still an appreciable amount of fragments can be seen. We have trapped ejected glass particles in a simple setup using two PMMA plates placed at forward and bottom of ground glass target assembly. Figure 2.5 shows particle size distribution for several kinds of diffusion plates with different surface roughness. Two peaks of trapped particle size are observed irrespective of glass surface roughness indicating a scaling law for the physical process of the destruction of brittle materials.

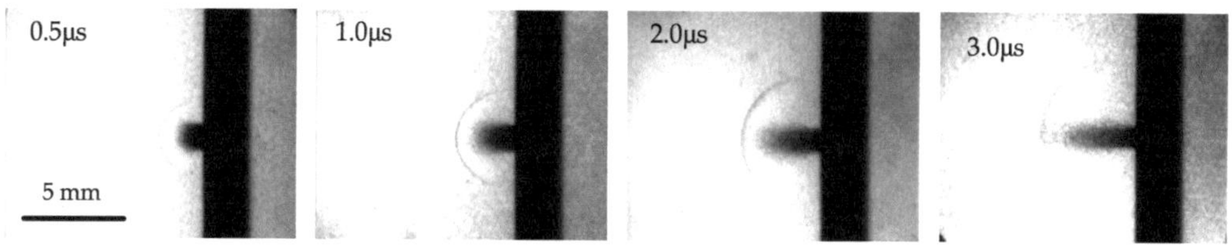

(a) (b) (c) (d)

Fig. 2.3 High speed imaging record of glass fragments ejected from the ground glass surface with #240 surface roughness by pulse laser beam focused through the glass plate from right. Delay time of photographs are shown in the figure. Laser energy, focused diameter and fluence are 0.39 J, 1.8 mm, and 15.5 J/cm^2, respectively.

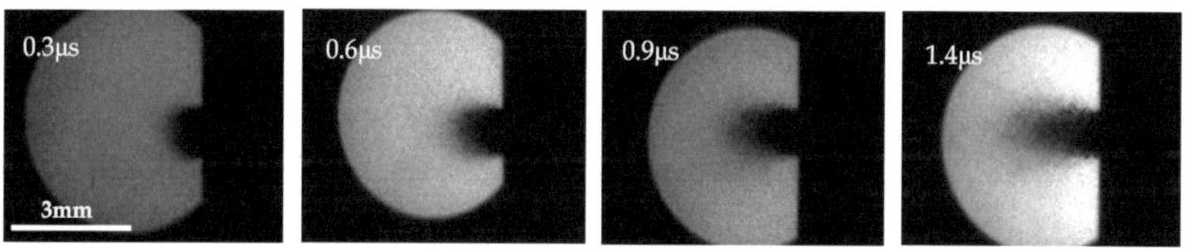

(a) (b) (c) (d)

Fig. 2.4 High speed imaging record of glass fragments ejected from the ground glass surface with #240 surface roughness by pulse laser beam focused through the glass plate in vacuum. Delay time of photographs are shown in the figure. Laser energy, focused diameter and fluence are 0.38 J, 1.4 mm, and 25 J/cm^2, respectively.

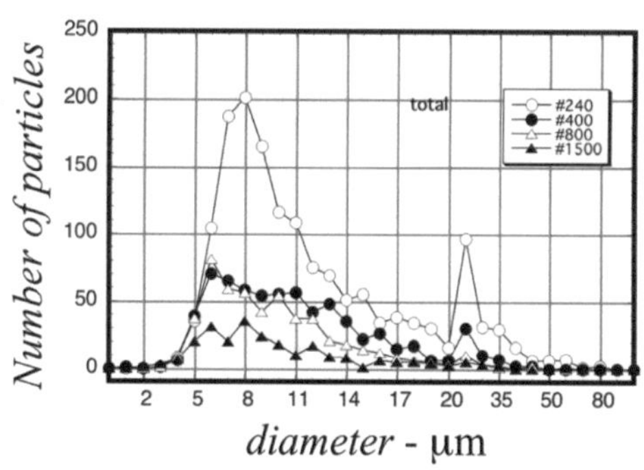

Fig. 2.5 Size distribution of trapped glass fragments from pulse laser ablation of diffusion plates with several different surface roughness (shown in the figure).

2.3 Initiation of PETN high explosive powder by glass ablation

Possibility of applying high velocity particle cloud generation by ground glass ablation to the initiation of high explosive powder has been pursued here. This is supposed to be realized due to the velocity of fragments of 1.5 km/s. We have observed the detonation of PETN powders in two ways. By putting thin PETN powder layers in contact with ground glass plates with PMMA block, we observed the detonation-induced strong stress wave in PMMA cube by a high-speed camera. (Hamamatsu C-7972-11) Figure 2.6 shows typical records with different delay time. Pulse laser ablation of ground glass induces weak precursor wave followed by higher stress wave in PMMA. They are clearly visible in the photographs in Fig. 2.6. One may note that laser ablation induced wave and detonation induced higher stress wave cannot be seen as one wave but separated. Finite time is seen to be required for the detonation reaction to occur after laser irradiation. This phenomenon is well known by several researchers. (Paisley, 1989, Watson et al, 2000) As one of side evidences on the detonation reaction is the self-emission of detonation-induced plasma in air.

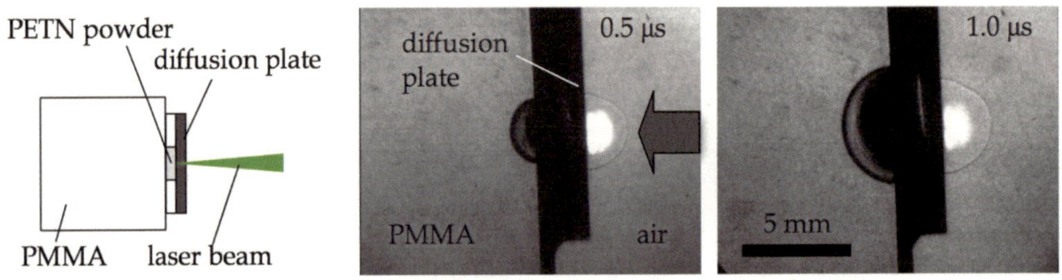

Fig. 2.6 Detonation induced stress wave in PMMA slab by pulse laser ablation of ground glass plate (diffusion plate) with surface roughness of #800 in contact with PETN high explosive powder of initial density of about 1.0 g/cm^3. Delay time of photographs is shown in the figure.

Another kind of experiments was planned to investigate the retardation of detonation reaction after laser irradiation. Since the detonation of PETN powder emits light that can be seen through PMMA cube. Figure 2.7 shows the high speed streak record of self emission by detonation reaction and its propagation to radial direction. It is convenient that pulse laser ablation itself emits light at that instant, this flash is also recorded in streak record. This flash must be the time 0 of the event, and one may see the delay in detonation from ablation.

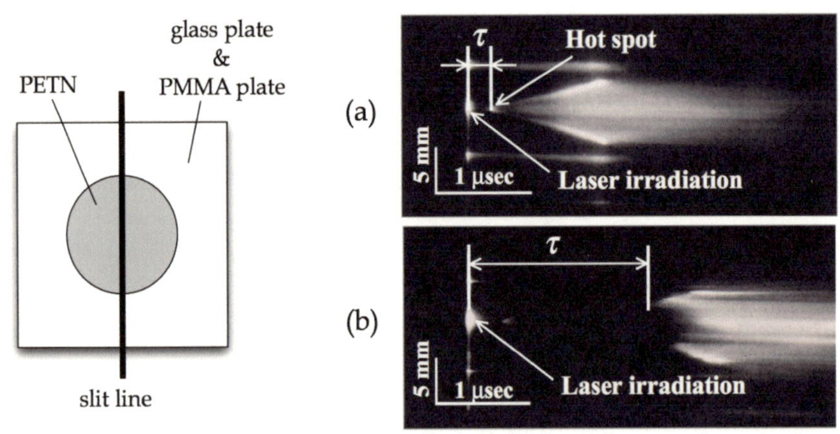

Fig. 2.7 Time sequence of PETN powder initiation experiment. Streak record of self emission by laser ablation and by reaction was obtained by a high-speed streak camera. (Imacon 790) Diffusion plate of #400 was used. Laser fluence is set to (a) 15.1J/cm^2, and (b) 9.1 J/cm^2, respectively. Ignition delay time τ is defined as in the figure.

Shortest delay time from ablation to detonation is around 200-300 ns, which is almost in harmony with the high speed imaging results as in Fig. 2.6. It is also possible to estimate the detonation velocity by the slope of streak record. Estimated detonation velocity from streak photographs agrees with available data for the detonation velocity of PETN powder of relevant initial density. Streak record in Fig. 2.7 (b) shows the case of incomplete reaction. In such cases, delay time is quite long, and detonation may not be completed.

3. Pressure wave production at material interface by the reflection of focused laser beam

3.1 Possible mechanism of reduction of laser ablation threshold for roughened surface

The key to understand the above-explained phenomena is the fact that they have direction dependence. That is, phenomena expected by laser irradiation are quite different depending upon the direction from which medium to focus through the material interface. Experimental results apparently show that laser beam must be focused through the material of higher refractive index to the interface with that of lower one.

The fact that present phenomena have very weak dependence on the surface roughness of the material suggests that the phenomena is regarded as not a microscopic effect but a macroscopic one. It is plausible that laser beam reflection at the interface between transparent material and adjacent medium must be the key phenomena. That is, difference in refractive index of the transparent material and that of the adjacent medium may lead to the present phenomena. Reduction of ablation threshold in this study can be explained at least qualitatively by Snell's law. (Nagayama et al, 2007, Utsunomiya et al, 2009)

For the reflection of light wave with moderate energy density, Fresnel's equations for the coefficient of reflection r and of transmission t at an interface between two media with refractive indices, n_1 and n_2 are given by

$$
\begin{aligned}
r_{\perp} &= \frac{n_1\cos\theta_1 - n_2\cos\theta_2}{n_1\cos\theta_1 + n_2\cos\theta_2}, \quad t_{\perp} = \frac{2n_1\cos\theta_1}{n_1\cos\theta_1 + n_2\cos\theta_2}, \\
r_{\parallel} &= \frac{n_2\cos\theta_1 - n_1\cos\theta_2}{n_1\cos\theta_2 + n_2\cos\theta_1}, \quad t_{\parallel} = \frac{2n_1\cos\theta_1}{n_1\cos\theta_2 + n_2\cos\theta_1}
\end{aligned}
\tag{1}
$$

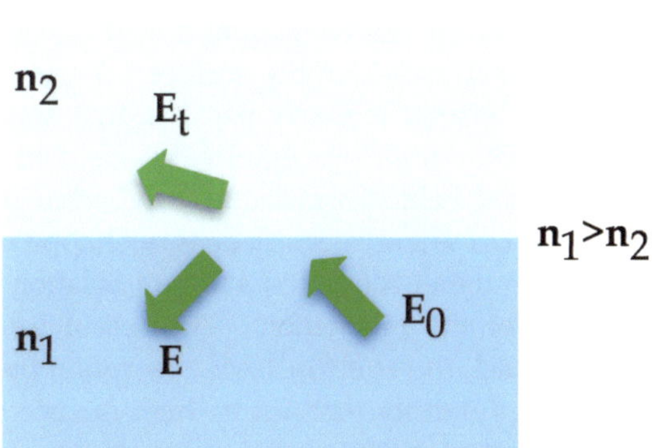

Fig. 3.1 Schematic illustration of the reflection of laser beam at material interface. E_0 and E denote the Electric field strength of the incident beam, and that of the reflected beam.

where suffices $\perp$ and $\|$ denote light waves whose polarization is perpendicular and parallel to the reflection plane, respectively. Within the light intensities treated here, it is assumed that medium cannot absorb light energy. As for all the variables and notations used here, please refer to the textbook of optics. (Hecht, 1989) By considering the above two kinds of polarizations, one may describe light waves with arbitrary polarization direction. Figure 3.1 shows schematic illustration of electromagnetic wave reflection at a interface of two media with different refractive index.

One may note that how the light reflection at glass-air interface depends on the laser irradiation direction, i.e., from glass to air or from air to glass. Only difference within these two is phase shift by reflection. Important point is the fact that phase shift of reflection at the interface for transmission into medium of lower refractive index is 0. Therefore, the interference between the incident radiation and reflected radiation is **constructive**. In the present case, we have

$$\begin{aligned} E_{\perp} &= E_0 + \frac{n_1\cos\theta_1 - n_2\cos\theta_2}{n_1\cos\theta_1 + n_2\cos\theta_2}E_0 = \frac{2n_1\cos\theta_1}{n_1\cos\theta_1 + n_2\cos\theta_2}E_0, \\ E_{\|} &= E_0 + \frac{n_2\cos\theta_1 - n_1\cos\theta_2}{n_1\cos\theta_2 + n_2\cos\theta_1}E_0 = \frac{2n_1\cos\theta_1}{n_1\cos\theta_2 + n_2\cos\theta_1}E_0, \end{aligned} \tag{2}$$

By using Snell's law,

$$n_1 = n_2\frac{\sin\theta_1}{\sin\theta_2} \tag{3}$$

Eq.(2) can be rewritten to

$$\begin{aligned} E_{\perp} &= \frac{2\sin\theta_1\cos\theta_1}{\sin\theta_1\cos\theta_1 + \sin\theta_2\cos\theta_2}E_0, \\ E_{\|} &= \frac{2\sin\theta_1\cos\theta_1}{\sin\theta_1\cos\theta_2 + \sin\theta_2\cos\theta_1}E_0, \end{aligned} \tag{4}$$

From Eq. (4), electric field intensity at the material surface has dependence on the incident angle. Maximum field intensity enhancement is obtained for the incident angle of critical value for total internal reflection. In this case, maximum field intensity is two times that of incident field strength. We have to consider the surface energy density at the material interface, so that electromagnetic energy intensity per unit area must be four times that of initial one. On the contrary, it is shown that for light waves incident from the material with lower refractive index, superposed field of incident and reflected wave is **destructive**, so that enhancement of electromagnetic energy density does not happen.

These considerations can explain the effective reduction of ablation threshold fluence and also the direction dependence of the phenomena. Theoretical treatments based on the Fresnell's equations apparently lack the effect of laser absorption by the medium and also non-linear electromagnetic energy density realized by high fluence of focused laser beam. Quantitative explanation of the effect must be made by a theory containing the nonlinear effects of the phenomena and laser energy absorption by the material. Almost same discussion has been made by Greenway et al. (Greenway et al, 2002) Their discussion is limited to the case of normal incidence and glass-air interface due to their research topics on energy transmission by an optical fiber.

3.2 Evidence on reduction of laser ablation threshold for oblique incidence on material interface

If the above discussion is the case, one can observe the reduction of ablation threshold for the oblique surface of the transparent material prism, if the laser beam is focused through the transparent prism as shown schematically in Fig. 3.2(a). To show the ablation threshold reduction at the interface, we made an experiment using glass prism specimen. Laser fluence at oblique glass surface is set smaller than known ablation threshold. Focused laser beam is incident as shown in Fig. 3.2(a) so as to have a focus at a point inside the glass prism but after reflected at the oblique surface.

As shown in the pulse laser shadowgraph picture in Fig. 3.2(b), one observes air shock front emanated from the point of laser beam reflection. This experimental result and further similar experiments showed that reduction of ablation threshold really takes place at the material interface of two media with different value of refractive index. (Nakahara et al, 2008) Laser beam must be incident from the material of high refractive index for the present effect to occur. In Fig. 3.2(b), incident beam fluence at prism surface is 17 J/cm^2, and that at reflected prism surface is estimated to be 12 J/cm^2. One may note that incident beam is focusing from the incident surface to the oblique surface, but the cross section of the reflected area is larger than the incident area due to elliptical shape of the reflected area. Fluence of the reflected area is appreciably lower than the ablation threshold of this glass in the present pulse laser system of around 20 J/cm^2. We have observed similar phenomena for two materials combination of glass-air, glass-water, PMMA-air and PMMA-water.

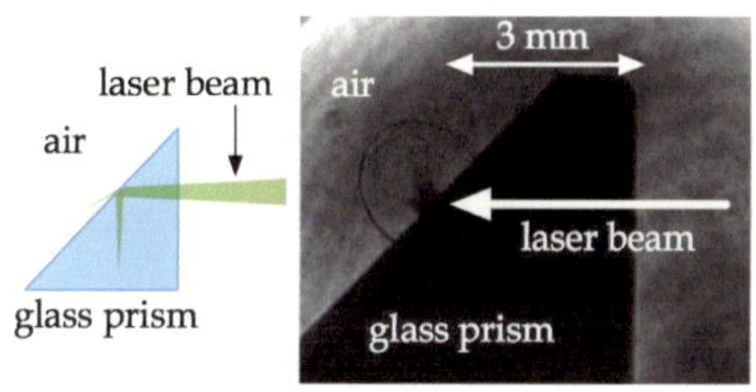

(a) (b)

Fig. 3.2 Demonstration experiment showing the reduction of laser ablation threshold by the reflection of laser beam at oblique material interface of glass and air. Estimated laser fluence at oblique surface is appreciably smaller than the measured value of ablation threshold.

In order to obtain further evidences on thershold fluence reduction, we have performed two different experiments. Figure 3.3 shows the experimental observation of pressure wave production at glass prism-water interface by using experimental assembly of Fig. 3.3 (a). In this case, estimated laser fluence at the reflected area is about 7 J/cm^2, which is smaller than threshold fluence of glass. One may note superposition of tiny waves emanated from the beam reflection area. In this case, incident angle of the beam onto the interface is less than the critical angle of total internal reflection between glass and water, transmitted wave induces cavitation bubbles on its way to the focused point. Even so, one may see generation of many waves at reflected interface.

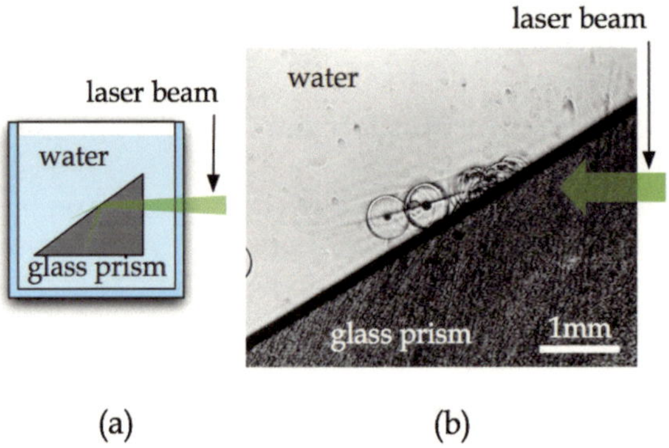

Fig. 3.3 Demonstration experiment showing the reduction of laser ablation threshold by the reflection of laser beam at oblique material interface of glass immersed in water. Estimated laser fluence at oblique surface is smaller than the measured value of ablation threshold.

Reproduction of evidences for the reflection induced ablation is shown in Fig. 3.4. In this case, laser beam is incident on the material interface perpendicularly. Therefore, the field enhancement is not very large, but threshold fluence reduction for ablation can be seen in the figure. This is the case where Greenway et al observed for their optical fiber experiment. These observations lead us to the conclusion that the present effect can be extended to various combination of transparent materials.

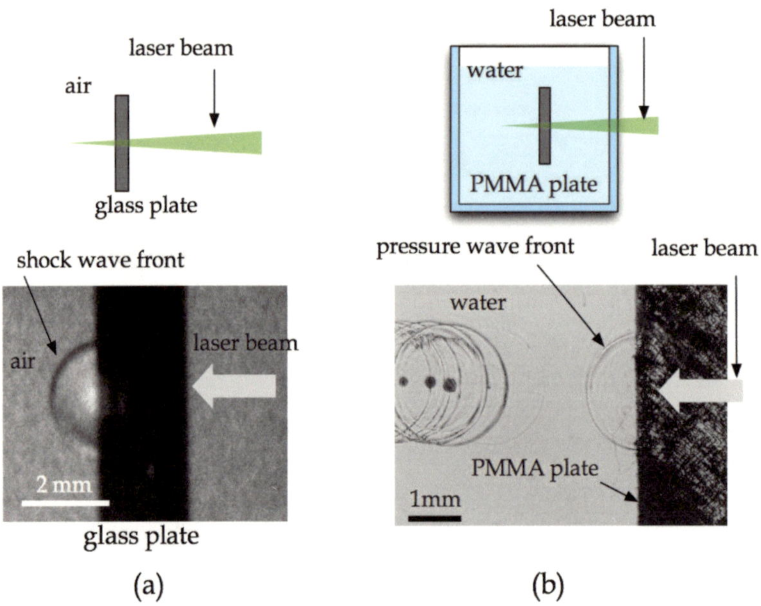

Fig. 3.4 Demonstration experiment showing the reduction of laser ablation threshold by the reflection of laser beam at material interface of glass-air and PMMA immersed in water. Estimated laser fluence at reflected surface is smaller than the measured value of ablation threshold.

4. Liquid jet production at liquid-air interface

4.1 Motivation of experiments on fluid media combination

For medical applications, various activities have reported in the breakdown or ablation of liquids. (Vögel and Venugopalan, 2000, Sigrist, 1986) Natural extension of the present study leads us to the combination of two transparent fluid media. (Utsunomiya et al, 2009, 2010, and Utsunomiya, 2010a) It is expected that at least ablation threshold reduction must be realized even in the case of fluid combinations. We also expect further events of dynamic flows induced by ablation inherent to the properties of fluids. There must also be a general rule of events for any fluid combination irrespective of various material properties such as reactivity, viscosity, etc.

As one of fundamental fluid combinations, we have made extensive studies on the water-air interface ablation by using a very simple assembly of Fig. 4.1. Point of focus of the laser beam is chosen away from the water surface in order to avoid mixing of the breakdown flash at the laser focus and the expected flash from ablation at the water surface. Also in this case, laser beam is lead so as to the water surface and reflected back to the focal point as shown in Fig. 4.1.

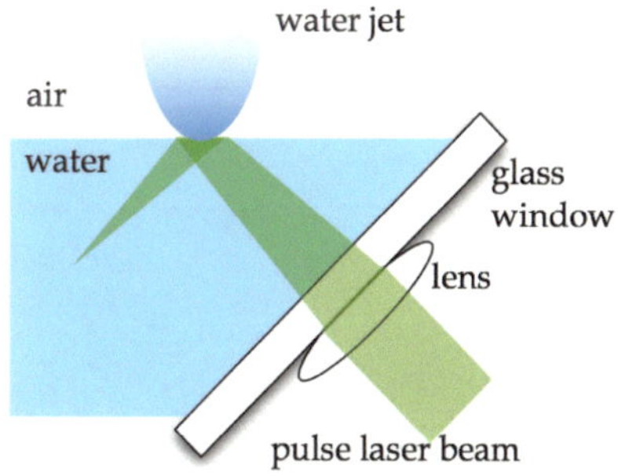

Fig. 4.1 Concept of experimental assembly for fluid experiment.

4.2 Precision pulse laser photography by using high-resolution film with pulse laser light source

We have started the experiments using high speed imaging system by an intensified CCD system. Soon, we have noticed that the phenomena can be recorded only by the system capable of recoding the phenomena with much higher spatial resolution. For this purpose, we have chosen high-resolution film together with pulse laser shadowgraphy. Photographic film we have used is the Minicopy film supplied by FUJI Film Co., Japan. This black and white film has the specifications of 850 lp/mm for 35 mm frame with ASA 25. The only disadvantage of the film is its high contrast of the recorded images. Recorded images are stored in computer by using the film scanner with maximum reading resolution of 12,800 dpi.

We needed at least two pulse laser systems for the ablation energy source and for the light source of photography. We used three Nd:YAG laser systems with 4-10 ns duration and with fundamental frequency. SHG is used only in the case of light source. Figure 4.2 shows diagram of photographic assembly. An aperture is inserted to the objective lens system for the film box to eliminate the flash from ablation and to record the image as shadowgraphs.

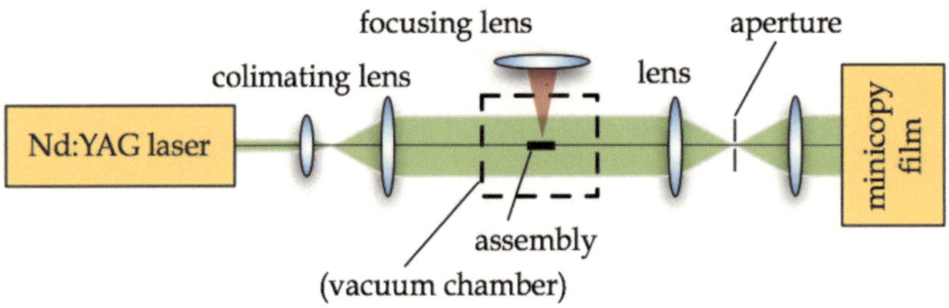

Fig. 4.2 Pulse laser shadowgraphy schematic.

4.3 Water jet generation and evolution

Figure 4.3(a) shows typical pulse laser shadowgraphs representing generation of liquid jet induced by laser ablation of water surface using the assembly of Fig. 4.1. In Fig. 4.3(b), one may also see laser induced cavitation bubble in water by focusing laser pulse from air to inside water.

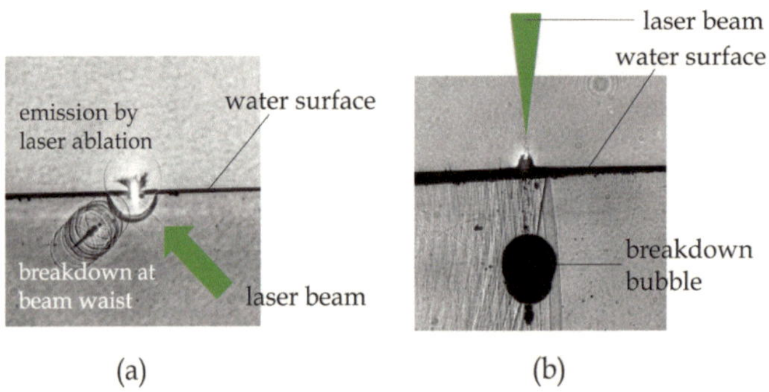

Fig. 4.3 Reproduction of pulse laser shadowgraphs of (a) laser ablation of water surface by using target assembly, Fig. 4.1, and (b) laser ablation induced bubble in water focused from air. Laser fluence on water surface is (a) 25 J/cm^2 and (b) 190 J/cm^2. To observe waves induced by laser ablation or breakdown at water surface or in water, a small aperture in Fig. 4.2 is used.

One may note that the value of fluence at water surface in (a), 25 J/cm^2 and in (b), 190 J/cm^2 is quite different almost one order of magnitude. This is another example of clear dependence of laser ablation threshold on the laser irradiation direction. That is, laser ablation is found to be much easier only in the case of laser irradiation from inside water to air interface. As is seen in Fig. 4.3(a), one may see that pressure waves are produced both at water surface and at beam waist of focal region after laser reflection at water surface. These waves are separated spatially and at least at this delay time, they do not interact with each other, indicating that these waves are the results of independent events. Phenomenon after laser ablation of liquid surface is also laser direction dependent. The value of laser ablation fluence, 25 J/cm^2 is almost two times larger than those obtained in cases of solid materials. It is emphasized that laser ablation of liquid materials is, in our experience, more difficult than of solid materials.

Laser ablation induced dynamic phenomena for liquids are quite different from those for solid materials. Temporal behaviour after ablation is shown in Fig. 4.4. In this sequential display of photographs, we have used three different laser conditions with different beam patterns, which are (a) tophat, (b) Gaussian, and (c) torus shape. Laser #1 and 2 in Table 1 are used in these experiments. As is seen in the figure, liquid jet and its evolution strongly depend on beam profile. Unfortunately, laser #2 has smaller output energy and this makes the phenomena smaller and faster. We noticed that the phenomena depend on (i) laser fluence, (ii) beam profile, and (iii) spot size at water surface. Since the product of laser fluence and spot size gives incident laser energy, parameter (iii) can also be replaced by laser energy.

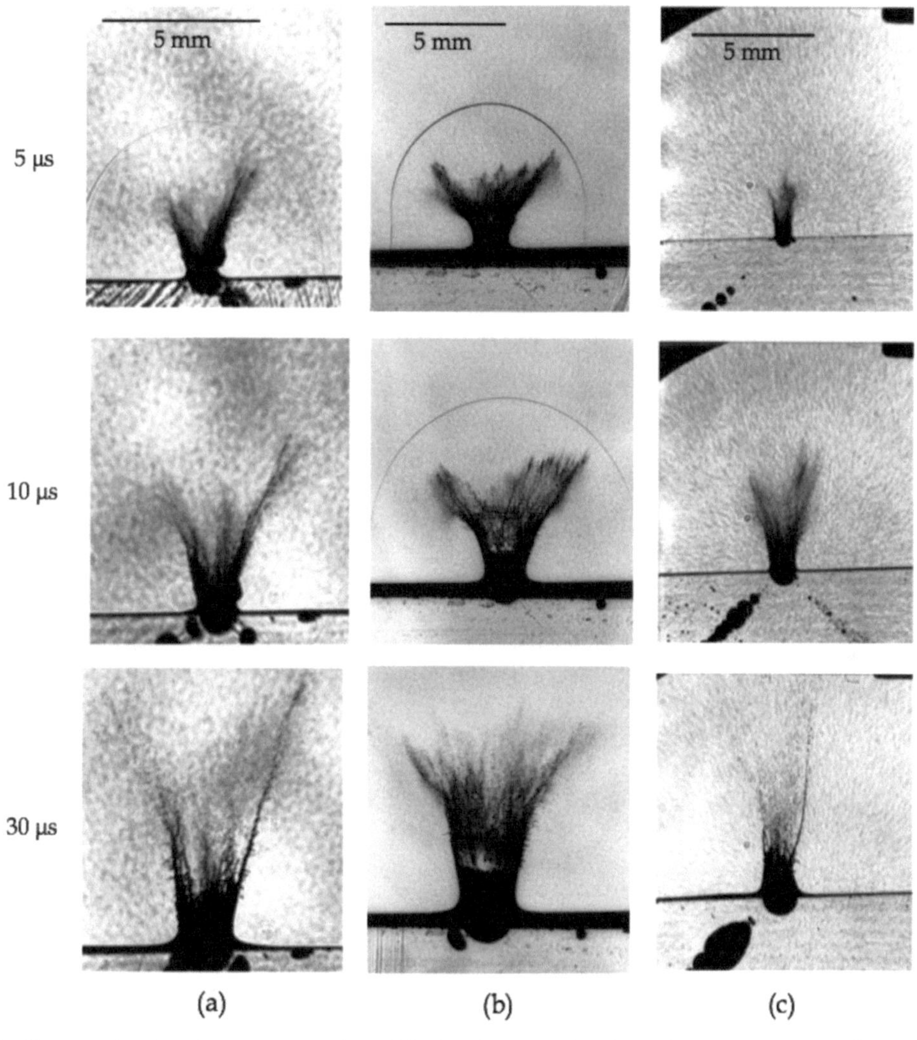

Fig. 4.4 Jet production by water surface ablation by focusing laser beam in water reflected at the surface and to a focus in water. Delay time of these pictures are given left side of pictures. Beam pattern used are (a) tophat, (b) Gaussian, and (c) torus shape, respectively. Laser fluence and energy of each series of experiments are (a) 30 J/cm^2, 300 mJ/pulse, (b) 20 J/cm^2, 300 mJ/pulse, and (c) 25 J/cm^2, 62 mJ/pulse, respectively.

From the photographs, one may see that air shock is followed by a bunch of slender water jet as a form of ligaments from the water surface. By taking successive photographs of jet propagation, one may see that ligament will elongate with time to disintegrate at the tip into droplets. Atomized water particles can be recorded even in these photographs in Fig. 4.4.
It is shown that liquid jet has the velocity of around 1 km/s at the initial stage but slows down very rapidly associated with atomization. Jet production behaviour is found to be dependent strongly on the laser beam profile as well as laser fluence. (Utsunomiya et al, 2010) Tip of liquid ligaments is found to slow down very rapidly. This result can be explained by very fast disintegration of ligaments into small liquid droplets. It is plausible that small droplets themselves will slow down very rapidly due to air drag. In this sense, liquid jet does reach finite distance depending on experimental parameters. In the end, small liquid droplets will evaporate to liquid vapour, and vapour molecules may diffuse into air. These considerations lead to the conclusion that after the liquid jet reaches its maximum height, very slow process of jet and liquid droplets and liquid vapour will be expected.

4.4 Dependence of water jet generation on various parameters

Most of the pulse laser ablation process is governed by so-called laser fluence on the focused target surface. Above the threshold fluence, explosive burst of electrons, ions, atoms, molecules and clusters of the target materials takes place. In the case of liquid surface ablation, however, further dynamics will follow after these material burst. They are dynamic liquid flow induced by instantaneous high plasma pressure of the ablated liquid. Air shock precursor wave, high-speed liquid ligament extension, disintegration into droplets, evaporation, etc. All these dynamics except for the initial burst may be scale dependent, i.e., they will depend on the focused beam size on the liquid surface, since plasma volume is given by this size.
As noted earlier, liquid surface ablation in this study is found to be dependent not only on laser fluence but also on laser energy with keeping fluence constant. We then planned to make experiments with varying fluence, energy and beam pattern systematically. We chose the tophat beam pattern as a standard, and compare the results of other profiles with that of tophat configuration. As shown in Fig. 4.4, it is found that water surface ablation takes place with laser fluence larger than 20 J/cm^2. We have performed systematic experiments for three beam patterns by varying fluence in order to determine the threshold fluence for ablation. Threshold fluence of ablation is almost independent of beam pattern and is 20-25 J/cm^2. The value is also independent of incident laser energy in the interval of 50 mJ/pulse to 400 mJ/pulse.
Increase in laser fluence leads to the occurrence of breakdown along incident beam path toward water surface. Appreciable incident laser energy is absorbed until beam reaches the water surface. As a result, part of the incident laser energy is used to produce cavitation bubbles along incident beam path.
Dependence of ablation behaviour on the laser beam pattern can be seen in Fig. 4.4. In fact, the fastest velocity of ligament tip is obtained in the case of Gaussian beam, and the slowest one is in the case of torus beam. Difference in the ablation behaviour on the beam pattern is attributed to the spatial distribution of local fluence on the water surface due to beam profile. It is also plausible that the difference in the coefficient of absorption of laser energy above the threshold fluence may not be linear to the difference of fluence from the threshold value.

Ablation behaviour depends on the incident laser energy itself as well as laser fluence, i.e., surface energy density. This is a peculiar characteristic of fluid ablation, since almost all of the solid surface ablation must be determined solely by the fluence. As is stated earlier, larger spot size on the water surface creates larger number of ligaments, thereby larger and longer dynamic flow of water.

4.5 Water surface ablation in various geometries

We have also tried water surface ablation experiments in various geometries. Figure 4.5 shows some of the results of these kinds of experiments. We first tried slant water surface by a glass plate with small opening and is ablated by a laser pulse focused in horizontal direction as shown in Fig. 4.5(a). Next, we tried to ablate a part of air bubble created in water, and is ablated to generate slender jet inside the bubble as shown in Fig. 4.5(b). In both cases, water surface is curved by surface tension. The last example is the behaviour of two bubbles by the pulse laser ablation as shown in Fig. 4.5(c). These two bubbles are near enough to interact with each other after ablation. Bubble breakup, or smaller bubble production, is observed depending on the number of bubbles, and their spatial position relative to the focused laser beam.

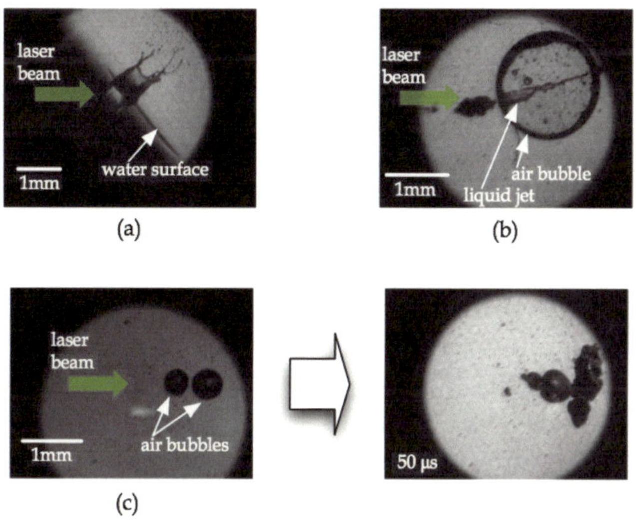

Fig. 4.5 Surface ablation in various geometries. (a) oblique water surface, (b) liquid jet inside air bubble, and (c) bubble collapse by laser ablation.

4.6 Ablation of various liquid samples

We have investigated laser ablation behaviour of several liquid samples other than water. Sample materials tested in this study include ethanol and vacuum oil. Further flammable materials were also tested, since the ablation produces cloud of small droplets or vapour that can be ignited by some means. All of the flammable materials can be ablated and are possible to ignite by a heated wire. Ignition, however, requires large delay time probably due to the fact that atomized or evaporated liquid vapour may reach the place of heated wire with very slow velocity.

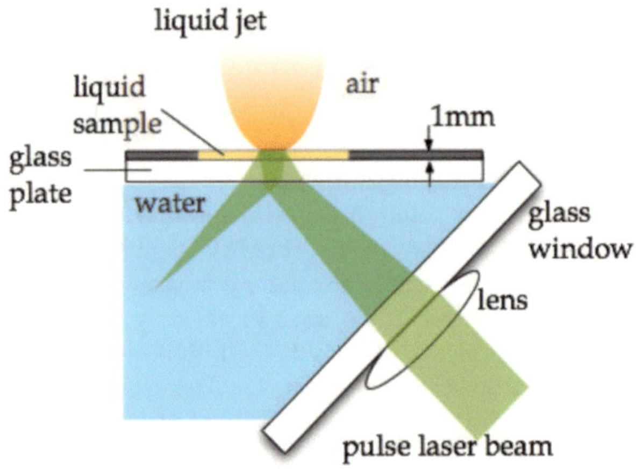

Fig. 4.6 Thin liquid film assembly for energetic liquid samples.

We have used a different sample assembly from that used for water. Schematic illustration of the assembly is shown in Fig. 4.6. Liquid sample is used in the form of a thin layer on glass substrate, which is placed on a water vessel. Laser beam is led from bottom right to the water vessel through the glass plate and to the thin sample. This assembly structure is adopted by the following reason. Since precise specification of the exact laser fluence at the liquid sample surface is essential in this study, the distance from the focusing lens to the sample surface must be precisely controlled and fixed. At least for ethanol sample, constant evaporation from the ethanol surface prevents from precise control of the above distance in the previous assembly in Fig. 3.1. By the use of the assembly in Fig. 4.6, only small amount of sample is set to the sample section and pulse laser is shot before they are about to dry out. We also found that if the laser beam is incident to the assembly from the angle for total internal reflection for water and not for sample, angle of reflection on the sample surface is automatically equal to the critical angle of total internal reflection between the sample and air.

In order to check the usefulness of this setup, we have also observed the ablation behaviour of water samples and compared the results with those for other liquid samples. Figure 4.7 shows reproduction of observed ablation and jet behaviour of water, ethanol and vacuum oil. These three liquid samples are quite different in properties of viscosity and vapour pressure. Very high viscosity and very low vapour pressure for vacuum oil is the reason why we chose this material as a liquid specimen. Table 2 summarizes typical material properties of these materials.

As stated in the previous section, ablation threshold for these materials are almost the same, so that the same experimental conditions are used for each experiment. Comparison of the photographs of Fig. 4.7 shows that among three liquids, ethanol ablation induces mist-like particle cloud and no appreciable generation of ligaments, while ligament-like structure is kept for a long time in the case of vacuum oil. These differences are attributed to the difference in their material properties. Compared with water jet, extension of ligaments for vacuum oil is shorter and the disintegration or atomization delays. One may note a hump just right of jet especially in the later frame of the photographs indicates an influence of thin film sample structure in this assembly.

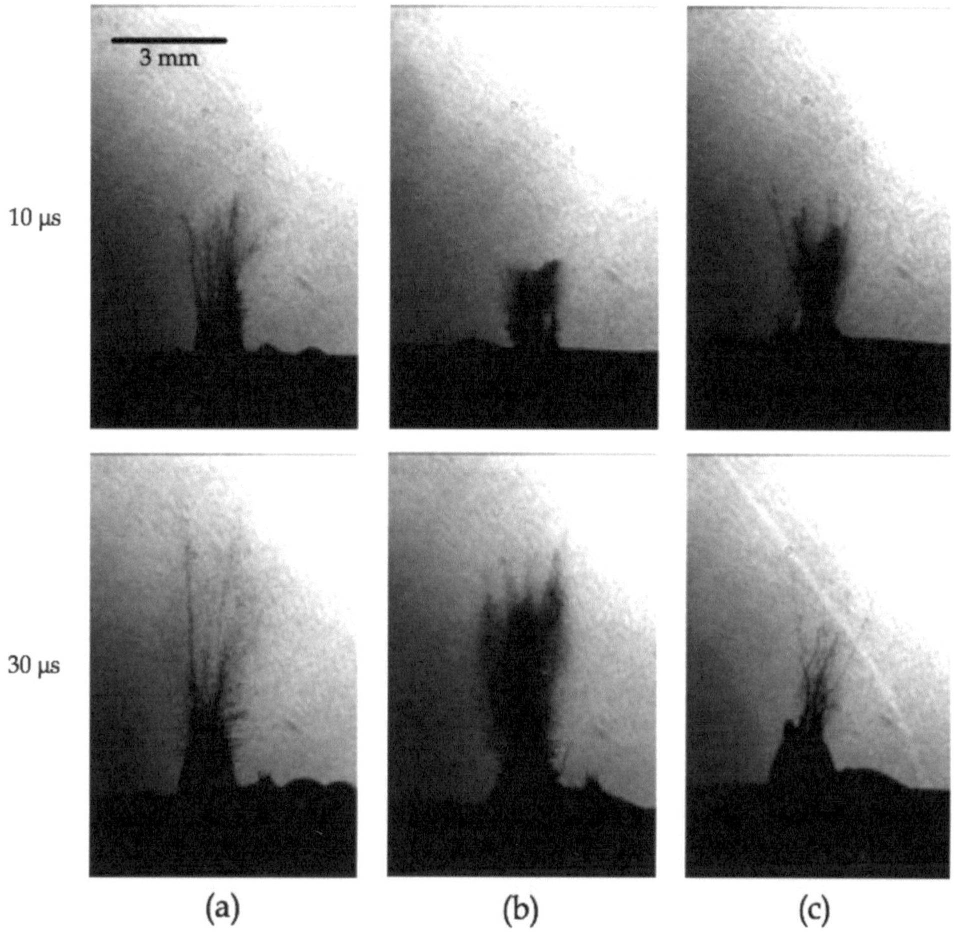

Fig. 4.7 Jet behavior of liquid specimens using assembly of Fig. 4.6. Liquid samples used are (a) : water, (b) : ethanol and (c) : vacuum oil.

sample	initial density (g/cm³)	viscosity (mPa•s)	vapour pressure (mmHg)
water	1.0	0.89	24
ethanol	0.8	1.2	59
vacuum oil	0.9	23	$7.5X10^{-9}$

Table 2. Material properties of liquid samples

5. Summary and conclusion

Present study shows two extremely different threshold fluence for pulse laser ablation for transparent materials depending on the direction of laser incidence on the material interface. Let us summarize these two threshold fluences. Threshold fluence for transparent materials must be very high easily inferred from the very low absorption of light energy at least for low laser intensity. Experience showed that threshold fluence for ground glass surface through the glass is almost one order of magnitude lower than that for direct laser focus from air. Smallest laser fluence for ground glass surface in this study is about a few joules per square centimeter. Typical and minimum fluence for pulse laser ablation of various material surface are summarized as

2-5 J/cm^2	: normal incidence, air, roughened PMMA surface
10 J/cm^2	: normal incidence, air, smooth PMMA surface
1-3 J/cm^2	: normal incidence, air, roughened glass surface
20 J/cm^2	: normal incidence, air, smooth glass surface
6-10 J/cm^2	: oblique incidence, water, smooth glass surface
12 J/cm^2	: oblique incidence, air, smooth glass surface
20-25 J/cm^2	: oblique incidence, air, surface of all liquid samples

These values may vary depending on the laser system. That is, pulse duration and beam profile, in other words, instantaneous irradiance and its spatial distribution over the focused area. For the case of transparent fluid media, little information for ablation threshold has been known. Threshold fluence for fluids is found to be almost two times larger than that for solid transparent materials.

Pulse laser ablation of transparent materials is found to be special in that from which direction the focused laser beam is incident determines the behaviour. As explained earlier, they are relevantly but only qualitatively described by the optics equations. Gross reduction of ablation threshold fluence for roughened surface of transparent materials can also be described by the same physics. Surface ablation of transparent solids and furthermore, that of transparent liquids can also be described by the same theoretical basis.

Ground glass ablation produces high velocity glass fragments cloud that can be successfully applied to the initiation of small amount of high explosive powder. Ablation of roughened PMMA plate in water produces pressure pulse, and that can be produced repetitively. This phenomenon will be applied to controllable medical tools in microsurgeries. Ablation of liquid surface produces high-speed liquid jet and eventually produces small droplet cloud. High-speed liquid jet can be used as an injector in medical area. In order to realize usable liquid jet, jet must be shaped as a convenient straight slender jet. Artificial laser beam profile must be used for this purpose, as seen in the dependence of the phenomena on beam profile. We observed very slender jet in air bubble due to the curvature of bubble surface, which has similar effect as the beam profile.

6. Acknowledgment

Authors wish to thank graduate students and undergraduate students involved in these projects especially in experimental activities. They are also indebted to the support of Mr. Tamio Iwasaki of Nobby Tech. Ltd. for high speed imaging, and Ms. Mariko Murakami for arranging laser instrumentation. Part of the present work was supported financially from the Ministry of Education, Culture and Science, Japan.

7. References

Ben-Yakar,A., Byer,R.L., Harkin,A., Ashmore,J., Stone,H.A., Shen, M., Mazur, E. (2003), *Morphology of femotosecond-laser-ablated borosilicate glass surfaces*. Appl. Phys. Lett. Vol. 83, pp. 3030–3032

Chrisey, D.B., and Hubler, G.K. (1994), *Pulse Laser Deposition of Thin Films* (Wiley, New York)

Greenway, M.W.,Proud, W.G.,Field, J.E.(2002), *The development and study of a fiber delivery system for beam shaping*. Rev. Sci. Instrum., Vol. 73, (2002), pp. 2185–2189.

Hecht, E. (1989), *Optics*, (Addison-Wesley), p.94

Kajiwara, T., Utsunomiya, Y., Nishiyama, T., Nagayama, K. (2009), *Generation of Energetic Liquid Jet and Atomization by Pulse Laser Reflection at Inclined Surface of High Refractive Index Material*, Sci. Tech. Energetic Materials, Vol.70, No.4, (2009), pp. 105-108.

Kane, D.M., Halfpenny, D.R.: *Reduced threshold ultraviolet laser ablation of glass substrates with surface particle coverage: a mechanism for systematic surface laser damage*, J. Appl. Phys. 87, 4548–4552 (2000)

Nagayama, K.; Kotsuka, Y., Nakahara, M., and Kubota, S., (2005), *Pulse laser ablation of ground glass surface and initiation of PETN*, Sci. Tech. Energetic Materials, Vol.66, No.6, (2005), pp. 416-420.

Nagayama, K.; Kotsuka, Y., Kajiwara, T., Nishiyama, T., Kubota, S., Nakahara, M. (2007), *Pulse laser ablation characteristics of quartz diffusion plate and initiation of PETN*, Sci. Tech. Energetic Materials, Vol.68, No.3, (2007), pp. 65-72.

Nagayama, K., Kotsuka, Y., Kajiwara, T., Nishiyama, T., Kubota, S., Nakahara, M. (2007a), *Pulse Laser Ablation of Ground Glass*, Shock Waves, Vol.17, (2007), pp. 171-183.

Nakahara, M., Nagayama, K., Kajiwara, T., and Nishiyama, T. (2008), *High-Speed Photographic Observation of Shock-Pressure Pulse in Water Induced by Laser Energy Absorption at Roughened Surface*, Material Science Forum, Vol. 566 (2008) pp. 47-52.

Paisley, D.L.(1989), *Prompt detonation of secondary explosives by laser*, Proc. 9th Int. Symp. Detonation, (1989) pp. 1110–1117.

Petr Chylek, M.A. Jarzembski, Chou, N.Y.(1986), *Effect of size and material of liquid spherical particles on laser-induced breakdown*. Appl. Phys. Lett. Vol. 49, pp. 1475–1477

Sigrist, M.W. (1986), *Laser generation of acoustic waves in liquids and gases*, J. Appl. Phys. Vol. 60, pp. R83-R122

Tominaga, T., Nakagawa, A., Hirano, T., Sato, J., Kato, K., Hosseini, S.H.R., and Takayama, K. (2006), *Application of underwater shock wave and laser-induced liquid jet to neurosurgery*, Shock Waves, Vol. 15, pp. 55-67

Utsunomiya, Y., Kajiwara, T., Nishiyama, T., Nagayama, K., Kubota, S., and Nakahara, M. (2009), *Liquid Atomization Induced by Pulse Laser Reflection underneath Liquid Surface*, Japn J. Appl. Phys., Vol. 48 (2009) 052501-1-5.

Utsunomiya, Y., Kajiwara, T., Nishiyama, T., Nagayama, K. (2010), *Pulse laser ablation at water–air interface*, Appl. Phys. A, Vol.99, (2010) pp. 641–649.

Utsunomiya, Y. (2010a), *Laser ablation phenomena at liquid surface* [in Japanese], Doctorial Thesis, Kyushu University, 2010.

Vogel, A., and Venugopalan, V. (2003), *Mechanisms of Pulsed Laser Ablation of Biological Tissues*, Chem. Rev. (Wash. DC) Vol. 103, pp. 577-644

Watson, S., Gifford, M.J. Field, I.E. (2000), *The initiation of fine grain pentaerythritol tetranitrate by laser-driven flyer plates*, J. Appl. Phys. Vol. 88, pp. 65-

8

Pulsed-Laser Ablation of Au Foil in Primary Alcohols Influenced by Direct Current

Karolína Šišková
Dept. of Physical Chemistry, RCPTM,
Palacky University in Olomouc
Czech Republic

1. Introduction

Almost two decades ago, Henglein pioneered the application of laser pulses for the synthesis of nanoparticles (Amendola & Meneghetti, 2009, as cited in Henglein et al., 1993). Since that the pulsed-laser ablation process of a foil performed in liquids is one of the top-down processes of nanomaterials generation. In a nutshell, laser pulses are focused into a metallic target immersed in a particular liquid producing thus nanoparticles dispersions (Amendola & Meneghetti, 2009; Georgiou & Koubenakis 2003; Zhigilei & Garrison, 1999). Noble metal nanoparticles are frequently formed by this approach because of a well-known fact that the as-prepared nanoparticle solutions do not contain any by-products and chemicals remaining from usual bottom-up approaches such as chemical syntheses. Hence, pulsed-laser ablation constitutes a "green" technique of nanoparticles formation.

There are several other benefits which make pulsed-laser ablation process attractive. One of them lies in the choice of ablation medium which is usually determined by a further usage of noble metal nanoparticles. So far, numerous papers have been written about pulsed-laser ablation performed in water and in aqueous solutions of simple ions (e.g. Procházka et al, 1997; Srnová et al, 1998; Šišková et al, 2008), surfactants (e.g. Fong et al, 2010), organic molecules (e.g. Darroudi et al, 2011; Kabashin et al, 2003; Mafune et al, 2002; Šišková et al, 2007, 2008, 2011), or even DNA (Takeda et al, 2005). In the literature, there have also been published pulsed-laser ablation processes of metallic foils performed in ionic liquids (Wender et al., 2011), or in a wide range of organic solvents, such as chloroform (Compagninni et al., 2002; Mortier et al, 2003; Šišková et al, 2010), toluene (Amendola et al., 2005), tetrahydrofurane (Amendola et al., 2007), dimethylsulfoxide (Amendola et al., 2007), N,N-dimethylformamid (Amendola et al., 2007), acetonitrile (Amendola et al., 2007), acetone (Burakov et al., 2005, 2010; Boyer et al., 2010; Tarasenko et al, 2005), primary alcohols (Burakov et al, 2010; Compagnini et al, 2002; Simakin et al, 2004; Werner et al, 2008).

Another substantial advantage of pulsed-laser ablation process is the possibility to choose (at least in principle) laser wavelength, pulse duration (ns, ps, fs), energy per pulse, and fluence (energy per area). All these parameters distinctly influence the final nanoparticles size, shape, uniformity, and their production efficiency. The reader is referred to the appropriate literature for more details, namely concerning the other advantages and disadvantages of the pulsed-laser ablation process in conjunction with the parameters (e.g. Amendola & Meneghetti, 2009; Franklin & Thareja 2004; Semerok et al., 1999; Sobhan et al., 2010; Tsuji et al., 2004).

Laser pulses can be applied not only for the generation, but also for the size reduction and reshaping of noble metal nanoparticles, the process known as nanoparticles fragmentation (Dammer et al, 2007; Kamat et al., 1998; Kurita et al., 1998; Link et al., 1999; Mafune et al, 2001, 2002; Peng et al., 2005; Shoji et al., 2008; Šmejkal et al., 2003, 2004; Takami et al., 1999; Werner et al., 2010; Yamada et al, 2006, 2007). Laser-pulses induced nanoparticles fragmentation has been described by two possible mechanisms so far: (i) coulomb explosion due to the sequential photo-ejection of electrons during the absorption of a single laser pulse (Link & El-Sayed, 2003; Yamada et al., 2006), and/or (ii) vaporization of particles due to the heating, induced by photon absorption, to a temperature higher than the boiling threshold (Franklin & Thareja 2004; Inasawa et al., 2006; Kurita et al., 1998; Takami et al., 1999). Similarly as in the case of pulsed-laser ablation process, particles fragmentation strongly depends on laser wavelength, pulse duration (ns, ps, fs), energy per pulse, and fluence. For instance, Au nanoparticles with the maximum of extinction at 520 nm can be efficiently fragmented by using the nanosecond laser pulses of 532 nm wavelength using reasonable values of fluence (Amendola & Meneghetti, 2009).

In the past three decades, nanoparticles have gained an increasing attention due to their unique optical, electrical, and magnetic properties which differ from bulk materials (Roduner, 2006). In particular, it has been demonstrated that noble metal nanoparticles (Ag, Au, Cu) possess surface plasmons which are responsible for enhanced light scattering and absorption (Le Ru & Etchegoin, 2008). This characteristic property of noble metal nanoparticles is fully exploited in surface-enhanced Raman scattering (SERS) spectroscopy. Recently, noble metal nanoparticles have also been employed in cancer diagnosis and therapy (Jain et al., 2007) as well as in photovoltaic devices (Atwater, H.A. & Polman A., 2010; Kim et al., 2008; Morfa et al., 2008; Tong et al. 2008).

According to a particular exploitation, either liquid dispersions of nanoparticles, or nanoparticles deposited on a substrate are preferentially required. Noble metal nanoparticles can be deposited on a particular substrate by several different ways depending on the force which is responsible for nanoparticles assembling. Roughly divided, nanoparticles assembling can be directed by molecular interactions, or by external fields as reviewed in more details in (Grzelczak et al., 2010). An elegant method is to allow self-assembling of nanoparticles exploiting spontaneous processes (Rabani et al., 2003; Siskova et al., 2011).

When molecular interactions are intended to be exploited for nanoparticles assembling, either substrate or nanoparticles have to be suitably modified by a surface modifier which enables the mutual interaction between nanoparticles and substrates. As an excellent example, the modification by amino- and/or mercapto-alkylsiloxane, or porphyrins can be referenced (Buining et al., 1997; Doron et al., 1995; Grabar et al., 1996; Šloufová-Srnová & Vlčková, 2002; Sládková et al., 2006). Obviously, surface modifications may be useful in or, on the contrary, disable some applications because they change electrical and optical properties of nanoparticles as well as of substrates (Carrara et al., 2004; de Boer et al., 2005; Durston et al., 1998; Rotello, 2004; Schnippering et al., 2007; Wu et al., 2009). Therefore, many research groups look for other types of nanoparticles assembling. One of many possibilities is the electrophoretic deposition technique which is based on the fact that charged nanoparticles are driven to and deposit on a substrate`s surface when an electric field is applied perpendicular to the substrate (Zhitomirsky et al., 2002).

Recently, a few papers have appeared using electrophoresis for the deposition of noble metal nanoparticles on substrates intended for particular purposes. ZnO nanorod arrays have been decorated by electrophoretically deposited Au nanoparticles (He et al., 2010). Such Au nanoparticle-ZnO nanorod arrays have exhibited an excellent surface-enhanced Raman scattering performance and enabled the detection at a single molecule level (He et al., 2010). Another electrophoretic deposition of Au nanoparticles performed in acetone has been motivated by the effort to prepare a SALDI (surface-assisted laser desorption ionization) substrate (Tsuji et al., 2011). Kim and co. used electrodeposited Au nanoparticles for electrochromic coloration (Nah et al., 2007). In another study (Yang et al., 2009), it has turned out that electrophoresis carried out for a long time (14 hours) can even lead to a preferential growth of nanoparticles on a substrate resulting in nanoplates. By changing the parameters of electrophoresis, namely the current density, the morphologies and structures of the obtained films can be easily controlled and tuned (Yang et al., 2009).

This chapter deals with Au nanoparticles prepared by pulsed-laser ablation process exploiting nanosecond laser pulses of 532 nm wavelength, performed in primary alcohols while direct current passes simultaneously through the ablation medium. Due to the charges present on the surface of arising Au nanoparticles, they are moved toward electrodes where they deposit. We assume the impact of simultaneous electrophoresis on the outcomes of pulsed-laser ablation, i.e., on the resulting nanoparticles dispersions. This point has never been addressed yet. Although electrophoresis of nanoparticles formed by pulsed-laser ablation process, however, using femtosecond laser pulses and aqueous environment, have been investigated by Barcikowski group (Menendez-Manjon et al., 2009), the authors focussed mainly on the velocities of nanoparticles using laser scattering velocimetry and on the surface patterning of metal target induced by the impact of a train of femtosecond laser pulses. In contrast, a complete characterization of Au nanoparticles solutions gets attention in this chapter.

Moreover, the chapter reports brand new results concerning not only the as-prepared solutions of Au nanoparticles influenced by direct current, but also microscopic and spectroscopic characteristics of three selected types of substrates which Au nanoparticles are deposited on due to electrophoresis.

Last, but not least, a possible elucidation of the influence of direct current value on the mechanism of Au nanoparticles generation by the pulsed-laser ablation process combined with electrophoretic deposition and performed in primary alcohols is suggested.

2. Experimental

2.1 Materials

Ethanol and butanol of UV-spectroscopy grade purchased from Fluka were used. Cleaning of a pure Au foil (99.99%, Aldrich) and ablation cell by washing in piranha solution (H_2O_2:H_2SO_4, 1:1) was carried out. The latter was also washed with aqua regia (HNO_3:HCl, 1:3) in order to remove any residual Au nanoparticles from the previous experiments. Indium-tin-oxide (ITO) and fluorine-tin-oxide (FTO) coated glass substrates purchased from Aldrich were ultrasonicated in acetone (p.a., Penta) and dried by nitrogen flow prior to their use as electrodes in the course of the simultaneous pulsed-laser ablation and electrophoretic deposition process. Alternatively, freshly cleaved highly oriented pyrolytic graphite plates (HOPG, purchased from RMI, Lazne Bohdanec, Czech Republic) were employed as electrodes.

2.2 Simultaneous pulsed-laser ablation and electrophoretic deposition

Homemade experimental setup for the simultaneous pulsed-laser ablation and electrophoretic deposition process is depicted in Scheme 1. Cylindrical glass ablation cell with a teflon cover was equipped with (i) two glass tubes allowing inert gas (Ar, 99.999%) to come in and leave, (ii) two electrode holders connected with a power supply, and (iii) a Au foil holder. Inert atmosphere is employed in order to increase the yield of nanoparticles which has been demonstrated in the literature (Werner et al., 2008). Laser pulses provided by Q-switched Nd/YAG laser system (Continuum Surellite I), wavelength 532 nm (the second harmonic) with the repetition rate of 10 Hz, effective diameter of a pseudo-Gaussian spot of 5 mm, and pulse width (FWHM) of 6 ns were used for the pulsed-laser ablation of the Au foil immersed in primary alcohols (100 mL). Pulsed-laser beam passed through ca. 8 mm column of a primary alcohol solution before hitting the Au target. Lenses (plano-convex, BK7, 25 mm in diameter) of 250 mm focal length were used to focus the pulsed-laser beam. The Au foil was irradiated for 6 min by a train of laser pulses of the 105 mJ/pulse energy as determined by a volume absorber powermeter PS-V-103 (Gentec Inc.). Simultaneously with the pulsed-laser ablation, electrophoresis took place, i.e. direct electric current (controlled by an ampere-meter) passed through the ablation medium due to the immersed electrodes (3 cm distant). Two values of direct current were employed, 10 μA and 17 μA (the applied voltage was set accordingly). The experiments have been performed at least 3 times.

2.3 Instrumentation

UV-visible extinction spectra of Au nanoparticle solutions in a 1 cm cuvette as well as of the selected substrates with electrophoretically deposited Au nanoparticles were recorded on a

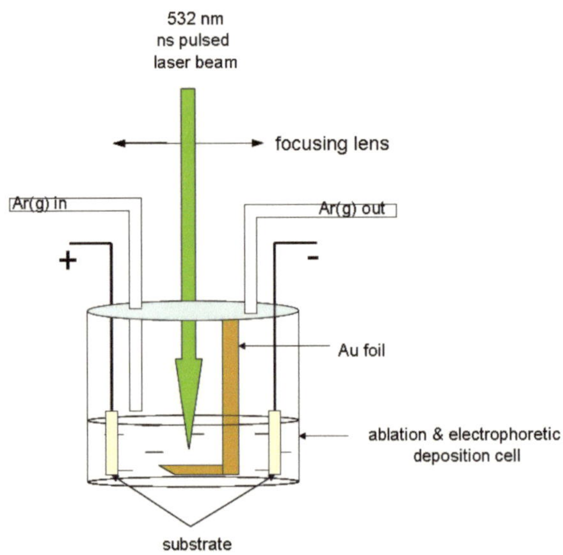

Scheme 1. Depiction of experimental setup for simultaneous pulsed-laser ablation and electrophoresis.

double-beam spectrophotometer (Perkin-Elmer Lambda 950). Zeta-potentials were measured by means of Zetasizer Nano series (Malvern Instruments). Transmission electron microscopy (TEM) was used for the characterization of sizes of Au nanoparticles dispersed in alcoholic solutions after the simultaneous pulsed-laser ablation and electrophoresis. TEM imaging of dried drops of the Au nanoparticle solutions deposited on a carbon-coated Cu grid was performed using a JEOL-JEM200CX microscope. Scanning electron microscopy (SEM) was employed for the characterization of ITO- and/or FTO-coated glass substrates. SEM images were recorded on a SEM microscope Quanta 200 FEI. HOPG substrates were measured on Ntegra scanning tunnelling microscope (STM). Mechanically clipped Pt/Ir tip was approached toward a sample until a set tunnel current was detected. All STM experiments were done under ambient conditions. STM images were recorded and treated by using Nova 1.0.26 software provided by NT-MDT.

3. Results and discussion

Our choice of Au target, primary alcohols, and the other parameters for the combined pulsed-laser ablation and electrophoretic deposition (PLA+EPD) process has been influenced by several good reasons. First of all, Au nanoparticles are preferred by many applications as it has been well documented in Introduction. Furthermore, they do not undergo surface oxidation as easily as Ag and/or Cu nanoparticles (Muto et al., 2007). Primary alcohols as ablation medium have been chosen because of a good stability of Au nanoparticles in ethanol and other aliphatic alcohols as reported in the literature many times (Amendola et al., 2006, 2007; Amendola & Meneghetti, 2009; Compagnini et al., 2002, 2003). Laser pulses of nanosecond time duration have been rather used because the occurrence of explosive boiling or other photomechanical ablation mechanisms is suppressed in comparison to the situation when using femtosecond pulses (Amendola & Meneghetti, 2009). The 532 nm wavelength has been employed in our study owing to the fact that a narrow particle size distribution can be obtained due to an efficient Au nanoparticles fragmentation accompanying their generation (by pulsed-laser ablation) at this wavelength. The selection of substrate types serving as electrodes is given by possible applications of Au nanoparticles-modified substrates in photovoltaic devices. Therefore, indium-tin-oxide (ITO) and/or fluorine-tin-oxide (FTO) coated glass substrates have been used. On the contrary, highly oriented pyrolytic graphite (HOPG) plates serving as electrodes in the PLA+EPD process have been employed with the aim to investigate the influence of the surface roughness on the character of electrodeposited Au nanoparticles, thus, HOPG has been chosen for a purely scientific reason.

3.1 Au nanoparticles solutions resulting from PLA+EPD process

In general, Au nanoparticles posses surface plasmon (collective oscillations of free electrons) resonances in the visible region of the electromagnetic spectrum. The position of the maximum of surface plasmon extinction (i.e., absorption + scattering) strongly depends on the nanoparticle size, shape, surrounding, and aggregation state (Rotello, 2004). Thus, measurements of extinction spectra of Au nanoparticle dispersions can serve as a first tool of their characterization. However, this characterization is insufficient since it does not report solely about one feature of nanoparticles. Therefore, transmission electron microscopy (TEM) has to be used as well in order to visualize Au nanoparticles and to distinguish between influences of shape and/or size on extinction spectrum for instance. Another important feature of nanoparticles in solutions is their zeta-potential which enables to predict their stability in solutions, their aggregation state. Obviously, the combination of all three measurements can fully characterize the Au nanoparticle alcoholic solutions resulting from the PLA+EPD process.

Figure 1 shows UV-visible extinction spectra of Au nanoparticles generated by the PLA+EPD process in ethanol. Two distinct values of direct current, 10 and 17 μA, have been allowed to pass through the ethanolic ablation medium. These Au nanoparticles solutions are labelled from now on as Au10 and Au17 according to the passing direct current values.

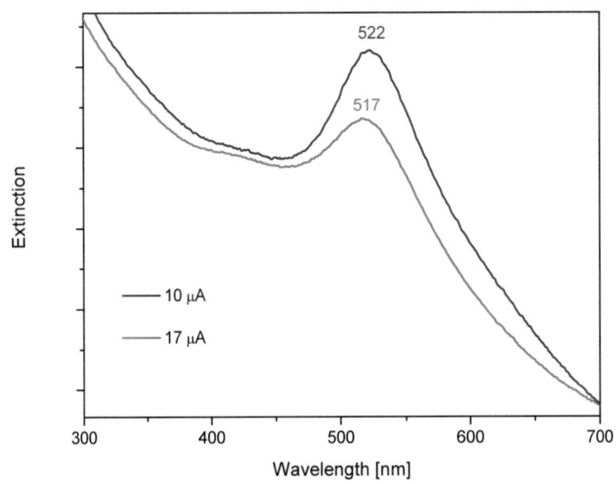

Fig. 1. UV-vis extinction spectra of Au nanoparticles generated by PLA+EPD process in ethanol while direct current of 10 μA and/or 17 μA passed through.

The maximum of surface plasmon extinction of Au10 is located at 522 nm, while that of Au17 is placed at 517 nm - Figure 1. Considering that all the other conditions, except for the direct current value, are the same (duration of PLA+EPD, laser fluence, experimental setup, etc.), and taking into account Mie theory (Rotello, 2004), the average nanoparticle size of Au17 could be smaller than that of Au10. This assumption is corroborated by particle size distribution (PSD) determined on the basis of TEM imaging – Figure 2. While Au10 contains the nanoparticles of 7.3 ± 3.1 nm in diameter (Figures 2A,B), nanoparticles of 4.0 ± 0.9 nm in diameter are encountered in Au17 (Figures 2C,D).

Interestingly, the optical density of Au10 is slightly higher than that obtained for Au17 (Figure 1) which can be related to a lower concentration of nanoparticles in Au17 solution. The decrease of Au nanoparticles concentration in Au17 solution is most probably caused by a higher amount of electro-deposited Au nanoparticles on electrode surface when the direct current of 17 μA is passed through the ablation medium. This hypothesis will be discussed in the next section.

Zeta potentials of Au nanoparticles ethanolic solutions have been measured and are presented in Table 1. Both types of solutions, Au10 and Au17, reveal values below -30 mV which indicates stable nanoparticle dispersions.

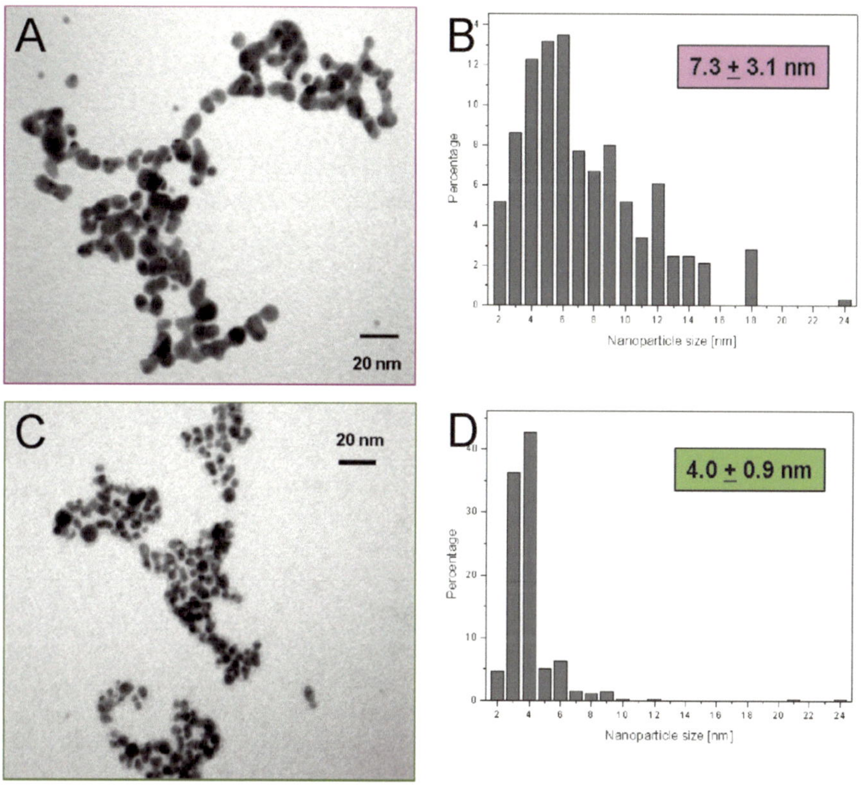

Fig. 2. TEM images (A, C) and appropriate PSD (B, D) of Au nanoparticles formed by PLA+EPD process while 10 μA (A, B) and/or 17 μA (C, D) passing through ethanolic ablation medium.

System label	Solvent	DC [μA]	Zeta potential [mV]
Au10	Ethanol	10	-37.4 ± 2.0
Au17	Ethanol	17	-42.6 ± 0.8
Au10B	Butanol	10	-10.8 ± 1.1
Au17B	Butanol	17	-12.9 ± 1.2

Table 1. Zeta potentials determined for ethanolic as well as butanolic Au nanoparticle solutions. DC means direct current.

In the next step, the PLA+EPD process has been performed in butanol. The resulting Au nanoparticles solutions are entitled as Au10B and Au17B when direct current of 10 μA and 17 μA passed through the butanolic ablation medium, respectively. The values of zeta potentials of these systems are presented in Table 1. They indicate rather unstable Au nanoparticles solutions since the values are above -30 mV and below 30 mV. The differences in zeta potential values of Au10, Au17, and Au10B, Au17B can be ascribed to different dielectric constants of solvents: ethanol possess the value of 24.3, while butanol 17.1 (Sýkora, 1976).

UV-visible extinction spectra of Au10B and Au17B solutions are shown in Figure 3. Both systems manifest themselves by a well pronounced surface plasmon extinction band with the maximum located at 526 nm indicating thus similar sizes of Au nanoparticles. This idea has been confirmed by PSD based on TEM imaging, presented in Figure 4. Au nanoparticles in Au10B solution reveal sizes of 4.9 ± 1.2 nm and in Au17B sizes of 5.2 ± 1.7 nm in diameter.

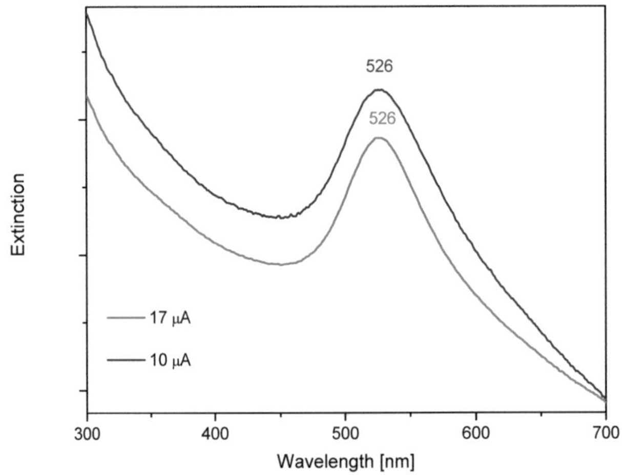

Fig. 3. UV-vis extinction spectra of Au nanoparticles generated by PLA+EPD process in butanol while direct current of 10 μA and/or 17 μA allowed to pass through.

Similarly as in the case of ethanolic Au nanoparticle solutions, the concentrations of Au nanoparticles appear to be slightly higher in Au10B than in Au17B solution. The reason will be discussed in the next section.

To sum up, it can be concluded that Au nanoparticles of controlled sizes dispersed in ethanol can be prepared by changing the direct current passing through the ethanolic ablation medium during the PLA+EPD process. In contrast, the same factor (direct current value) does not induce any changes in the average size of Au nanoparticles when formed by the PLA+EPD process in butanol. Considering the zeta potential values of ethanolic and butanolic Au nanoparticles solutions, this result is fully understandable since the higher the zeta potential value, the stronger effect of applied electric field on the generated nanoparticles. The longer aliphatic chain of butanol induces smaller zeta potential values of generated Au nanoparticles and, consequently, the effect of direct current passing during the PLA+EPD process is decreased.

Furthermore, ethanolic Au nanoparticles solutions can be prepared with a narrower particle size distribution when the direct current of 17 μA instead of 10 μA employed. On the contrary, the dispersity of butanolic Au nanoparticles solutions is almost negligibly influenced. Obviously, the length of primary alcohols has a distinct effect on the average size of Au nanoparticles generated by the PLA+EPD process.

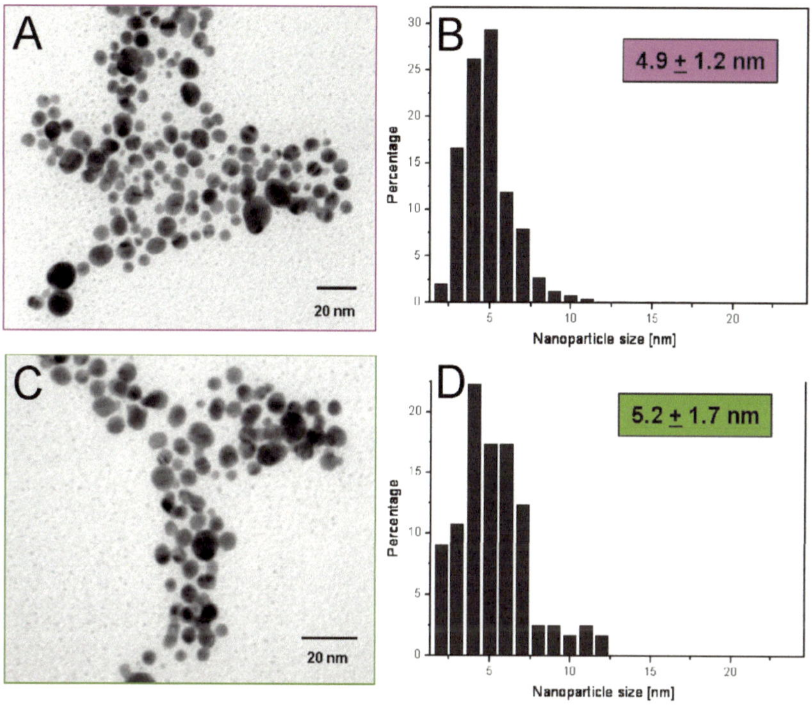

Fig. 4. TEM images (A, C) and appropriate PSD (B, D) of Au nanoparticles formed by PLA+EPD process while 10 μA (A, B) and/or 17 μA (C, D) passing through butanolic ablation medium.

3.2 Substrates with electrophoretically-deposited Au nanoparticles

In this section, three types of substrates serving as electrodes in the PLA+EPD process will be characterized by means of microscopic techniques and visible absorption spectroscopy. With respect to the negative values of zeta potential of generated Au nanoparticles in both primary alcohols, they are preferentially deposited on anodes.

3.2.1 ITO-coated glass substrates

SEM images of the ITO-coated glass substrates modified by electrodeposited Au nanoparticles during the PLA+EPD process performed in ethanol are shown in Figure 5. Comparing the SEM images of substrates in Figure 5A (10 μA direct current) and 5B (17 μA direct current), a higher surface coverage of substrates by Au nanoparticles is observed at higher current values than at the lower one. This microscopic observation goes hand in hand with the fact deduced from the UV-visible extinction spectra of Au nanoparticles solutions (discussed in the previous section): the final concentration of Au17 solution is lower than that of Au10 solution. The reason for this difference lies in a larger amount of Au nanoparticles being deposited under the higher than the lower current value and, as a consequence, a decrease of Au nanoparticles concentration in Au17 solution being determined.

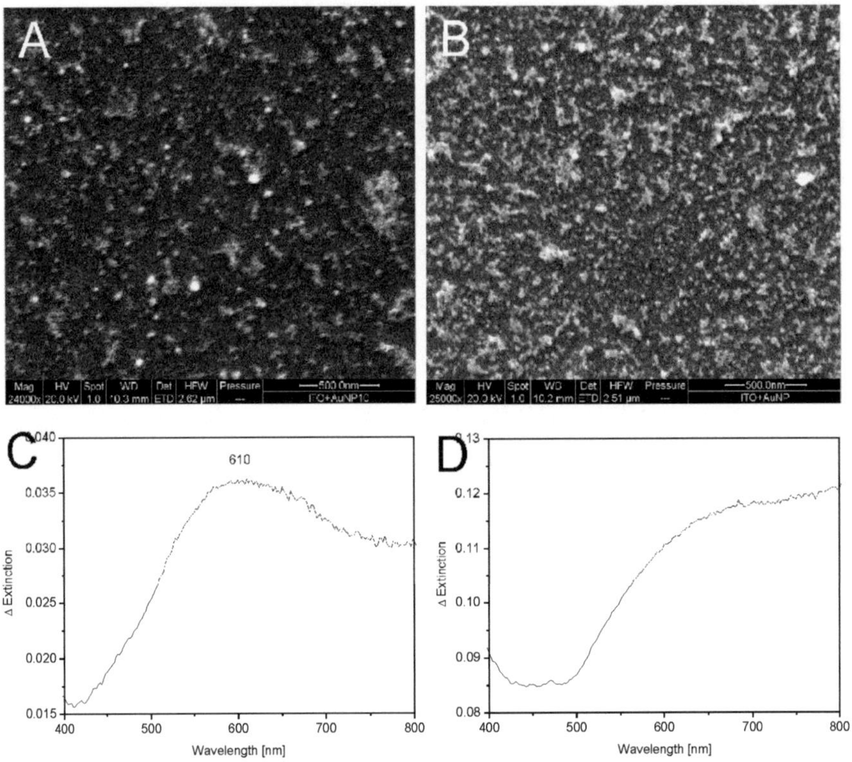

Fig. 5. SEM images (A, B) and particular differential visible extinction spectra (C, D) of ITO-coated glass substrates modified by Au nanoparticles electrodeposited at 10 μA (A, C) and/or 17 μA (B, D) during PLA+EPD process performed in ethanol.

Furthermore, Au nanoparticle aggregates are frequently encountered under both direct current values (Figures 5A and 5B). The aggregation can be also derived from the measured visible extinction spectra of the two discussed substrate samples, presented in Figure 5C and 5D. The differential extinction spectra have been obtained by the subtraction of the extinction spectrum of an unmodified ITO-coated glass substrate from that of a nanoparticles-modified ITO-coated glass substrate. The position of the maximum located at around 610 nm (Figure 5C) reports about aggregated Au nanoparticles on the substrates modified under 10 μA. In the case of Au nanoparticles deposited on ITO-coated glass substrates under 17 μA, there is even no distinct maximum of extinction band (Figure 5D) indicating thus an extensive aggregation of Au nanoparticles.

The same type of experiments using ITO-coated glass substrates as electrodes in the PLA+EPD process has been performed in butanol. The resulting SEM morphologies and differential visible extinction spectra are shown in Figure 6. Comparing Figures 6A (10 μA direct current) and 6B (17 μA direct current), a slightly higher amount of Au nanoparticles can be seen on ITO-coated glass substrates when a higher current value used. This is quite similar result to that observed in ethanolic systems. However, regarding the absolute counts of electro-deposited Au nanoparticles, the substrates from ethanolic solutions are generally more covered by Au nanoparticles than that obtained in butanolic solutions. As it has been already stated above, the higher zeta potential values of Au nanoparticles in ethanolic solutions are most probably responsible for this result. Furthermore, Au nanoparticles are more evenly dispersed on ITO-coated glass substrates immersed in butanolic than in ethanolic solutions. This can be related to the effect of aliphatic chain length.

The differential extinction spectra, shown in Figures 6C and 6D, reveal a distinct band with the maximum positioned at 575 nm when the lower, and at 615 nm when the higher current values employed. The positions of the maxima of surface plasmon extinction bands correlate with the microscopic observation presented in Figures 6A and 6B. Indeed, the higher the surface coverage of substrates by Au nanoparticles, the more intense and red-shifted surface plasmon extinction observed. In comparison to the extinction spectra of substrates immersed in ethanolic solutions during the PLA+EPD process, the surface plasmon extinction band is well-developed at both current values exploited for the PLA+EPD process performed in butanol. Thus, regarding the aggregation of Au nanoparticles electro-deposited on ITO-coated glass substrates, it is less pronounced in butanolic than in ethanolic samples. Again, the same result evidenced by two independent methods, microscopic and spectroscopic one.

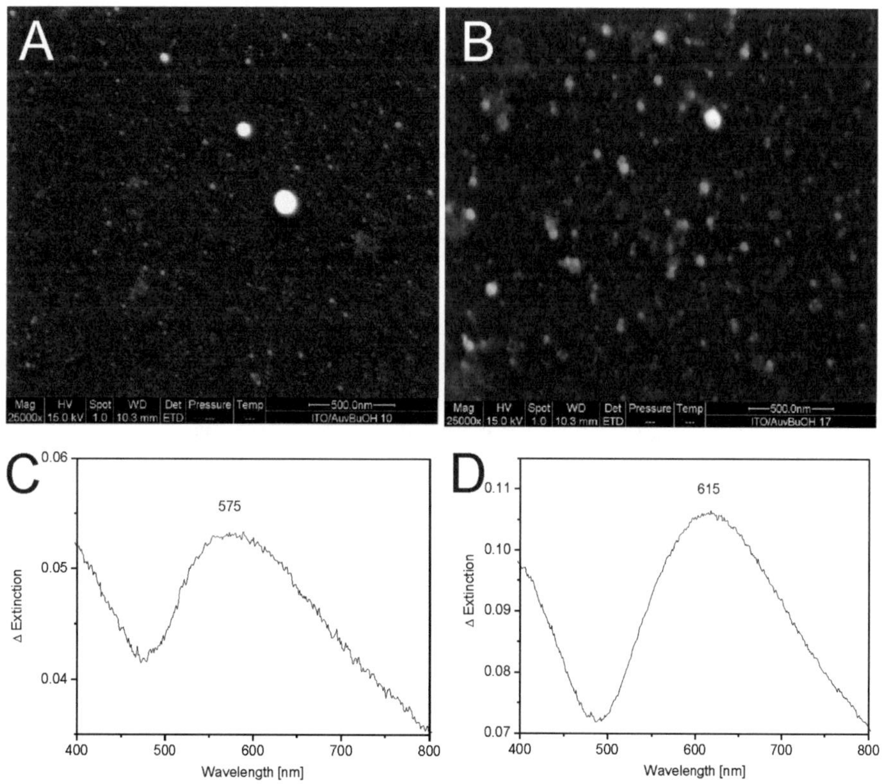

Fig. 6. SEM images (A, B) and particular differential visible extinction spectra (C, D) of ITO-coated glass substrates modified by Au nanoparticles electro-deposited at 10 μA (A, C) and/or 17 μA (B, D) during PLA+EPD process performed in butanol.

3.2.2 FTO-coated glass substrates

As it has been already evidenced in the previous section, there are strong effects of direct current value and the type of alcohol on the final coverage of a substrate by Au nanoparticles. In order to investigate if there is any additional influence of substrate roughness, FTO-coated glass substrates have been used as electrodes during the PLA+EPD process performed in ethanol at both values of direct current, 10 μA as well as 17 μA.

In Figure 7A, the SEM image of a cleaned bare FTO-coated glass substrate surface is shown. Obviously, the surface of a FTO-coated glass substrate is very rough with plates and crystals being of sizes of hundreds of nanometers. Taking into account that the local current density can be very different on the edges of a plate and/or crystal, an inhomogeneous distribution of Au nanoparticles and their aggregates on FTO-coated glass substrate can be awaited. Figures 7B and 7C depict the SEM images of Au nanoparticles-modified FTO-coated glass substrates when the lower and the higher electric field applied, respectively. Mutually

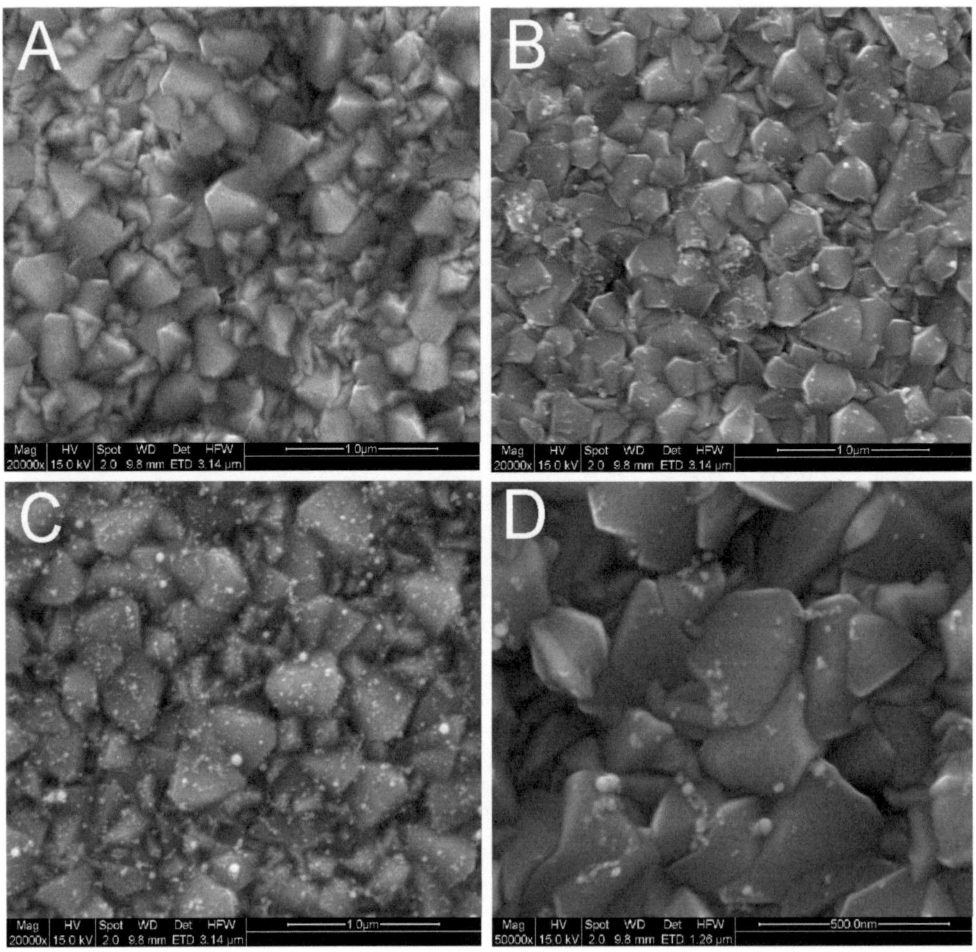

Fig. 7. SEM images of (A) cleaned bare FTO-coated glass substrate, (B, D) Au nanoparticles-modified FTO-coated glass substrate when 10 μA allowed to pass through the ethanolic ablation medium, and (C) Au nanoparticles-modified FTO-coated glass substrate when 17 μA used. Higher magnification is intentionally shown in (D).

compared, there is seen the effect of the direct current value, i.e., with an increasing current a higher surface coverage by Au nanoparticles is observed. When compared to ITO-coated glass substrates serving as electrodes under otherwise the same experimental conditions (SEM images presented in Figures 5A and 5B), the surface of FTO-coated glass substrates is covered even more randomly by Au nanoparticles and their aggregates. These results indicate that the surface roughness of a substrate does play an important role in the course of electrophoretic deposition of Au nanoparticles. The assumption about the inhomogenity of electric field, made a few lines above, is well documented by a characteristic SEM image in Figure 7D revealing a preferential deposition of Au nanoparticles on the edges of plates and crystals of a FTO-coated glass substrate.

3.2.3 HOPG substrates

Considering the results of the two previous sections, it can be hypothesized that a substrate with a very smooth surface, such as HOPG for instance, could lead to homogeneously dispersed electrodeposited Au nanoparticles since the current density will be homogeneous everywhere on the substrate surface. In order to prove this hypothesis, the PLA+EPD process performed in ethanol while 17 μA passed through has been chosen because under these conditions, the highest degree of aggregation of electrodeposited Au nanoparticles and inhomogenity in surface coverage were observed as shown in the two previous sections. With respect to the fact that Au nanoparticles in the selected system are tiny (around 4 nm in diameter), another microscopic technique than SEM has to be employed in order to visualize isolated nanoparticles on HOPG substrates. Scanning electron microscopy (STM) can fulfil this task when appropriate measuring conditions met (Durston et al., 1998; Wang et al., 2000).

Figure 8 shows topographic as well as tunnelling current images of a freshly cleaved HOPG surface without and with electrodeposited Au nanoparticles. Smoothness of HOPG surface is well evidenced in Figure 8A where the value along z axis (perpendicular to the plane of the image) stays well below 1 nm. The values of tunnelling current below 0.2 pA have been recorded on a freshly cleaved HOPG substrate measured under ambient conditions, in air and at the room temperature – Figure 8B. Under the same conditions, STM measurements of a HOPG substrate which served as the anode during the PLA+EPD process have been undertaken and one of the resulting topographic images together with its tunnelling current values are shown in Figures 8C and 8D, respectively. Evidently, isolated Au nanoparticles are randomly, however, quite homogeneously dispersed on the surface of a HOPG plate (Figure 8C), the value of 6 nm along z axis is not surpassed. It is worth noting that tunnelling current exceeds 0.4 nA (Figure 8D), which is the value of more than three orders of magnitude higher than on a bare HOPG substrate (Figure 8B). This can be related to the presence of Au nanoparticles.

It is known that HOPG substrate can contain terraces and steps as observed in Figure 8C. Hypothetically, the edges of these terraces and steps could be the places of a locally higher electrical density, hence, more electrodeposited Au nanoparticles could be awaited to occur on these edges. However, this was not the case as evidenced in Figure 8C.

Figure 9 shows topography and tunnelling current image of a smaller flat surface area (200 x 200 nm^2) on a HOPG substrate decorated with electrodeposited Au nanoparticles. At this place it should be noticed that the bias voltage of +0.1 V has been applied between the measured HOPG substrate and the tip during STM imaging. This is a sufficiently low value to suppress any unwanted manipulation of Au nanoparticles (Durston et al., 1998).

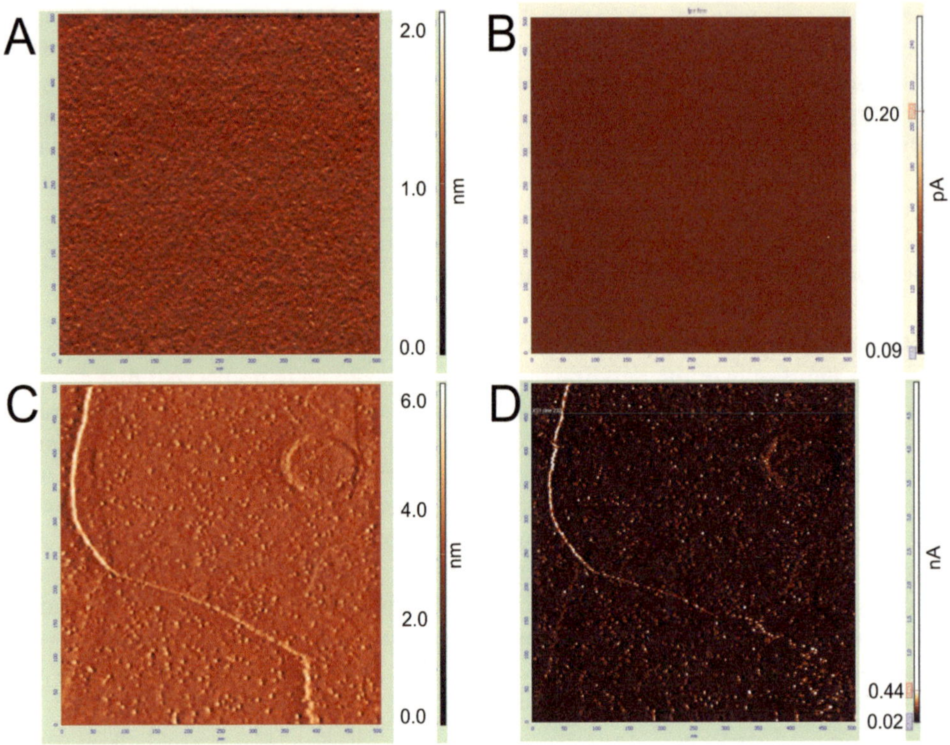

Fig. 8. Topographic (A, C) and tunnelling current (B, D) images of HOPG substrate serving as cathode (A, B) or anode (C, D) in PLA+EPD process while 17 μA passed through ethanolic ablation medium. Dimensions of scans are 500x500 nm^2.

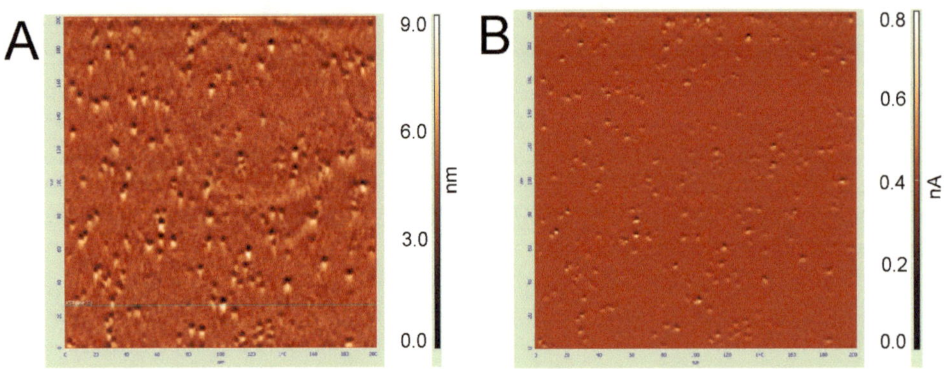

Fig. 9. 200 x 200 nm^2 scan of HOPG substrate with electrodeposited Au nanoparticles due to PLA+EPD process while 17 μA passed through ethanolic ablation medium:
(A) Topography, (B) tunnelling current image.

Therefore, on the basis of these results, it can be concluded that the substrate roughness in the range of hundreds of nanometers (the case of FTO-coated glass substrate) distinctly impedes the homogeneous coverage of substrate by electrodeposited Au nanoparticles. On the contrary, the surface roughness being below 1 nm does not hamper a homogeneous distribution of electrodeposited Au nanoparticles.

3.3 Possible elucidation of direct current value influence on mechanism of Au nanoparticles generation during PLA+EPD process in primary alcohols

Since *the final stages* of Au nanoparticles solutions and/or the electrodeposited Au nanoparticles on different substrates have been investigated, it cannot be unambiguously stated the exact formation mechanism of Au nanoparticles during the PLA+EPD process. However, it can be hypothesized the influence of direct current value on the mechanism of Au nanoparticles generation by the PLA+EPD process in comparison to a generally adopted mechanism of pulsed-laser ablation itself.

The prevailing formation mechanism of nanoparticles by a classical pulsed-laser ablation process implies the generation of a plasma plume followed by its cooling (Amendola & Meneghetti, 2009; Tsuji et al., 2004). The former step is nothing else than the vaporization of the part of a target which was attacked by the focused beam of laser pulses. During the second step (plasma plume cooling), the formation of nanoparticles nuclei starts. The driving force for the nucleation is the supersaturation in the plasma plume (Amendola &. Meneghetti, 2009). Subsequently, the nuclei grow and coalesce into the sizes of resulting nanoparticles. This last step strongly depends on the polarity of solvents, the presence and/or the absence of simple ions or adsorbing species which may stabilize nanoparticles of a particular size.

Under the assumption that our pulsed-laser ablation process in the selected solvent (e.g. ethanol or butanol) is repeatedly performed in the same way and under otherwise the same experimental conditions, the value of the applied electric field can induce changes rather in the step of nuclei growth and coalescence than during the plasma plume generation and/or the nucleation process. As it has been pointed out in section 3.1, ethanol possesses a higher dielectric constant than butanol which means that ethanol is more easily polarized by an increasing electric field than butanol. This implies that a further nuclei growth and coalescence is possibly hindered in the case of 17 μA direct current value passing through the ethanolic ablation medium when pulsed-laser ablation takes place. Basically, charged nanoparticles of smaller sizes in diameter can be efficiently stabilized in a more polarized solvent, i.e., in our case, at a higher current value passing through the ethanolic ablation medium. Thus, a smaller average particle size is observed in Au17 than in Au10 solutions. Nevertheless, this hypothesis needs a further experimental support which is beyond the scope of this chapter.

4. Conclusions

The application of nanosecond laser pulses of 532 nm laser wavelength for the generation of Au nanoparticles in primary alcohols (ethanol and butanol) has been discussed and the impact of a direct current passing simultaneously through the ablation medium during the pulsed-laser ablation process has been determined. On the basis of a complete characterization of Au nanoparticles solutions, it has been concluded that the average size of

Au nanoparticles can be influenced by the type of an alcoholic ablation medium as well as by the direct current value (the latter induces changes only in the case of ethanol). Moreover, the length of aliphatic chains in the two selected alcohols affects the character of coverage of ITO-coated glass substrates by Au nanoparticles; more evenly dispersed electro-deposited Au nanoparticles have been encountered in the butanolic ablation medium. In contrast, aggregates of Au nanoparticles have been observed when the ethanolic ablation medium used. The amount of electrodeposited Au nanoparticles is generally higher in ethanol than in butanol which can be related to the differences in zeta potential values of Au nanoparticles. The surface roughness of substrates has appeared to be another very important parameter influencing the final characteristic coverage of substrates by Au nanoparticles generated by the PLA+EPD process. An excellent correlation between microscopic and spectroscopic results has been demonstrated. Finally, a possible explanation of the influence of direct current value on the mechanism of Au nanoparticles generation during the PLA+EPD process has been proposed.

5. Acknowledgment

The author thanks to Mrs. Jiřina Hromádková for SEM and TEM imaging. Financial support by GAČR P108/11/P657 is gratefully acknowledged.

6. References

Amendola, V., Rizzi, G.A., Polizzi, S. & Meneghetti, M. (2005). Synthesis of gold nanoparticles by laser ablation in toluene: quenching and recovery of the surface plasmon absorption. *The Journal of Physical Chemistry B*, Vol. 109, No. 49, pp. 23125-23128, ISSN 1520-6106 print / ISSN 1520-5207 online

Amendola, V., Polizzi, S. & Meneghetti, M. (2007). Free silver nanoparticles synthesized by laser ablation in organic solvents and their easy functionalization. *Langmuir*, Vol. 23, No. 12, pp. 6766-6770, ISSN 0743-7463 print / ISSN 1520-5827 online

Amendola, V. & Meneghetti, M. (2009). Laser ablation synthesis in solution and size manipulation of noble metal nanoparticles. *Physical Chemistry Chemical Physics*, Vol. 11, pp. 3805-3821, ISSN 1463-9076 print / ISSN 1463-9084 online

Atwater, H.A. & Polman A. (2010). Plasmonics for improved photovoltaic devices. *Nature Materials*, Vol. 9, pp. 205-213, ISSN 1476-1122 print / ISSN 1476-4660 online

Boyer, P., Menard, D. & Meunier, M. (2010). Nanoclustered Co-Au particles fabricated by femtosecond laser fragmentation in liquids. *J. Phys. Chem. C*, Vol. 114, No. 32, pp. 13497-13500, ISSN 1932-7447 print / ISSN 1932-7455 online

Buining, PA, Humbel, BM, Phillipse, AP & Verkleij, AJ. (1997). Preparation of functional silane-stabilized gold colloids in the (sub)nanometer size range. *Langmuir*, Vol. 13, No. 15, pp. 3921-3926, ISSN 0743-7463 print / ISSN 1520-5827 online

Burakov, V.S., Tarasenko, N.V., Butsen, A.V., Rozantsev, V.A. & Nedelko, M.I. (2005). Formation of nanoparticles during double-pulse laser ablation of metals in liquids. *Eur. Phys. J. Appl. Phys.*, Vol. 30, pp. 107-112, ISSN 0021-8979 print / ISSN 1089-7550 online

Burakov, V.S., Butsen, A.V. & Tarasenko, N.V. (2010). Laser-induced plasmas in liquids for nanoparticle synthesis. *Journal of Applied Spectroscopy*, Vol. 77, No. 3, pp. 386-393, ISSN 0021-9037 print / ISSN 1573-8647 online

Carrara, M., Kakkassery, J.J., Abid, J.P. & Fermin, D.J. (2004). Modulation of the work function in layer-by-layer assembly of metal nanoparticles and poly-L-lysine on modified Au surfaces. *Chem. Phys. Chem.*, Vol. 5, pp. 571-575, ISSN 1439-7641

Compagninni, G., Scalisi, A.A. & Puglisi, O. (2002). Ablation of noble metals in liquids: a method to obtain nanoparticles in a thin polymeric film. *Phys. Chem. Chem. Phys.*, Vol. 4, pp. 2787-2791, ISSN 1463-9076 print / ISSN 1463-9084 online

Dammer, O., Vlckova, B., Slouf, M. & Pfleger, J. (2007). Interaction of high-power laser pulses with monodisperse gold particles. *Materials Science and Engineering B*, Vol. 140, pp. 138-146, ISSN 0921-5107

Darroudi, M., Ahmad, M. B., Zamiri, R., Abdullah, A.H., Ibrahim, N.A., Shameli, K. & Husin, M.S. (2011). Preparation and characterization of gelatine mediated silver nanoparticles by laser ablation. *Journal of Alloys and Compounds*, Vol. 509, pp. 1301-1304, ISSN 0925-8388

de Boer, B., Hadipour, A., Mandoc, M.M., van Woudenbergh, T. & Blom, P.W.M. (2005). Tuning of metal work functions with self-assembled monolayers. *Adv.Mater.*,Vol. 17, No. 5, pp. 621-625, ISSN 0935-9648 print / ISSN 1521-4095 online

Doron, A., Katz, E. & Willner, I. (1995). Organization of Au colloids as monolayer films onto ITO glass surfaces: application of the metal colloid films as base interfaces to construct redox-active monolayers. *Langmuir*, Vol. 11, No. 4, 1313-1317, ISSN 0743-7463 print / ISSN 1520-5827 online

Durston, P.J., Palmer, R.E. & Wilcoxon, J.P. (1998). Manipulation of passivated gold clusters on graphite with the scanning tunneling microscope. *Appl. Phys. Lett.*, Vol. 72, No. 2, pp. 176-178, ISSN 0003-6951 print / ISSN 1077-3118 online

Fong, Y.Y., Gascooke, J. R., Visser, B.R., Metha, G.F. & Buntine, M.A. (2010). Laser-based formation and properties of gold nanoparticles in aqueous solution: formation kinetics and surfactant-modified particle size distributions. *J. Phys. Chem. C*, Vol. 114, No. 38, pp. 15931-15940, ISSN 1932-7447 print / ISSN 1932-7455 online

Franklin, S.R. & Thareja, R.K. (2004). Simplified model to account for dependence of ablation parameters on temperature and phase of the ablated material. *Applied Surface Science*, Vol. 222, pp. 293-306, ISSN 0169-4332

Georgiou, S.& Koubenakis, A. (2003). Laser-induced material ejection from model molecular solids and liquids: mechanisms, implications, and applications. *Chemical Reviews*, Vol. 103, No. 2, pp.349-393, ISSN 0009-2665 print / ISSN 1520-6890 online

Grabar, K.C., Allison, K.J., Baker, B.E., Bright, R.M., Brown, K.R., Freeman, R.G., Fox, A.P., Keating, Ch.D., Musick, M.D. & Natan, M.J. (1996). Two-dimensional arrays of colloidal gold particles: a flexible approach to macroscopic metal surfaces. *Langmuir*, Vol. 12, No. 10, 2353-2361, ISSN 0743-7463 print / ISSN 1520-5827 online

Grzelczak, M., Vermant, J., Furst, E.M. & Liz-Marzan, L.M. (2010). Directed self-assembly of nanoparticles. *ACS Nano*, Vol. 4, No. 7, pp. 3591-3605, ISSN 1936-0851 print / ISSN 1936-086X online

He, H., Cai, W., Lin, Y. & Chen, B. Surface decoration of ZnO nanorod arrays by electrophoresis in the Au colloidal solution prepared by laser ablation in water. *Langmuir*, Vol. 26, No. 11, 8925-8932, ISSN 0743-7463 print / ISSN 1520-5827 online

Inasawa, S., Sugiyama, M., Noda, S. & Yamaguchi, Y. (2006). Spectroscopic study of laser-induced phase transition of gold nanoparticles on nanosecond time scales and longer. *J. Phys. Chem. B*, Vol. 110, No. 7, pp. 3114-3119, ISSN 1520-6106 print / ISSN 1520-5207 online

Jain, P.K., El-Sayed, I.H. & El-Sayed M.A. (2007). Au nanoparticles target cancer. *Nanotoday*, Vol. 2, No. 1, pp. 18-28, ISSN 1748-0132

Kabashin, A.V., Meunier, M., Kingston, Ch. & Luong, J.H.T. (2003). Fabrication and characterization of gold nanoparticles by femtosecond laser ablation in an aqueous solution of cyclodextrins. *J.Phys.Chem. B*, Vol. 107, No. 19, pp. 4527-4531, ISSN 1520-6106 print / ISSN 1520-5207 online

Kamat, P.V., Flumiani, M. & Hartland, G.V. (1998). Picosecond dynamics of silver nanoclusters. Photoejection of electrons and fragmentation. *J. Phys. Chem. B*, Vol. 102, No. 17, pp. 3123-3128, ISSN 1520-6106 print / ISSN 1520-5207 online

Kim, S.S., Na, S.I., Kim, D.Y. & Nah, Y.Ch. (2008). Plasmon enhanced performance of organic solar cells using electrodeposited Ag nanoparticles. *Appl. Phys. Lett.*, Vol. 93, 073307, ISSN 0003-6951 print / ISSN 1077-3118 online

Kurita, H., Takami, A. & Koda, S. (1998). Size reduction of gold particles in aqueous solution by pulsed laser. *Appl.Phys.Lett.*, vol. 72, No. 7, pp. 789-791, ISSN 0003-6951 print / ISSN 1077-3118 online

Le Ru, E. & Etchegoin, P. (2008). *Principles of Suface-Enhanced Raman Spectroscopy and related plasmonic effects,* Elsevier Science, ISBN 0444527796, Amsterdam, The Netherlands

Link, S., Burda, C. Mohamed, M.B., Nikoobakht, B. & El-Sayed, M.A. (1999). Laser photothermal melting and fragmentation of gold nanorods: energy and laser pulse-width dependence. *J. Phys. Chem. A*,Vol. 103, No. 9, pp. 1165-1170, ISSN 1089-5639 print / ISSN 1520-5215 online

Link, S & El-Sayed, MA. (2003). Otpical properties and ultrafast dynamics in metallic nanocrystals. *Annu. Rev. Phys. Chem.*, Vol. 54, pp. 331-366, ISSN 0066-426X

Mafune, F., Kohno, Jy., Takeda, Y. & Kondow, T. (2001). Dissociation and aggregation of gold nanoparticles under laser irradiation. *J. Phys. Chem. B*, Vol. 105, No. 38, pp. 9050-9056, ISSN 1520-6106 print / ISSN 1520-5207 online

Mafune, F., Kohno, Jy., Takeda, Y. & Kondow, T. (2002). Full physical preparation of sized-selected gold nanoparticles in solution: laser ablation and laser-induced size control. *J. Phys. Chem. B*, Vol. 106, No. 31, pp. 7575-7577, ISSN 1520-6106 print / ISSN 1520-5207 online

Menendez-Manjon, A., Jakobi, J., Schwabe, K., Krauss, J.K. & Barcikowski, S. (2009). Mobility of nanoparticles generated by femtosecond laser ablation in liquids and its application to surface patterning. *JLMN-Journal of Laser Micro/Nanoengineering,* Vol. 4, No. 2, pp. 95-99, ISSN 1880-0688

Morfa, A.J., Rowlen, K.L., Reilly III, T.H., Romero, M.J. & van de Lagemaat, J. (2008). Plasmon-enhanced solar energy conversion in organic bulk heterojunction photovoltaics. *Appl. Phys. Lett.*, Vol. 92, 013504, ISSN 0003-6951 print / ISSN 1077-3118 online

Mortier, T., Verbiest, T. & Persoons, A. (2003). Laser ablation of gold in chloroform solutions of cetyltrimethylammoniumbromide. *Chem.Phys.Lett.*, Vol. 382, pp. 650-653, ISSN 0009-2614

Muto, H., Yamada, K., Miyajima, K. & Mafune, F. (2007). Estimation of surface oxide on surfactant-free gold nanoparticles laser-ablated in water. *J.Phys.Chem. C*, Vol. 111, No. 46, pp. 17221-17226, ISSN 1932-7447 print / ISSN 1932-7455 online

Nah, Y.Ch., Kim, S.S., Park, J.H. & Kim, D.Y. (2007). Electrochromic coloration of MEH-PPV films by electrodeposited Au nanoparticles. *Electrochemical and Solid-State Letters*, Vol. 10, No. 1, pp. J12-J14, ISSN 1099-0062

Peng, Z., Walther, T. & Kleinermanns, K. (2005). Photofragmentation of phase-transferred gold nanoparticles by intense pulsed laser light. *J.Phys.Chem.B*, Vol. 109, No. 33, pp. 15735-15740, ISSN 1520-6106 print / ISSN 1520-5207 online

Procházka, M., Mojzeš, M., Štěpánek, J., Vlčková, B. & Turpin, P.Y. (1997). Probing applications of laser-ablated Ag colloids in SERS spectroscopy: improvement of ablation procedure and SERS spectral testing. *Anal. Chem.*, Vol. 69, No. 24, pp. 5103-5108, ISSN 0003-2700 print / ISSN 1520-6882 online

Roduner, E. (2006). *Nanoscopic Materials: Size-dependent Phenomena*, The Royal Society of Chemistry, ISBN-13: 978-0-85404-857-1, Dorchester, Dorset, UK

Semerok, A., Chaleard, C., Detalle, V., Lacour, J.L., Mauchien, P., Meynadier, P., Nouvellon, C., Salle, B., Palianov, P., Perdrix, M. & Petite, G. (1999). Experimental investigations of laser ablation efficiency of pure metals with femto, pico and nanosecond pulses. *Applied Surface Science*, Vol. 138-139, pp. 311-314, ISSN 0169-4332

Shoji, M., Miyajima, K. & Mafune, F. (2008). Ionization of gold nanoparticles in solution by pulse laser excitation as studied by mass spectrometric detection of gold cluster ions. *J. Phys. Chem. C*, Vol. 112, No. 6, pp. 1929-1932, ISSN 1932-7447 print / ISSN 1932-7455 online

Simakin, A.V., Voronov, V.V., Kirichenko, N.A. & Shafeev, G.A. (2004). Nanoparticles produced by laser ablation of solids in liquid environment. *Appl. Phys. A*, Vol. 79, pp. 1127-1132, ISSN 0947-8396 print / ISSN 1432-0630 online

Sládková, M., Vlčková, B., Mojzeš, P., Šlouf, M., Naudin & C. LeBourdon, G. (2006). Probing strong optical fields in compact aggregates of silver nanoparticles by SERRS of protoporphyrin IX. *Faraday Discuss.*, Vol. 132, pp. 121-134, ISSN 0301-7249 print / ISSN 1364-5498 online

Sobhan, MA, Ams, M, Withford, MJ & Goldys, EM. (2010). Ultrafast laser ablative generation of gold nanoparticles: the influence of pulse energy, repetition frequency and spot size. *J. Nanopart. Res.*, Vol. 12, pp. 2831-2842, ISSN 1388-0764 print / ISSN 1572-896X online

Srnová, I., Procházka, M., Vlčková, B., Štěpánek, J. & Malý, P. (1998). Surface-enhanced Raman scattering-active systems prepared from Ag colloids laser-ablated in chemically modified aqueous media. *Langmuir*, Vol. 14, No. 16, 4666-4670, ISSN 0743-7463 print / ISSN 1520-5827 online

Sýkora, V. (1976). *Chemicko-analyticke tabulky*. SNTL, ISBN 80-03-00049-1, Prague, Czech Republic

Sylvestre, J.P., Poulin, S., Kabashin, A.V., Sacher, E., Meunier, M. & Luong, J.H.T. (2004). Surface chemistry of gold nanoparticles produced by laser ablation in aqueous media. *J.Phys.Chem. B*, Vol. 108, pp. 16864-16869, ISSN 1520-6106 print / ISSN 1520-5207 online

Rabani, E., Reichman, D.R., Geissler, P.L. & Brus, L.E. (2003). Drying-mediated self-assembly of nanoparticles *Nature*, vol. 426, pp. 271-274, , ISSN 1545-0740

Rotello, V. (Ed.). (2004). *Nanoparticles: Building blocks for nanotechnology.* Kluwer Academi/Plenum Publishers, ISBN 0-306-48287-8, New York, USA

Schnippering, M., Carrara, M., Foelske, A., Kotz, R. & Fermin, D.J. (2007). Electronic properties of Ag nanoparticle arrays. A Kelvin probe and high resolution XPS study. *Phys.Chem.Chem.Phys.*, Vol. 9, pp. 725-730, ISSN 1463-9076 print / ISSN 1463-9084 online

Šišková, K., Vlčková, B., Turpin, P.Y., Fayet, C. Hromádková, J. & Šlouf, M. (2007). Effect of citrate ions on laser ablation of Ag foil in aqueous medium. *Journal of Physics: Conference Series*, Vol. 59, pp. 202-205, ISSN 1742-6588 print / ISSN 1742-6596 online

Šišková, K., Vlčková, B., Turpin, P.Y., Thorel, A. & Grosjean, A. (2008). Porphyrins as SERRS spectral probes of chemically functionalized Ag nanoparticles. *Vibrational Spectroscopy*, Vol. 48, pp. 44-52, ISSN 0924-2031

Šišková, K., Pfleger, J. & Procházka, M. (2010). Stabilization of Au nanoparticles prepared by laser ablation in chloroform with free-base porphyrin molecules. *Applied Surface Science*, Vol. 256, pp. 2979-2987, ISSN 0169-4332

Šišková, K., Vlčková, B., Turpin, P.Y., Thorel, A. & Procházka, M. (2011). Laser ablation of silver in aqueous solutions of organic species: probing Ag nanoparticle-adsorbate systems evolution by surface-enhanced Raman and surface plasmon extinction spectra. *J. Phys. Chem. C*, Vol. 115, pp. 5404-5412, ISSN 1932-7447 print / ISSN 1932-7455 online

Šišková, K., Šafářová, K., Seo, J.H., Zbořil, R. & Mashlan, M. (2011) Non-chemical approach toward 2D self-assemblies of Ag nanoparticles via cold plasma treatment of substrates. *Nanotechnology*, Vol. 22, 275601 (7pp) NANO/381585/PAP, ISSN 0957-4484 print / ISSN 1361-6528 online

Šloufová-Srnová, I. & Vlčková, B. (2002) Two-dimensional assembling of Au nanoparticles mediated by tetrapyridylporphine molecules. *NanoLetters*, Vol. 2, No. 2, 121-125, ISSN 1530-6984 print / ISSN 1530-6992 online

Šmejkal, P., Šišková, K., Vlčková, B., Pfleger, J., Šloufová, I., Šlouf, M. & Mojzeš, P. (2003). Characterization and surface-enhanced Raman spectral probing of silver hydrosols prepared by two-wavelength laser ablation and fragmentation. *Spectrochimica Acta A*, Vol. 59, pp. 2321-2329, ISSN 1386-1425

Šmejkal, P., Pfleger, J., Šišková, K., Vlčková, B., Dammer,O. & Šlouf, M. (2004). In-situ study of Ag nanoparticle hydrosol optical spectra evolution during laser ablation/fragmentation. *Appl. Phys. A*, Vol. 79, pp. 1307-1309, ISSN 0947-8396 print / ISSN 1432-0630 online

Takami, A., Kurita, H. & Koda, S. (1999). Laser-induced size reduction of noble metal particles. *J.Phys.Chem.B*, Vol. 103, No. 8, pp. 1226-1232, ISSN 1520-6106 print / ISSN 1520-5207 online

Takeda, Y., Kondow, T. & Mafune, F. (2005). Formation of Au(III)-DNA coordinate complex by laser ablation of Au nanoparticles in solution. *Nucleoside, Nucleotides, and Nucleic Acids*, Vol. 24, No. 8, pp. 1215-1225, ISSN 1525-7770 print / 1532-2335 online

Tarasenko, N.V., Butsen, A.V. & Nevar, E.A.(2005). Laser-induced modification of metal nanoparticles formed by laser ablation technique in liquids. *Applied Surface Science,* Vol. 247, pp. 418-422, ISSN 0169-4332

Tsuji, T., Tsuboi, Y., Kitamura, N. & Tsuji, M. (2004). Microsecond-resolved imaging of laser ablation at solid-liquid interface: investigation of formation process of nano-size metal colloids. *Applied Surface Science,* Vol. 229, pp. 365-371, ISSN 0169-4332

Tsuji, T., Mizuki, T., Yasutomo, M., Tsuji, M., Kawasaki, H., Yonezawa, T. & Mafune, F. (2011). Efficient fabrication of substrates for surface-assisted laser desorption/ionization mass spectrometry using laser ablation in liquids. *Applied Surface Science,* Vol. 257, pp. 2046-2050, ISSN 0169-4332

Wang, B., Xiao, X., Huang, X., Sheng, P. & Hou, J.G. (2000). Single-electron tunneling study of two-dimensional gold clusters. *Appl.Phys.Lett.,* Vol. 77, No. 8, pp. 1179-1181, ISSN 0003-6951 print / ISSN 1077-3118 online

Wender, D., Andreazza, M.L., Correia, R.R.B., Teixeira, S.R. & Dupont, J. (2011). Synthesis of gold nanoparticles by laser ablation of an Au foil inside and outside ionic liquids. *Nanoscale,* Vol. 3, pp. 1240-1245, ISSN 2040-3364 print + online / ISSN 2040-3372 online only

Werner, D., Hashimoto, S., Tomita, T., Matsuo, S. & Makita, Y. (2008). Examination of silver nanoparticle fabrication by pulsed-laser ablation of flakes in primary alcohols. *J.Phys. Chem. C,*, Vol. 112, No. 5, pp. 1321-1329, ISSN 1932-7447 print / ISSN 1932-7455 online

Werner, D., Hashimoto, S. & Uwada, T. (2010). Remarkable photothermal effect of interband excitation on nanosecond laser-induced reshaping and size reduction of pseudospherical gold nanoparticles in aqueous solution. *Langmuir,* Vol. 26, No. 12, pp. 9956-9963, ISSN 0743-7463 print / ISSN 1520-5827 online

Wu, K.Y., Yu, S.Y. & Tao, Y.T. (2009). Continuous modulation of electrode work function with mixed self-assembled monolayers and its effect in charge injection. *Langmuir,* Vol. 25, No.11, pp. 6232-6238, ISSN 0743-7463 print / ISSN 1520-5827 online

Yamada, K., Tokumoto, Y., Nagata, T. & Mafune, F. (2006). Mechanism of laser-induced size-reduction of gold nanoparticles as studied by nanosecond transient absorption spectroscopy. *J. Phys. Chem. B*, Vol. 110, No. 24, pp. 11751-11756, ISSN 1520-6106 print / ISSN 1520-5207 online

Yamada, K., Miyajima, K. & Mafune, F. (2007). Thermionic emisión of electrons from gold nanoparticles by nanosecond pulse-laser excitation of interband. *J. Phys. Chem. C,* Vol. 111, No. 30, pp. 11246-11251, ISSN 1932-7447 print / ISSN 1932-7455 online

Yang, S., Cai, W., Liu, G. & Zeng, H. (2009). From nanoparticles to nanoplates: preferential oriented connection of Ag colloids during electrophoretic deposition. *J. Phys. Chem. C*, Vol. 113, No. 18, 7692-7696, ISSN 1932-7447 print / ISSN 1932-7455 online

Zhigilei, L.V. & Garrison, B.J. (1999). Mechanisms of laser ablation from molecular dynamics simulations: dependence of the initial temperature and pulse duration. *Appl.Phys.A,* Vol. 69, pp. S75-S80, ISSN 0947-8396 print / ISSN 1432-0630 online

Zhitomirsky, I., Petric, A. & Niewczas, M. (2002). Nanostructured ceramic and hybrid materials via electrodeposition. *JOM,* Vol. September, pp. 31-35, ISSN 1047-4838 print / ISSN 1543-1851 online

9

Application of Pulsed Laser Fabrication in Localized Corrosion Research

M. Sakairi[1], K. Yanada[2], T. Kikuchi[1], Y. Oya[3] and Y. Kojima[3]
[1]*Faculty of Engineering, Hokkaido University, Kita-13, Nishi-8, Kita-ku, Sapporo*
[2]*Graduate School of Engineering, Hokkaido University, Kita-13, Nishi-8, Kita-ku, Sapporo*
[3]*Technical Research Division, Furukawa-Sky Aluminum Corp., Akihabara UDX, Sotokanda 4-chome, Chiyoda-ku, Tokyo*
Japan

1. Introduction

Aluminum and its alloys have been known as light metals because they are used to reduce the weight of automobiles and components. Aluminum is the second most used and produced metal in the world nowadays. It is well known that one of the typical corrosion morphologyies of aluminum alloys in chloride containing environments such as seawater is pitting corrosion. Many papers have been investigating pitting corrosion ((Ito et al., 1968), (Horibe et al., 1969), (Goto et al. 1970), (Blanc et al., 1997), (Kang et al., 2010)). Electrochemical techniques, such as model macro-pits (Itoi et al., 2003) and electrochemical noise analysis (Sakairi et al., 2005, 2006, 2007) have been applied to investigate pitting corrosion of aluminum alloys.

Details of the propagation of pitting corrosion (Fig. 1) are not fully understood, however, the aspect ratio of pits (pit depth/pit diameter) plays a very important role in the growth of corrosion pits (Toma et al., 1980). To understand this effect, an in-situ artificial pit fabrication technique with area selective dissolution measurements would be helpful. One such technique is pulsed laser fabrication, which uses focused pulsed Nd-YAG laser beam irradiation to remove material from the substrate, combined with anodizing. Some of the authors have reported on the electrochemical behavior of artificial pits fabricated on aluminum alloy ((Sakari et al., 2009), (Yanada et al., 2010)).

In this chapter, the results of the effect of aspect ratio on dissolution behavior of the artificial pits formed on aluminum alloys are explained, and the chapter also explains the rate of pit fabrication and how to activate only the bottom surface of the formed pits.

2. Artificial pit fabrication by pulsed laser

2.1 Experimental

Sheet specimens of 2024 (15 x 20 x 2.0 mm) and 1050 (15 x 20 x 1.1 mm) aluminum alloys were used. Table 1 shows the chemical composition of the used aluminum alloys. Specimens were cleaned in doubly distilled water and an ethanol ultrasonic bath, and then polished

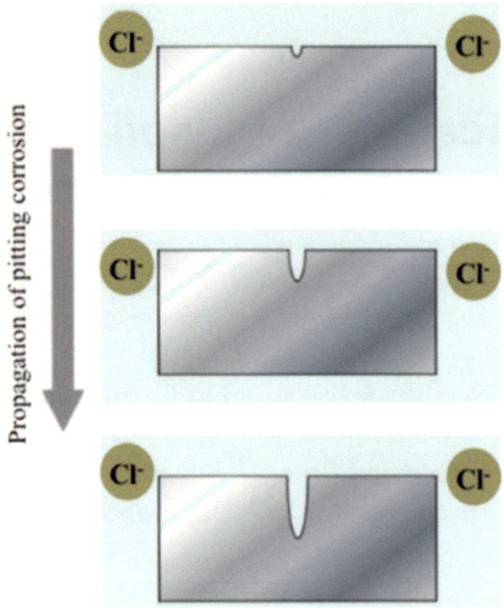

Fig. 1. Schematic representation of propagation of pitting corrosion in chloride ion containing solutions.

	Si	Fe	Cu	Mn	Mg	Zn	other	Al
2024	0.06	0.20	4.78	0.59	1.43	0.07	0.26	balance
1050	0.08	0.27	0.03	-	-	-	0.05	balance

Table 1. Chemical compositions of used aluminum alloys.

chemically in 0.1 kmol m^{-3} NaOH for 1800 s. A protective film is required to investigate the electrochemical reactions at only the laser beam irradiated area, and porous type anodic oxide films were formed at a constant current density of 100 A m^{-2} in 0.22 kmol m^{-3} $C_2H_2O_4$ at 293 K for 1800 s. Anodized specimens were dipped in boiling doubly distilled water for 900 s (pore sealing) to improve the protective performance of the formed anodic oxide films.

Specimens with protective films were irradiated by a focused Nd-YAG laser beam (Sepctra Physics GCR-130, wave duration 8 ns, frequency 10 s^{-1}, wave length 532 nm) for t_i = 0 to 30 s while immersed in 0.5 kmol m^{-3} H_3BO_3 - 0.05 kmol m^{-3} $Na_2B_4O_7$ (Borate). The laser beam power was adjusted to 30 mW in front of the lens. Fig. 2 shows a schematic outline of the laser irradiation and electrochemical measurement apparatus used in this chapter.

Surface and pit size observations: Specimen surfaces after the experiments were examined by an optical microscope, confocal scanning laser microscope (CSLM; 1SA21, LASERTEC Co.), and a scanning electron microscope (SEM; TM 1000, Hitachi Co.). The formed artificial pit depths and geometry were measured with the analysis function of the CSLM and X-ray Computed Tomography (X-ray CT; ELE SCAN mini, NS-ELEX Co. Ltd.).

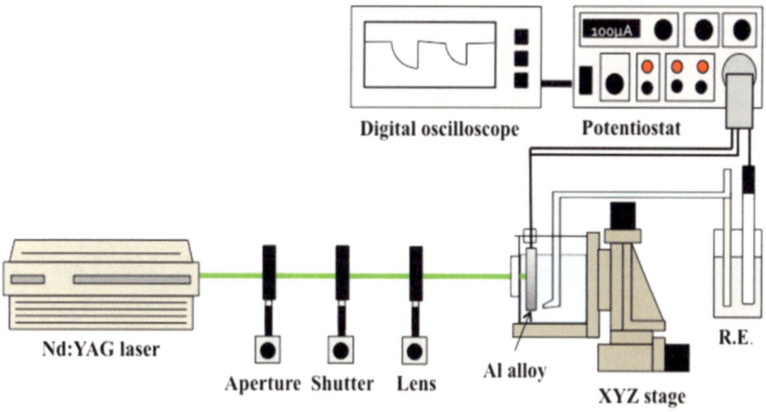

Fig. 2. Schematic representation of the laser irradiation and electrochemical measurement apparatus.

2.2 Results and discussion

2.2.1 Kinetics of pit fabrication

Artificial pit depth and morphology changes with continuous laser beam irradiation time were investigated in Borate. Fig. 3 shows SEM surface images of laser beam irradiated 2024 aluminum alloy specimens at t_i = 0.1 to 150 s. The anodic oxide film and aluminum alloy substrate can be removed at the irradiated area, and the shape of the area where oxide film was removed is almost circular. The center of the oxide film removed area is bright initially and then becomes dark with t_i, and it becomes larger with t_i.

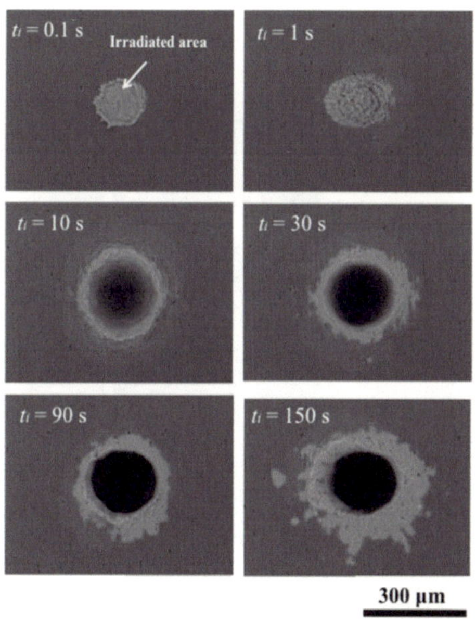

Fig. 3. SEM surface images of fabricated artificial pits on 2024 aluminum ally at different laser beam irradiation times.

Figure 4 shows X-ray CT vertical sectioning images of fabricated artificial pits on 2024 aluminum alloy. Fig. 5 shows horizontal section images of a t_i= 150 s pit and a schematic representation of the section positions. From Fig. 4, the depth of a fabricated pit becomes deeper with longer t_i. From the horizontal sectional images in Fig. 5, the shape of fabricated artificial pits are almost completely circular from the top to the bottom.

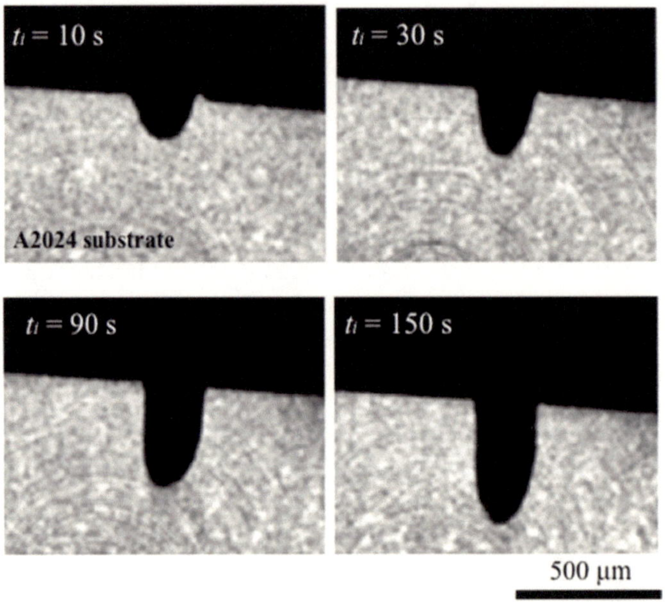

Fig. 4. X-ray computed tomography (X-ray CT) vertical section images of fabricated artificial pits on 2024 aluminum alloy.

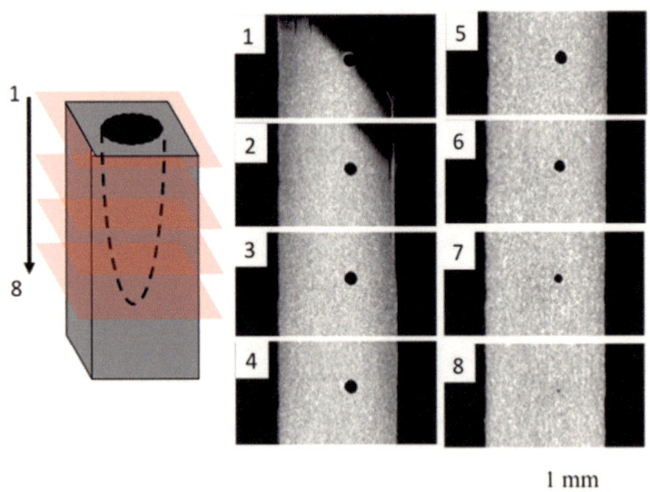

Fig. 5. X-ray CT horizontal section images of pits fabricated on 2024 aluminum alloy (t_i= 150 s) and schematic representation of section positions.

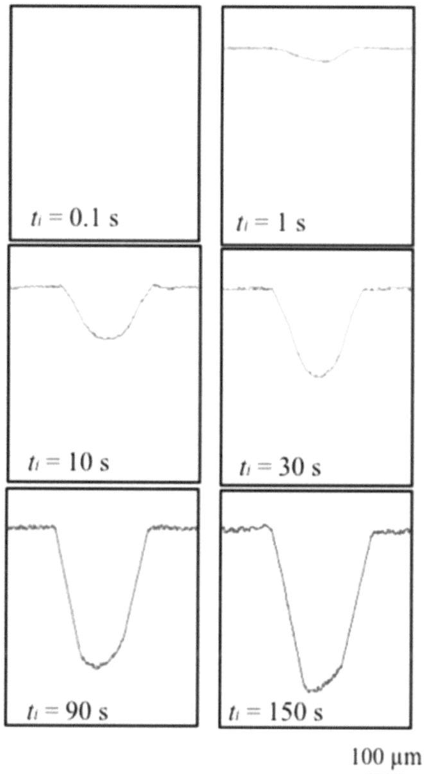

Fig. 6. CLSM depth profiles of fabricated artificial pits in 2024 aluminum alloy.

Figure 6 shows CSLM depth profile of fabricated pits with t_i. The CSLM depth profile also shows that the center area of the laser beam irradiated area is deeper than the other areas. The pit becomes deeper with t_i, however, the cross-sectional shape is independent of t_i.

Figure 7 shows changes in diameter (width) of artificial pits fabricated in both the 1050 and 2024 aluminum alloys with t_i. The pit diameter increases sharply at $t_i < 1$ s and the slope of the diameter change curve becomes flatter with longer t_i. The value of the diameter of 1050 aluminum alloy is about 20% lager than that of 2024 aluminum alloy. The laser beam used here has a Gaussian energy distribution and aluminum metal changes to gas or plasma only at the center of the irradiated area. However, the outer rim of the laser beam has sufficient energy to melt the aluminum substrate. This melted metal was ejected or flows by the effect of the evaporated gas or formed plasma (Fig. 8). If the irradiating conditions do not change during the experiments, then, after some time, the size of the melted area would not change with t_i.

Figure 9 shows the increases in depth of artificial pits fabricated on both the 1050 and 2024 aluminum alloys with t_i. The pit depth increases sharply at $t_i < 1$ s and the slope of the depth change curve becomes flatter with t_i. The specimen did not move during the laser beam irradiation, and therefore the distance between lens and irradiated surface (bottom of the pit) becomes longer with t_i. This distance change causes a decrease in the mean beam energy available for pit fabrication. This is a reason why the slope of the pit depth change curve becomes flatter with t_i. The pit formation rate of the 1050 aluminum alloy is about twice that of the 2024 aluminum alloy.

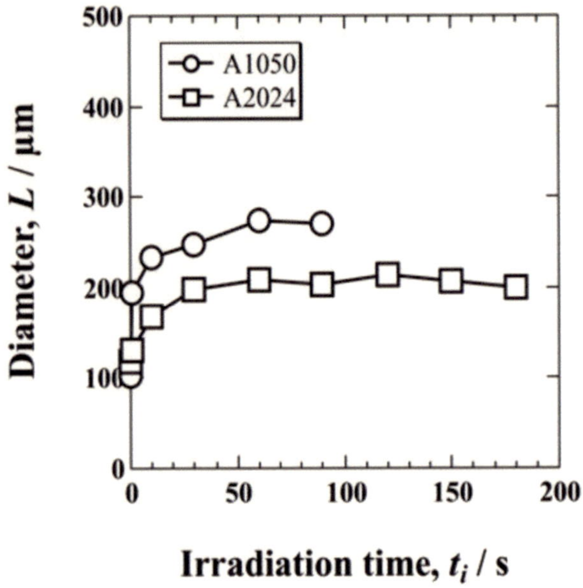

Fig. 7. Changes in the diameters of fabricated artificial pits on 1050 and 2024 aluminum alloys as a function of. irradiation time.

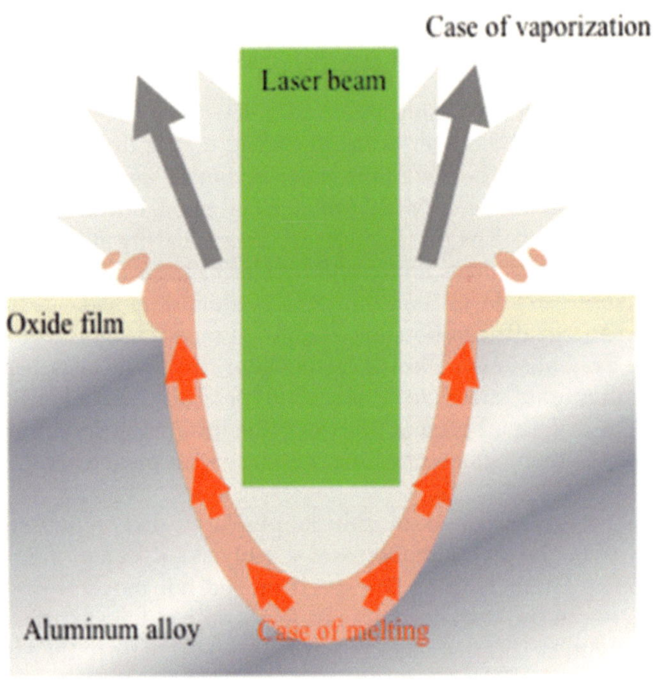

Fig. 8. Schematic representation of pit fabrication mechanism by continuous laser beam irradiation.

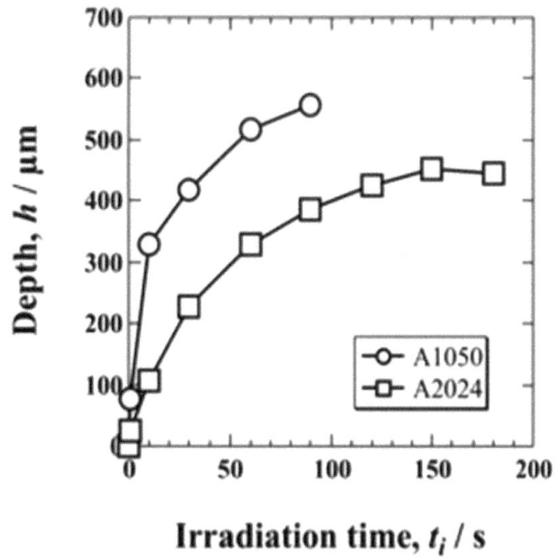

Fig. 9. Changes in the depths of fabricated artificial pits as a function of irradiation time.

These results shown here clearly substantiate that continuous focused pulse laser beam irradiation makes it possible to fabricate artificial pits on aluminum alloy in solutions, and the shape of the pits appear to be bell shaped. The detailed mechanism of pit fabrication is explained in section 2.2.2, but a possible mechanism is laser ablation.

Figure 10 shows changes in aspect ratio with t_i for both the 1050 and 2024 aluminum alloys. The aspect ratio of the formed pits on both aluminum alloys increases with t_i and the aspect ratio of both aluminum alloys at the same t_i are similar. This result shows that the proposed technique here makes it possible to fabricate artificial pits with different aspect ratios (0.13 - 2.3) on anodized aluminum in solutions.

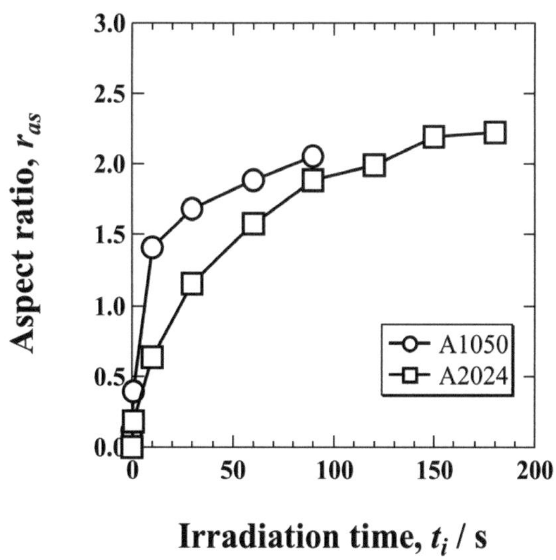

Fig. 10. Changes in the aspect ratios of fabricated artificial pits as a function of irradiation time.

2.2.2 Pit fabrication mechanism

The detailed explanation of laser ablation to remove oxide film or metals is shown in the literature (Sakairi et al., 2007).

Anodic oxide films formed on aluminum alloys are almost completely transparent at the laser frequency of 532nm used here. As continuous irradiation, oxide films are removed after several irradiation pulses by the laser beam. These situations indicate that almost all of the irradiated laser light energy reaches the metal-oxide interface or metal surface. It is not certain that the reflectivity of high energy density light is the same as low energy density beams, however, the reported reflectivity value of 0.82 at 532 nm (Waver, 1991-1992) is used to estimate the adsorbed power density here. The adsorbed power density in this experimental condition, with the wave duration 8 ns, frequency 10 s^{-1}, irradiated diameter 300 μm, and P = 3.0 mJ (30mW/10 Hz) becomes about 10^{12} W/m^2. According to the literature (Ready, 1971), the approximate expression of the minimum laser power density for ablation of aluminum (r = 2700 kg m^{-3}, L = 10778 kJ kg^{-1}, k = 1.0 x 10^{-4} m^2 s^{-1}) is about 0.7 x 10 ^{12} W / m^2. The value of the adsorbed power density in the present investigation is larger than that of ablation of aluminum. This suggests that laser ablation takes place beneath the area where the laser was irradiated. The ablation of metal produces pressure at the film/substrate or solution/substrate interfaces immediately after the irradiation. The pressure of laser ablation is simply calculated by using the laser power density for ablation, the specific thermal capacity of the aluminum, the initial, and the vaporization temperatures of the aluminum (Scruby, 1990). The estimated value is about 10^8 Pa. The deformation pressures of micro-filters made of porous type anodic oxide films with 45 μm thickness is about 2 x 10^8 Pa, and the pressures are proportional to the film thickness (Hoshino et al., 1997). The pressure estimated here is almost same as the deformation pressure of the thick porous type anodic oxide film. It may be concluded that the anodic oxide film and aluminum substrate can be destroyed and removed by the high pressure at the interface produced by the laser ablation of the aluminum substrate itself.

3. Corrosion behavior in formed artificial pits

3.1 Experimental

Borate with 0 to 0.01 kmol m^{-3} NaCl was used as electrolyte for the corrosion tests. An Ag/AgCl sat. KCl electrode was used as the reference electrode, and Pt plate (2x2 cm) was used as the counter electrode.

Polarization curves of chemically polished 2024 alloy specimens (un-anodized) were measured to determine the optimum applied potential and Cl^- concentration for investigation of the effect of the aspect ratio on the current transient in the artificial pits. In this experiment, the potential was swept at the constant rate of 0.83 mV/s from the rest potential to the anodic potential direction.

Two different types of electrochemical corrosion tests were carried out after fabrication of artificial pits with different aspect ratios on 2024 alloy, namely with the current transients at constant potential and with the rest potential changing.

Current transients: Artificial pits with two different depths formed by t_i = 1 s and 120 s were formed in Borate with 0.01 kmol m^{-3} NaCl, then a constant potential of -300 mV was applied. The current was measured to establish that no further dissolution or passivation was occurring in the pits, and after that one more pulse of laser light was applied to activate the bottom of the pits. The current transients after the activation were measured with a digital oscilloscope (Yokogawa Electric Co., DL708E).

Rest potential: The artificial pits with two different depths formed at t_i = 1 s to 120 s were formed in Borate with 0.001 to 0.01 kmol m^{-3} NaCl, then the bottom of the pits were activated by one more pulse of laser light. After the activation, the rest potential was measured with the digital oscilloscope.
One of the authors have reported on the effect of an aperture on the oxide film removed area (Sakairi et al., 1998). Using an aperture makes it possible to narrow the irradiated area at the focus point. Therefore, to activate only the bottom of the pits, an aperture placed in front of the lens was used. In the case of specimens with deep pits, t_i = 10 s to 120 s, the distance between lens and pit bottom was adjusted to focus the point of the used lens by using an XYZ stage (Fig. 1).
After the experiments, the specimen surfaces were examined by an optical microscope.

3.2 Results and discussion

3.2.1 Polarization results

The potentio-dynamic polarization curves of chemically polished 2024 aluminum alloys were measured in Borate with different concentrations of NaCl (Fig. 11) to determine the Cl^- concentration and applied potential for electrochemical corrosion tests of the artificial pits. The rest potential shifts to lower potentials with higher NaCl concentrations. At the low Cl^- concentration of 0.001 kmol m^{-3}, the relationship between current density and potential shows no changes due to the added Cl^-. No current fluctuations are observed reveals suggesting that no localized corrosion is taking place. However, at higher Cl^- concentrations (>0.002 kmol m^{-2}), the current shows sudden increases at the start of the polarization with further current fluctuations. This result at the higher Cl^- concentrations shows that when

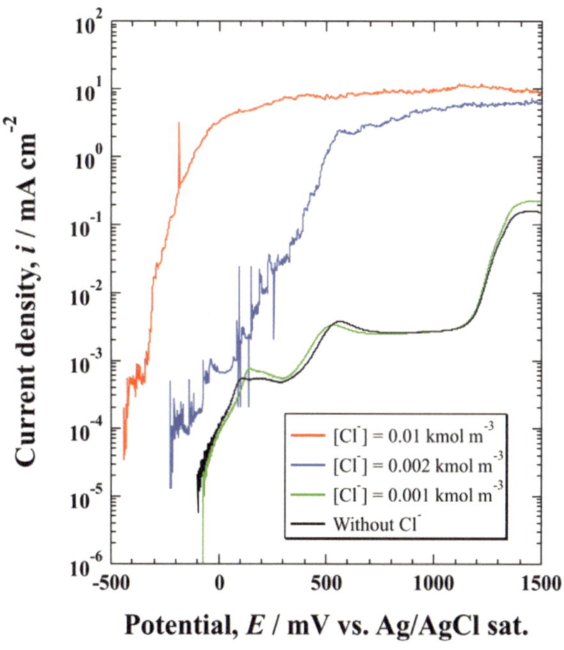

Fig. 11. Potentio-dynamic anodic polarization curves of chemically polished 2024 aluminum alloy in 0.5 kmol m^{-3} H_3BO_3 - 0.05 kmol m^{-3} $Na_2B_4O_7$ with 0 to 0.01 kmol m^{-3} NaCl.

aluminum substrate becomes exposed to the solution by laser beam irradiation, pitting corrosion tends to occur even at the open circuit condition. From these results, Borate with 0.01 kmol m^{-3} NaCl and an applied potential of -0.3 V was chosen for the electrochemical measurements of the artificial pits formed on the 2025 aluminum alloy. To investigate the effect of Cl^- concentration on pit propagation at open circuit condition (no potential applied), Borate with 0.001 and 0.01 kmol m^{-3} NaCl were chosen.

3.2.2 Current measurements

Figure 12 shows changes in the current of the pits formed on 2024 aluminum alloy with t_i = 1 s and 120 s after activation by 1 pulse of laser beam irradiation at -0.3 V in Borate with 0.01 kmol m^{-3} NaCl. After the laser beam irradiation, the current increased instantaneously through a maximum, then decreases with time in both pit conditions. After the test, white corrosion products can be seen on the specimen surfaces at the pit (Fig. 13).

Figure 14 shows changes in rest potential during and after pit formation on 2024 aluminum alloy in Borate with 0.01 kmol m^{-3} NaCl. Results with specimens without pits are also shown in the figure to evaluate the protectiveness of the anodic oxide film. The rest potential in both irradiated specimens show negative values during pit fabrication, and then there are increases after the pit fabrication. Fluctuations which relate to localized corrosion events are also observed in the rest potential changes. There are no very large potential fluctuations in the results for the anodized specimens here, indicating that anodic oxide film has good corrosion resistance for long times and that the measured potential fluctuations are related to events inside the formed pits.

Figure 15 shows changes in the rest potential during and after pit formation on 2024 aluminum alloy in Borate with 0.001 kmol m^{-3} NaCl. The changes in the rest potential at each t_i are very similar to those in Fig. 14, with no significant fluctuations observed. This means that the formed pits are repassivated after some time of pit formation, because of the low Cl^- concentration.

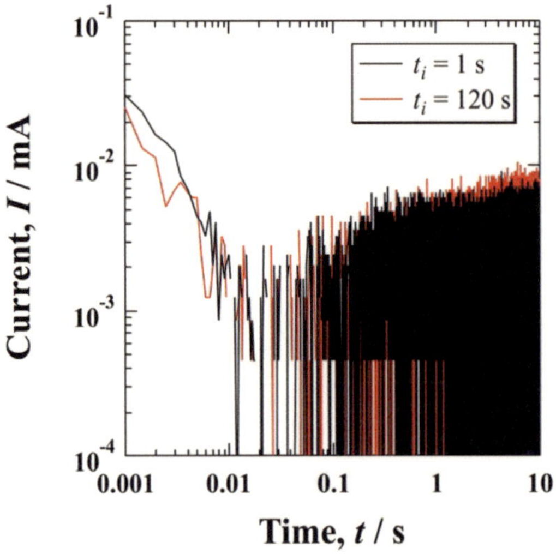

Fig. 12. Changes in the current of the pit formed on 2024 aluminum alloy after activation by one pulse of laser beam irradiation at -0.3 V in 0.5 kmol m^{-3} H_3BO_3 - 0.05 kmol m^{-3} $Na_2B_4O_7$ with 0.01 kmol m^{-3} NaCl.

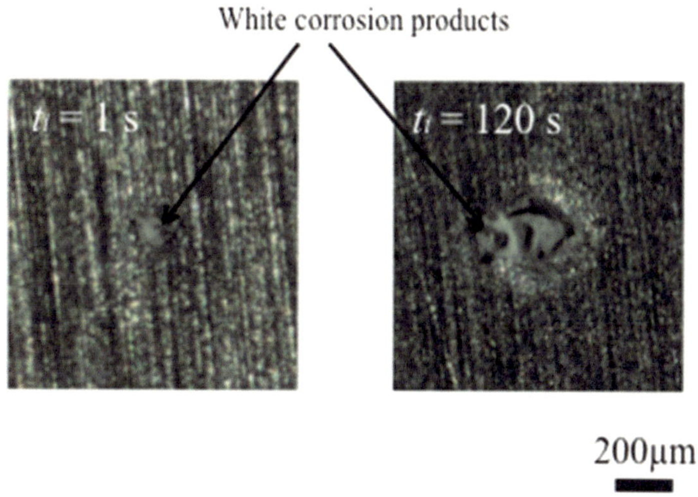

Fig. 13. Optical images of specimen surfaces after the test in Fig. 12.

Figure 16 shows rest potentials 2400 s after the finish of the pit fabrication, E_{2400}, in Fig. 15 as a function of t_i. The E_{2400} decreases with increasing t_i, suggesting pit depth or aspect ratio influence on the protective thin oxide film formation or differences in repassivation kinetics. Figure 17 shows different stages of rest potential changes by polarization curves at each step of the pit formation process. During the pit formation, the aluminum substrate is frequently activated causing increases in anodic currents and decreases in the rest potential. After the pit formation, the aluminum substrate is not further activated by the laser irradiation, and there is repassivation or further localized corrosion progresses in higher concentrations Cl^- solutions. If the surface repassivates, the anodic current decreases causing rest potential increases.

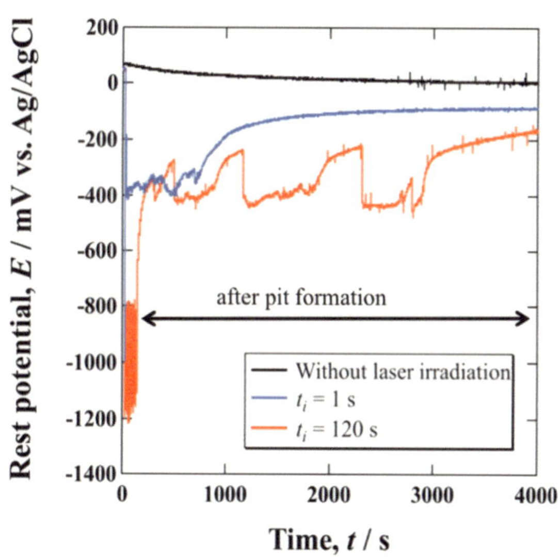

Fig. 14. Changes in rest potential during and after pit formation in 0.5 kmol m^{-3} H_3BO_3 - 0.05 kmol m^{-3} $Na_2B_4O_7$ with 0.01 kmol m^{-3} NaCl. The rest potential of specimens without pits is also shown in the figure.

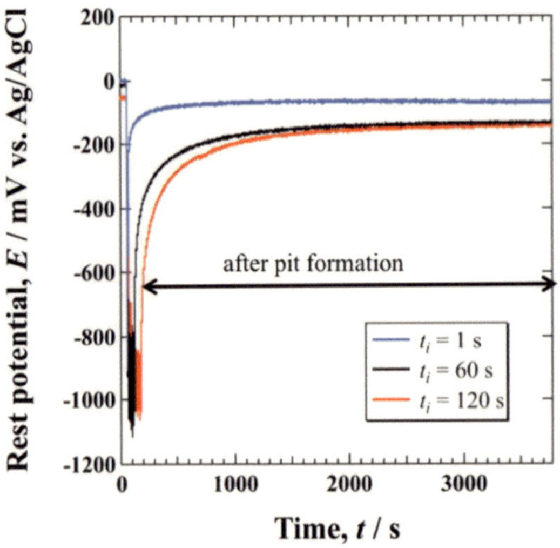

Fig. 15. Changes in rest potential during and after pit formation on 2024 aluminum alloy in 0.5 kmol m^{-3} H_3BO_3 - 0.05 kmol m^{-3} $Na_2B_4O_7$ with 0.001 kmol m^{-3} NaCl

To investigate the effect of pit aspect ratio on the repassivation kinetics, the bottom of the pit was re-activated by one pulse of laser beam irradiation. The re-activation was carried out 2400 s after completion of the pit formation. Fig. 18 shows the changes in the rest potential after this re-activation of 2024 aluminum alloy in Borate with 0.001 kmol m^{-3} NaCl. The potential changes to the negative direction in all specimens, shows a minimum value, and then shifts to the positive direction. This potential shift to the positive direction suggests repassivation of the re-activated surface at the bottom of the pit.

Figure 19 shows optical images after the re-activation tests. No corrosion products are observed in either the t_i = 1s or 120 s specimens. This result suggest that repassivation took place because of the low concentration of Cl^-, in good agreement with the potential changes in Fig. 18.

The lowest rest potential after the re-activation as a function of aspect ratio is shown in Fig. 20. Low aspect ratio samples show the lowest reached potential, while higher aspect ratio specimens show very similar values.

To clarify the effect of the aspect ratio on the repassivation kinetics, a repassivation ratio concept is introduced. The repassivation ratio, r_p, is explained as follows

$$r_P = \frac{\Delta E_p}{\Delta E_{ac}}$$
$$\Delta E_{ac} = E_{2400} - E_{ac}$$
$$\Delta E_p = E_p - E_{ac}$$

where E_{2400} is the rest potential 2400 s after pit formation, E_{ac} is the lowest rest potential after re-activation, and E_p is the average value of the rest potential around 0.01 s or 10 s after the re-activation. Therefore, a high value of r_p indicates that repassivation has progressed.

Changes in r_p at t_c = 0.01 s (Fig. 21) and 10 s (Fig. 22) with the aspect ratio of the pit in Borate with 0.001 kmol m^{-3} NaCl were established. The r_p of the 1050 aluminum alloy is also shown

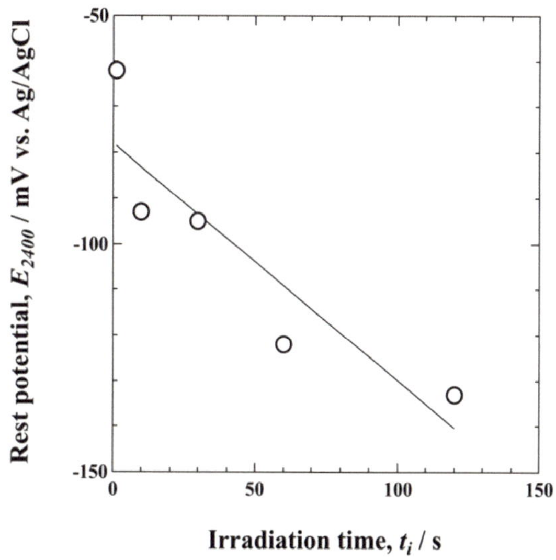

Fig. 16. Rest potential at 2400 s after the finish of the pit fabrication in Fig. 15 as a function of the aspect ratio of the formed pits.

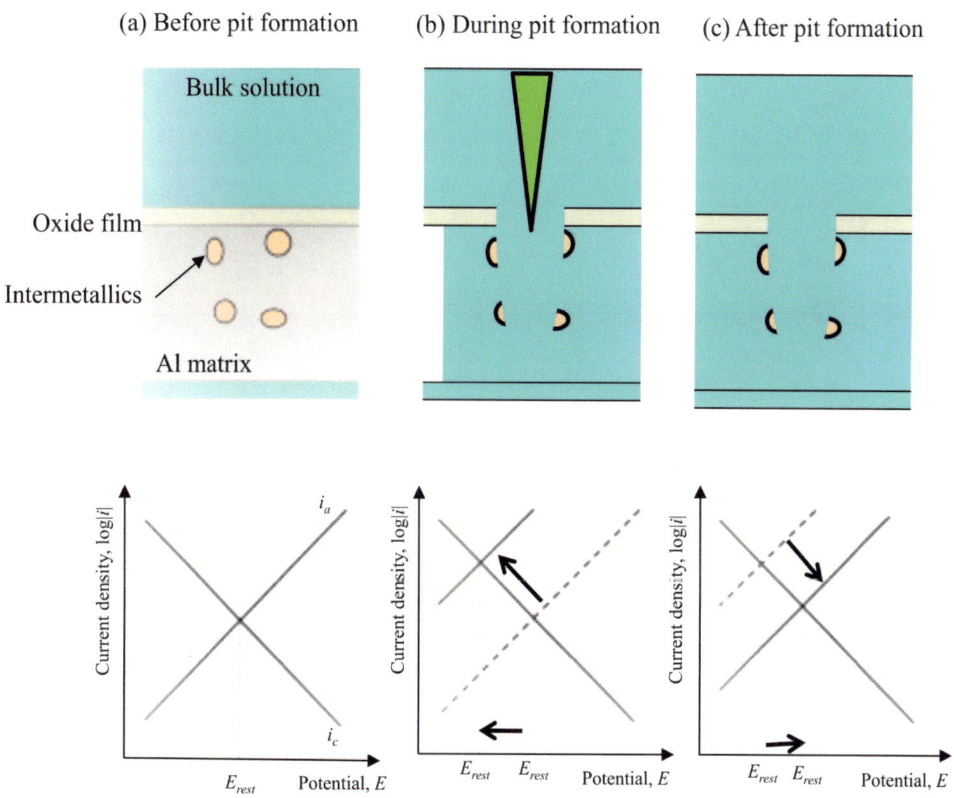

Fig. 17. Schematic representation of rest potential changes by polarization curves at each step of the pit formation process.

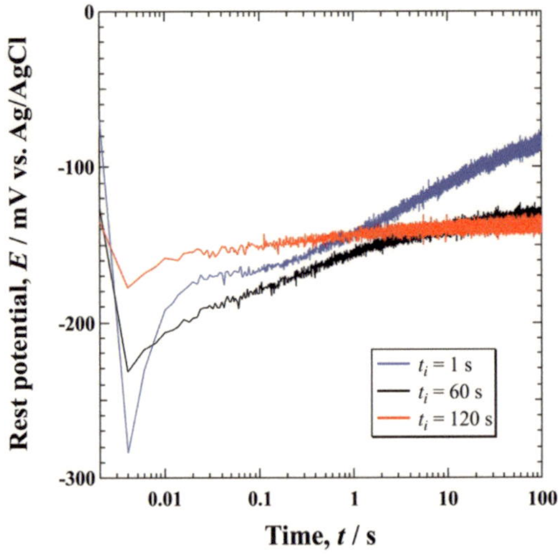

Fig. 18. Changes in rest potential after the re-activation of 2024 aluminum alloy in 0.5 kmol m^{-3} H_3BO_3 - 0.05 kmol m^{-3} $Na_2B_4O_7$ with 0.001 kmol m^{-3} NaCl. Re-activation was carried out 2400 s after the pit formation.

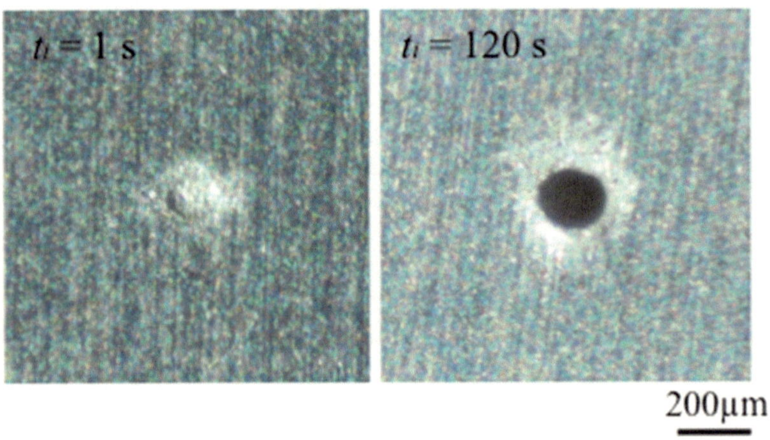

Fig. 19. Optical images after the re-activation tests in Fig. 18.

in the figures. It is clearly shown that r_p at 0.01 s decreases with the aspect ratio of the pit (Fig. 21) while at 10 s it increases with aspect ratio (Fig. 22). At both t_c, the r_p of the 2024 aluminum alloy is higher than that of the 1050 aluminum alloy. These results suggest that the repassivation rate of the 2024 aluminum alloy is faster than that of the 1050 aluminum alloy in low NaCl containing solutions.

Figure 23 shows a schematic representation of the situation after activation at the bottom of a pit, (a) transfer of oxygen and (b) selection of anodic and cathodic sites and transfer of hydrogen. With 2024 aluminum alloy, a number of copper rich intermetallics are present in the substrate. In the pit as formed by laser irradiation, these intremetallics may be exposed to the solution and act as cathodic reaction sites during immersion corrosion tests.

In the pit here, the main cathodic reaction is oxygen reduction because the solution pH is close to neutral and no nitrogen or argon gas was bubbled into the solution. The exposed area of the intermetallics in the pit wall may also increase with aspect ratio and cause the increasee in cathodic partial current, i_c, in Fig. 17. Here, as the anodic partial current, the dissolution of aluminum is increased by the re-activation, and there are no large rest potential changes. This also means an acceleration in the rate of passivation in low NaCl containing solutions. This may be concluded to be the reason why the r_p of the 2024 aluminum alloy is lager than that of the 1050 aluminum alloy in Figs 21 and 22.

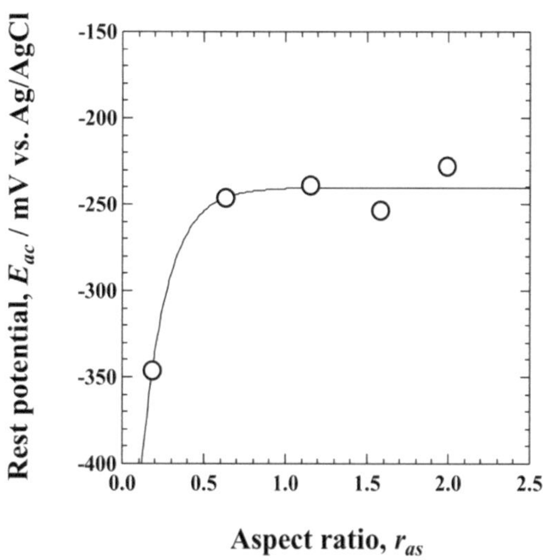

Fig. 20. Lowest rest potentials after the re-activation as a function of aspect ratio in 0.5 kmol m^{-3} H_3BO_3 - 0.05 kmol m^{-3} $Na_2B_4O_7$ with 0.001 kmol m^{-3} NaCl.

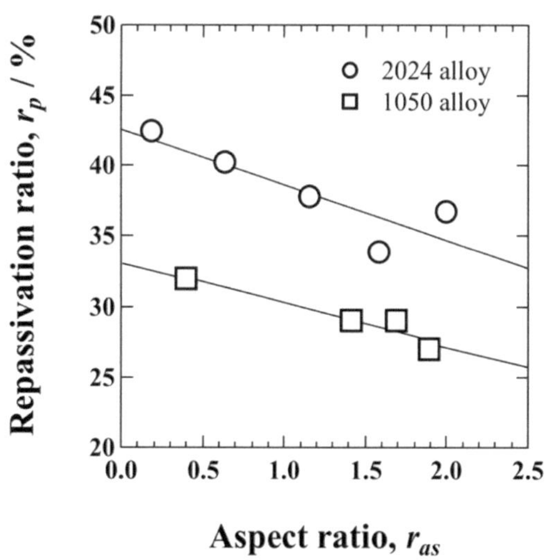

Fig. 21. Changes in repassivation ratios at 0.01 s with different aspect ratios of pits in 0.5 kmol m^{-3} H_3BO_3 - 0.05 kmol m^{-3} $Na_2B_4O_7$ with 0.001 kmol m^{-3} NaCl.

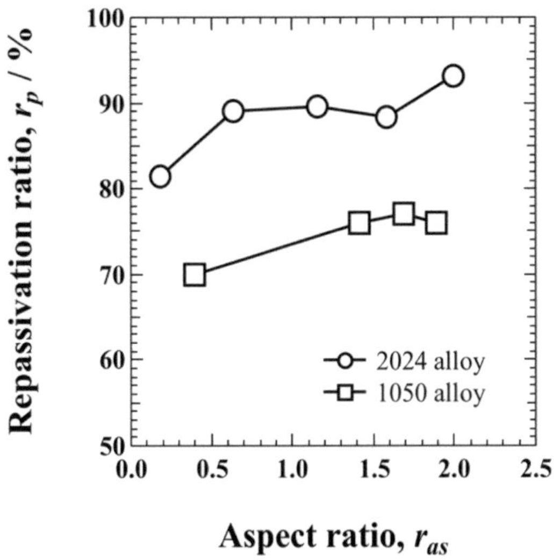

Fig. 22. Changes in repassivation ratios at 10 s with different aspect ratios of pits in 0.5 kmol m^{-3} H_3BO_3 - 0.05 kmol m^{-3} $Na_2B_4O_7$ with 0.001 kmol m^{-3} NaCl.

After some time, dissolved oxygen in the solution inside the pit may be consumed and oxygen diffuse from the bulk solution to the pit. In the high aspect ratio pit, the distribution of oxygen concentrations becomes dominant dividing the cathodic reaction (near the pit mouth) and anodic reaction areas (near the pit bottom) in the pit (Fig. 23 (b)). This means that the dissolution rate of aluminum at the bottom also increases. The dissolved Al^{3+} reacts with water to form H^+ and lowers the pH locally. The higher aspect ratio makes it difficult to dilute the H^+ ions at the pit bottom, and this is a possible reason why the r_p at 0.01 s decreases with increasing aspect ratio. At t_c= 10 s, the pH at the bottom of the pit may increase because of buffer reactions of the Borate and diffusion of H^+ ion into the bulk solution. The cathodic reaction rate of the high aspect ratio pit is still faster than that of the low aspect ratio pit. This fast cathodic reaction may make it easier to achieve repassivation at the bottom of the pit.

4. Summary

In this chapter, the application of a new in-situ artificial micro-pit formation method with an area selective electrochemical measurement technique was explained. The technique showed here uses focused pulsed Nd-YAG laser irradiation and anodizing. This technique was applied to investigate the effect of the geometry (aspect ratio) of artificially formed pits on the localized corrosion behavior of the formed artificial pits in aluminum alloys. The following conclusions may be drawn.

1. By controlling the laser irradiation time it becomes possible to form artificial micro-pits with different aspect ratios. An aspect ratio of about 2 is obtained by 120 s of laser irradiation.
2. The pit formation rate of the 2024 Al alloy is about four times slower than that of 1050 Al alloy.
3. The rest potential of the pits at 2400 s after completion of pit formation, E_{2400}, becomes lower with increasing aspect ratio.
4. The repassivation ratio at 0.01 s after activation becomes lower with increasing aspect ratio.

5. References

Ito, G. Ishida, S. Kato, M. Nakayama, T. and Mishima, R. (1968). Effect of minor impurities in water on the corrosion of aluminum, *Keikinzoku*, Vol. 18, No. 10, pp. 530-536, ISSN 0451-5994.

Horibe, K. (1969). Pit formation on aluminum immersed in artificial waters II, *Keikinzoku*, Vol. 19, No. 3, pp. 105-110, ISSN 0451-5994.

Goto, K. Ito, G. and Shimizu, Y. (1970). Effect of some oxidizing agents on pitting corrosion of aluminum in neutral water, *Keikinzoku*, Vol. 20, No. 2, pp. 88-94, ISSN 0451-5994.

Blanc, C. Lavelle, Mankowski, B. G. (1997). The role of precipitates enriched with copper on the susceptibility to pitting corrosion of the 2024 aluminium alloy, *Corrosion Science*, Vol. 39, No. 3, pp. 495-510, ISSN 0010-938X.

Kang, J. Fu, R. Luan, G. Dong, C. and He, H. (2010). In-situ investigation on the pitting corrosion behavior of friction stir welded joint of AA2024-T3 aluminium alloy, *Corrosion Science*, Vol. 52, No. 2, pp. 620-626, ISSN 0010-938X.

Ioti, Y. Take, S. and Okuyama, Y. (2003). Electrochemical noise in crevice corrosion of aluminum and possibility for its monitoring, *Ziryo-to-Kankyo*, Vol. 52, No. 9, pp. 471-476, ISSN 0917-0480.

Sakairi, M. Shimoyama, Y. and Takahashi, H. (2005). Electrochemical Noise Study on Galvanic Corrosion of Anodized Aluminum in Chloride Environments, *Proceedings of Electrochemical Society*, Volume 2004-14, pp. 265-272.

Sakairi, M. Shimoyama, Y. and Takahashi, H. (2006). Electrochemical Noise Study on Galvanic Corrosion of Aluminum Alloy in Chloride Environments- Effect of Oxide Film Structure, *ECS Transactions*, Vol. 1, No. 4, pp. 195-206, ISSN 1938-6737.

Sakairi, M. and Shimoyama, Y. (2007). Electrochemical Random Signal Analysis During Galvanic Corrosion of Anodized Aluminum Alloy, *Journal of Japan Society for Experimental Mechanics*, Special Issue, pp. 114-119 ISSN 1346-4930.

Tohma, K. and Yamada, K. (1980). Change of corrosion potentials of aluminum and aluminum alloys with pit growth, *Keikinzoku*, Vol. 30, No. 2, pp. 85-91, ISSN 0451-5994.

Sakairi, M. Kageyama, A. Kojima, Y. Oya, Y and Kikuchi, T. (2009). Effect of aspect ratio of artificial pits formed on Al by PRM on localized corrosion in chloride environments, *ECS Transactions*, Vol. 16, No. 43, pp. 19-21, ISSN 1938-6737.

Yanada, K. Sakairi, M. Kikuchi, T. Oya, Y. and Kojima, Y. (2010). Formation of artificial micro-pits on Al alloy with PRM and the localized corrosion behavior of the formed pits, *Surface and Interface Analysis*, Vol. 42, pp. 189-193, ISSN 1096-9918.

Sakairi, M. Uchida, Y. Itabashi, K and Takahashi, H. (2007). Re-passivation and initial stage of localized corrosion of metals by using photon rupture technique and electrochemistry, In: *Progress in Corrosion Research*, Emilio L. Bettini, pp. 133-157, Nova Science Publishers Inc., ISBN 1-60021-734-6, New York.

Weaver J. H. (1991-1992). Optical properties of metals, In: *CRC Handbook of Chemistry and Physics, A Ready-Reference Book of Chemical and Physical Data 72nd.*, Lide, D.R., CRC Press Inc., p. 12-101, ISBN 0-8493-0472-5, Boston.

Ready, J. F. (1971). Emission, In: *Effect of High Power Laser Radiation*, Academic Press, New York.

Scruby, C. B. Drain, L. E. (1990). Ultrasonic generation by laser, In: *Laser Ultrasonics - Techniques and Applications-*, Adam Hilger pp. 223-274, ISBN 0-7503-0050-7, New York.

Hoshino, S. Suzuki, K. and Nakane, K. (1997). Characteristics of microfilters made of anodic oxide films of aluminum, Transactions of the Institute of Metal Finishing, Vol. 75, No. 4, pp. 134 - 137, ISSN 0020-2967.

Sakairi, M. Wakabayashi, J. Takahashi, H. Abe, Y. and Katayama, N. (1998). *Journal of The Surface Finishing Society of Japan*, Vol. 49, No. 11, pp 1220-1232, ISSN 0915-1869.

Part 3

Biological Applications

10

Laser Pulse Application in IVF

Carrie Bedient, Pallavi Khanna and Nina Desai
Cleveland Clinic Foundation
U.S.A

1. Introduction

In-vitro fertilization (IVF) involves the culture and manipulation of gametes and embryos within a laboratory environment. IVF procedures are channeled towards enhancing fertilization and assisting the normal developmental physiology of the growing embryo to increase implantation potential, culminating in the birth of a healthy baby. Laser and its selective application to various steps in the IVF process is an area of growing interest.

In this chapter, we review the use of laser technology in the field of assisted reproduction as well as in stem cell research. The first step in the IVF process involves fertilization of the oocyte. For this to occur, sperm must penetrate the outer membrane known as the "zona pellucida" which surrounds the egg. This natural barrier prevents the entry of multiple sperm. Often it is necessary to assist fertilization by directly injecting a single sperm into the oocyte, a technique known as Intracytoplasmic Sperm Injection (ICSI). Laser pulse has been utilized to immobilize the human sperm tail before ICSI and in assisting the injection technique by creating a hole in the zona (laser assisted ICSI). Once successfully fertilized, the resulting embryo undergoes successive cell divisions. To implant on the uterine wall, the embryo must escape from the surrounding zona, a process known as hatching. Laser assisted hatching has been employed to create a controlled opening of the zona and facilitate embryo implantation after transfer to the patient's uterus. Zona opening through use of a laser pulse has also been used to extract a single cell from the growing embryo for preimplantation genetic diagnosis (PGD). Another application of the laser in reproductive biology has been cellular microsurgery. Embryonic stem cells can be isolated from a blastocyst stage embryo by selective ablation of trophectodermal cells, leaving behind the stem cell source material. More recently, laser has been used to induce fluid loss from the blastocyst stage embryo before cryopreservation. We discuss this novel application of laser and our own work with artificially collapsing blastocysts before freezing to reduce ice crystal damage.

This article also documents the evolution of laser pulse in IVF from the first generation of lasers with UV range wavelengths to the newer generation of lasers with emissions in the infrared range. Design characteristics for the ideal laser pulse for clinical IVF use are presented. Finally, safety considerations as regards laser usage at such early stages of development and potential risks to the newborn are discussed. The current FDA classification and approved devices are also reviewed.

Numerous engineering devices have been used in biomanipulation and a thorough understanding of both the disciplines of biology and engineering is imperative to develop an efficient system for handling biological materials. Lab procedures used during IVF involve some of the newest innnovations in medical technology, which may be attributed to the constant pressure to increase accuracy and efficiency in completing procedures. Among these innovations is laser technology. With the replacement of mechanical manipulation by laser pulse, interuser variability may be lessened and consistently high laboratory standards may be maintained.

In vitro fertilization (IVF) is one of several treatment options used in assisted reproduction. It involves an interplay of diagnostic tests, hormonal supplementation, surgery and laboratory techniques to help the subfertile couple achieve a pregnancy resulting in a healthy baby. When a couple approaches the physician with the issue of subfertility, they undergo a series of tests to determine the cause of subfertility and the optimal assisted reproductive technique for their clinical situation. Causes of infertility may include lack of eggs (oocytes), lack of sperm, inability of egg and sperm to meet due to blocked fallopian tubes, inability to grow or implant in the uterus, or an unknown etiology.

In a typical IVF procedure, oocytes are harvested from the ovary after hormonal ovarian stimulation. A sperm sample is collected from the male partner and washed from surrounding semen. Alternatively, sperm is surgically retrieved from the testis or epididymis. The oocytes are allowed to naturally fertilize in a Petri dish by co-incubation with sperm. If the sperm count or motility is compromised, the insemination step is carried out by direct injection of each oocyte with a single sperm using a glass needle. This specialized procedure is known as ICSI (Intracytoplasmic Sperm Injection). If fertilization occurs, a zygote forms. The zygote divides, undergoing cell cleavage, and forms an embryo. The cells within the embryo continue rapidly dividing over the 4-6 day culture interval, ultimately arranging in a distinct pattern to become a blastocyst. The blastocyst consists of a peripheral layer of cells called the "trophectoderm" and a discrete grouping of cells known as the inner cell mass (ICM) that will eventually form the fetus (Figure 1). The developing embryo is protected by an outer shell of protein called the "zona pellucida" until it is large enough to break free during a process known as "hatching", in preparation for implantation into the uterine wall.

Couples will have multiple embryos developing simultaneously in culture. Each embryo is evaluated throughout its growth process. On the day of transfer 1-3 embryos are selected from the laboratory dish and transferred to the patient's uterus. This transfer may occur on day 3 or day 5 after fertilization. Any additional embryos that are appropriately developed are frozen for possible later transfer. Selection of embryos most likely to implant and lead to a viable pregnancy is generally based on embryo morphology.

While some applications of lasers in IVF remain research topics, others have been successfully employed in clinical practice. Laser assisted ICSI is used to aid fertilization. Laser assisted hatching has been employed to create a controlled opening of the zona and facilitate embryo implantation after transfer to the patient's uterus. Zona opening through use of a laser pulse has also been used to extract a single cell from the growing embryo for preimplantation genetic diagnosis (PGD) and screen for genetic disorders prior to transfer. Another application of the laser in reproductive biology has been cellular microsurgery. Embryonic stem cells can be isolated from a blastocyst stage embryo by selective ablation of trophectodermal cells, leaving behind the stem cell source material.

When first approaching the application of lasers to reproductive medicine, concerns were raised as regards the safety profile and class of lasers to be used. Given the delicate stage of human development at the time of fertilization, the major concerns regarding the use of laser at earlier stages have been DNA damage, failed embryo development and possible congenital disorders. These concerns primarily centered on laser wavelength, heat generation and the amount of manipulation required of the fragile embryos. The primary aim of this review is to assimilate the significance and limitations of laser technology in the fast growing field of IVF and to outline the technical details to be considered when dealing with laser pulses in reproductive technology.

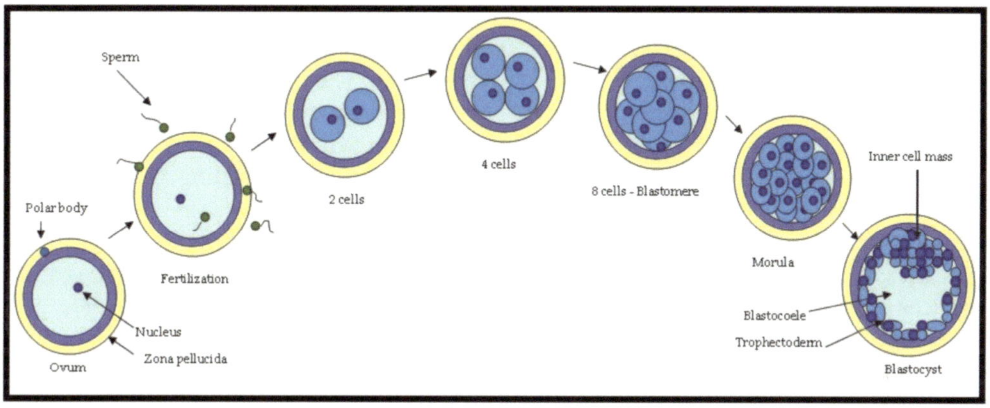

Fig. 1. Egg fertilization and development

2. History of lasers in IVF

Laser technology has been used in Assisted Reproductive Technology since the 1980s (Ebner *et al.,* 2005). Laser pulse has found wide application in IVF technology, particularly when efficient and precise manipulation is of paramount importance (Taylor *et al.,* 2010).

Two general types of laser systems exist: contact and noncontact. Noncontact lasers do not require additional physical manipulation of the embryo. Laser beams travel through the objective lenses and only microscope stage movement is required to adjust embryo position (Tadir *et al.,* 1989, 1990, 1991). In contrast, contact laser systems require direct contact between the laser and embryo, usually with either glass or an optical fiber (Neev *et al.,* 1992). This increases the likelihood of trauma to the embryo. Distance also affects damage - a greater distance from the embryo to the laser will result in a larger hole in the embryo, even if the difference in distance is only between the top and bottom of culture dish (Taylor *et al.,* 2010). Contact lasers also require use of a medium different than routine culture media in order to affect the most efficient energy transfer.

The first generation of lasers to be used in IVF included argon fluoride (ArF), Xenon chloride (XeCl), krypton fluoride (KrF), nitrogen and Nd:YAG lasers. The Nd:YAG laser (1064 nm) was the first non-contact laser used in reproductive technologies. Initial use was primarily for spermatozoa manipulation via optical trapping. Applications were then expanded to add a potassium-titanyl-phosphate crystal in order to create a hole in the zona pellucida to assist hatching (Tadir *et al.,*1989, 1990, 1991). Excimer lasers under development around the same time period function by temporarily exciting rare earth gasses. After comparing Nd:YAG lasers with the ArF (193 nm) excimer laser, the 193 nm was found to produce a more uniform, smooth tunnel in the zona pellucida (Palanker *et al.* 1991). Similar findings were noted with the XeCl (308 nm) excimer laser (Neev *et al.,* 1992). Many excimer

lasers, including KrF (248 nm), and nitrogen lasers (337 nm) function at a wavelength in the UV spectrum. Ultraviolet wavelengths are close to the absorption wavelength of DNA (260 nm). As a result, these lasers are minimally used in reproductive technologies due to concern for mutagenic effects (Green *et al.*, 1987; Hammadeh *et al.*, 2011; Kochevar *et al.*, 1989).
The next generation of lasers were designed to circumvent dangers of UV wavelength and cytotoxicity by emitting wavelengths in the infrared region (>800 nm) (Ebner *et al.*, 2005). The first of the newer generation of lasers to be used in IVF was the 2.9 um pulsed erbium:yttrium-aluminum-garnet laser (Er:YAG) (Feichtinger *et al.*, 1992). This device's use is limited by the need for constant contact with the embryo, as well as limitations due to interactions with the liquid media (Rink *et al.*, 1996). The next development was the holmium:yttrium-scandium-gallium-garnet laser (Ho:YSGG) with 2.1 um emission. In order to retain the beneficial effect of the infrared emission wavelength with this laser, the embryos require additional manipulation on a quartz slide, offsetting the advantages obtained by a safer wavelength (Schiewe *et al.*, 1995).
Currently, the 1.48 um diode wavelength indium-gallium-arsenic-phosphorus (InGaAsP) semiconductor laser is used in IVF. It is a non contact laser, has a safer wavelength and produces consistent results in the form of uniform, smooth edged tunnels (Rink *et al.*, 1996). This diode laser is delivered through a complex arrangement, requiring 3 mirrors and 3 lenses. A continuous laser beam is emitted and collimated by a microscope objective, and then paired with a visible beam. These pass through a mirror which reflects the invisible beam and is partially transparent to the 670 nm wavelength. Both beams are then directed through the primary microscope objective lens and to the desired object. The variability is less than 1 um, showing excellent reproducibility. Use of this laser does not require additional manipulation of the embryo or pose threat to DNA integrity by damaging radiation (Rink *et al.*, 1996).

3. Laser characteristics for IVF

Lasers in IVF have a wide variety of applications, however, the desirable characteristics of the laser used are similar across those applications. During laser targeting, the embryo's unique culture environment must remain consistent at all times to optimize the potential for a viable pregnancy. To that end, any laser used in the IVF laboratory must be very precise, extremely consistent with reproducible results and integrate well into the equipment required for routine IVF. In addition, it must not pose any additional threat to the integrity of the embryo. This includes an infrared wavelength to avoid direct chromosomal damage. It also helps when a non-contact mode is employed to avoid any unnecessary manipulation of the fragile embryo. Contact mode lasers requiring glass pipettes (UV wavelength) or quartz fibers (infrared wavelengths) add a layer of complexity with respect to additional manipulation of the embryo (Hammadeh *et al*, 2011). Similarly, no additional changes or alternations of media should be made to avoid undue stress on the embryo's environment, which should be kept at a physiologic pH of 7.2 and at 37 degrees Celsius at all times to optimize growth (Douglas-Hamilton & Conia, 2001 as cited in Al-Katanani *et al.*, 2002). This limits use to lasers which will not produce a thermal effect on the media containing the embryo, which is impacted by the laser's power, number of shots required, pulse length and irradiation time. Ease of use and speed of a technique also contribute to maintaining an appropriate environment for the embryo in that a faster procedure exposes the embryo to a hostile environment for a much shorter period of time.

Lasers have three characteristics directly impacting embryos: wavelength, power and pulse length. Wavelengths used in IVF tend to remain above 750 nm, in the infrared region, to avoid mutagenic effects on DNA (Kochevar *et al.,* 1989; Taylor *et al.,* 2010). The amount of power in a single laser remains constant but impacts the diameter of the hole created as well as the amount of heat emitted in the process, with higher power translating to larger diameter and increased heat (Taylor *et al.,* 2010). Different lasers may each have a different power. A similar scenario exists with pulse length, which can vary from 20 ms to >1,000 us. A longer pulse length also correlates with a larger hole (Rink *et al.,* 1996). Focusing the beam waist on a target provides a larger diameter of tunnel as well (Neev *et al.,*1992).
Beyond the physical characteristics of the laser itself are secondary characteristics and limitations impacting embryo use. For example, the mineral oil overlay may adhere to optical fibers in a contact mode laser, absorb additional heat and thus expand, moving the embryo and disrupting the path of the laser beam (Neev *et al.* 1992). The optical fibers used must be sterilized, as well as the micropipette tips, expensive disposable equipment leading to increased costs. Additional instruments used for manipulation introduce increased cost and possible damage to the embryo in the form of contamination and constant physical contact.

4. Applications of laser in IVF

Since the discovery of laser in 1960s, it has found application in many fields. The accuracy, versatility and spatial focusing potential have helped it to find a wide application in the

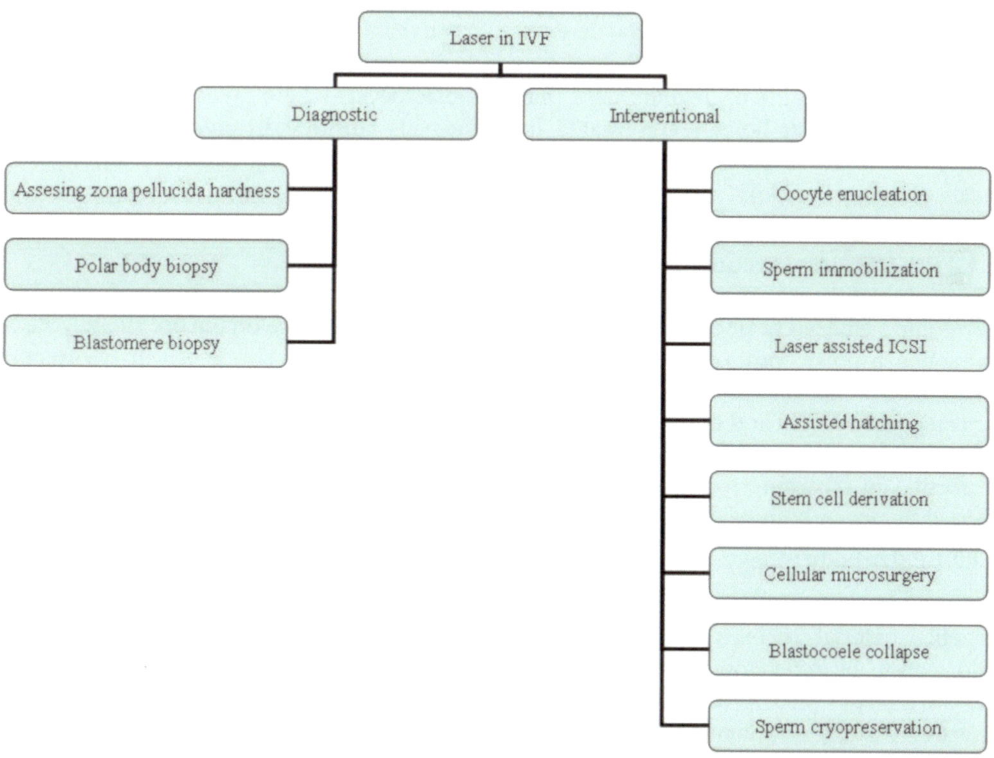

Fig. 2. Applications of lasers in IVF

medical arena. The applications of laser in IVF may be classified into diagnostic and interventional use for the ease of discussion (Figure 2). Diagnostic techniques include assessing the strength of the zona pellucida and pre-implantation genetic diagnosis. Interventional or therapeutic techniques involve manipulating individual gametes with oocyte enucleation and sperm immobilization, aiding fertilization and development with laser assisted ICSI and assisted hatching. Additional material may be obtained with stem cell derivation and cellular microsurgery. Embryos are optimized for freezing with blastocoele collapse. Regardless of the specific procedure, lasers provide an excellent method for precise intracellular surgery (Raabe *et al.*, 2009).

4.1 Diagnostic techniques

4.1.1 Assessing the zona pellucida

The zona pellucida is the hard protein coat surrounding and protecting the genetic material carried within the egg. This layer is approximately 15-20 um thick and must be breached in order for the sperm to make contact with the egg. In vivo, entry of the sperm initiates a reaction to ensure no other sperm obtains access to the egg and further hardens the protein layer to protect the zygote as it travels to the uterus. The proteinaceous coating must ultimately thin to allow the embryo to break out of the shell and implant in the uterine lining, or endometrium. Studies using laser pulses have determined the extent to which the zona hardens during the period from oocyte to blastocyst (Montag *et al.*, 2000b) and further identify which embryos may need assistance with sperm entry or hatching. Zona hardness is greater during in vitro culture as compared with in vivo growth. Montag et al. (2000b) and Inoue & Wolf (1975a) have shown that identical laser pulses create larger holes ranging from 13-17 um in the zona at earlier stages (oocyte, zygote) as compared to more advanced stages of development (morula, blastocyst) where holes are smaller at 10-13 um. Also, larger holes were created in blastocysts cultured in vivo when compared with in vitro grown blastocysts, suggesting zona hardening during culture (Montag *et al., 2000b*; Rink *et al.*, 1996).

4.1.2 Pre-implantation genetic diagnosis

Pre-implantation genetic diagnosis (PGD) is the analysis of genetic material from the developing embryo prior to transfer to the uterus. This can be done on the oocyte/zygote by extracting a polar body or on the 8-cell embryo by extracting a single cell or blastomere. Once genetic material has been obtained it may be analyzed for genetic abnormalities. Screening of oocytes and embryos for common chromosome abnormalities, such as trisomy 21, can improve pregnancy rates and reduce miscarriage rates. Some couples may be interested in screening for specific genetic problems typically severe or lethal conditions, carried by one or both partners, in order to avoid having an affected child.

4.1.2.1 Polar body biopsy

During oocyte maturation to the metaphase II stage and also after fertilization, duplicated genetic material is extruded as polar bodies. The polar body can provide helpful information by reflecting the maternal genetic material contained in that egg. (Clement-Sengewald *et al.*, 2002; Verlinsky *et al.*, 1990). Abnormal oocytes with genetic defects can be selectively excluded (Clement-Sengewald *et al.*, 2002). Genetic assessment of the unfertilized egg permits women who would not consider discarding an affected embryo due to personal beliefs to be screened for age related aneuploidy or hereditary chromosomal defects. It may

also be performed in countries where it is illegal to perform blastomere biopsy to genetically screen embryos (Dawson *et al.*, 2005; Clement-Sengewald *et al.*, 2002; Montag *et al.*, 2004).
The polar body is located in the perivitelline space directly under the zona pellucida and outside of the oocyte. It can be extracted by traversing the zona. Prior to the introduction of lasers, biopsy was typically done by degradation of the zona pellucida with Tyrode's acid, after which a capillary tube would be used to aspirate the polar body. This technique was highly variable, led to inconsistent opening size and could easily lead to further damage or loss of cells. It also requires changing culture media and increasing the risk of contamination. Alternatively to acid, mechanical biopsy could be performed with sharp glass instruments, again introducing possibility for structural damage or alteration during the manipulations (Clement-Sengewald *et al.*, 2002; Dawson *et al.*, 2005; Ebner *et al.*, 2005). Regardless of the method used, the oocyte must remain intact to continue development and the polar body must allow adequate, undamaged material for genetic analysis.
When polar body biopsy is performed using lasers, a pulse is directed at the region of zona pellucida nearest the polar body. In a description by Montag *et al.* (1998) two pulses of 14 ms are given by a 1.48 um non contact laser, creating an opening of approximately 14-20 um. The material is then extracted with a blunt capillary, avoiding potential damage to the oocyte with a sharp instrument, and the entire procedure is completed in just a few minutes (Montag *et al.*, 1998). A similar procedure has been described by Clement-Sengewald *et al.* using a nitrogen 337 nm laser and a Nd:YAG laser (Clement-Sengewald *et al.*, 2002). That same group described extraction of the polar body using optical tweezers (Nd:YAG, 1064 nm) and laser (nitrogen, 337 nm) pressure catapulting to collect the polar body, further eliminating a source of contamination by introduction of another pipette. To catapult the polar body, it was mounted to a membrane on a slide with the inner cap of a microfuge tube placed next to it. One pulse of the laser was aimed at the membrane, freeing it to catapult onto the nearby tube cap (Clement-Sengewald *et al.*, 2002; Schutze & Lahr, 1998). Oocyte recovery rates were only 67% in humans following this complete laser extraction method. An improved blastocyst survival rate was noted when access was obtained via laser as compared with acid solution, further strengthening the argument for laser use (Dawson *et al.*, 2005).

4.1.2.2 Blastomere biopsy

Blastomere biopsy is similar to polar body biopsy in that both techniques require careful extraction of genetic material from a very delicate structure followed by genetic screening. This procedure is also performed to facilitate selection of the embryo most likely to establish a viable pregnancy with healthy offspring. Blastomere biopsy becomes relevant at a later stage in development, after fertilization. Couples opt for this technique typically when one or both parents carry a hereditary genetic defect they want to avoid passing to children (Vela *et al.*, 2009) or in cases of advanced maternal age to screen against aneuploid embryos.
Until the introduction of laser assisted opening of the zona, blastomere biopsy was performed by zona drilling with an acid tyrodes solution (Talansky & Gordon, 1986, as cited in Malter & Cohen, 1989). The embryo is immobilized and held in place while acid in a microcapillary tube is gently blown against the zona until it starts to dissolve. The acid is then aspirated and the embryo is quickly rinsed to remove traces of acid. The technique requires speed and expertise so as not to injure the embryo. The hole size can often be variable.

The procedure for a blastomere biopsy using laser is similar to PGD with a polar body. Laser pulse(s) are utilized to create a hole in the zona pellucida, through which a blastomere is removed (Taylor *et al.,* 2010). Analysis of laser pulse length in generating a hole for blastomere extraction showed longer pulse duration (0.604 ms vs. 1.010 ms) produced larger hole sizes (10.5 nm vs. 16.5 nm, respectively) (Taylor *et al.,* 2010). However, Taylor *et al.* found no difference in number of blastomeres lysed for a given pulse duration. They did find a difference in number of blastomeres required to be obtained in each group. The longer pulse duration group was noted to require additional blastomere biopsy. These results were impacted by half of the affected embryos originating from the same patient with poor quality embryos and cannot clearly be attributed to laser use.

Studies comparing embryos after laser assisted biopsy to untreated embryos showed no adverse effects of treatment and similar hatching and development rates (Joris *et al.,* 2003). When performed with human embryos, pregnancy rates after laser blastomere biopsy are comparable to mechanical blastomere biopsy (Schopper *et al.,* 1999). Comparison of blastomeres obtained during acid and laser mediated biopsies showed laser biopsy generated more intact blastomeres (Joris *et al.,* 2003).

4.2 Interventional techniques

4.2.1 Laser assisted ICSI

With male factor infertility, it is often necessary to assist fertilization by directly injecting a single sperm in to the oocyte, a technique known as Intracytoplasmic Sperm Injection (ICSI). The limited number of viable or motile sperm decreases chances of fertilization and a successful pregnancy using the conventional oocyte insemination technique. ICSI is performed by aspirating a sperm into a sharp glass needle (5 um in diameter), perforating the oocyte's zona and depositing the sperm into the ooplasm (Palermo *et al.,* 1992). Deformation of the oocyte during the injection process can trigger oocyte degeneration either as a result of egg fragility or due to force required to traverse the membrane (Rienzi *et al.,* 2001, 2004; Abdelmassih *et al.,* 2002; Palermo *et al.,* 1996). Damage to the oocyte also occurs by disturbing the spindle apparatus, damaging the oocyte cytoskeleton, introducing harmful materials or by removal of cytoplasm during the injection procedure (Moser *et al.,* 2004; Hardarson *et al.,* 2000; Tsai *et al,* 2000; Dumoulin *et al.,* 2001).

Laser assisted zona drilling prior to ICSI can be used to increase the likelihood of successful fertilization (Palanker *et al.,* 1991). This may be done with a 193 nm ArF laser, which was shown to drill very precise holes without undesired damage to the zona pellucida (Palanker *et al.,* 1991). A 1.48 um diode laser can also be used to assist with ICSI (Rienzi *et al.,* 2001, 2004). A small channel of 5-10 um in diameter is drilled using low energy pulses of less than 2 milliseconds duration, taking care to leave the innermost layer of zona intact. The ICSI injection pipette is introduced through this channel to deliver the previously immobilized sperm (Rienzi *et al.,* 2001, 2004; Abdelmassih *et al.,* 2002). Prior to laser assistance, this technique was limited by operator skill and a non standardized tunnel size, potentially leading to polyspermy or loss of genetic material (Rink *et al,* 1996). Laser assisted ICSI provides a less traumatic method to create an opening in the zona pellucida for the purpose of sperm microinjection, leading to decreased breakdown of oocyte membrane (5% vs. 37%, Abdelmassih *et al.,* 2002) and increased oocyte preservation, 97% vs. 85%, after ICSI (Rienzi *et al.,* 2004). The type of laser used is in infrared range and is not absorbed by nucleotides and is considered safer than its counterparts (Ebner *et al.,* 2005; Kochevar *et al.,* 1989). The decreased force necessary in penetrating the egg with the ICSI needle in entry may also

preserve embryo quality (Rienzi *et al.*, 2001; Nagy *et al.*, 2001) and has been shown to improve embryo quality and survival, even when using poor quality oocytes (Abdelmassih *et al.*, 2002). To ensure even less traumatic manipulation, sperm may be injected into the oocyte through a laser drilled hole using optical tweezers to achieve fertilization (Clement-Sengewald *et al.*, 1996, 2002).
Ultimately, to establish a pregnancy the embryo must "hatch" out of zona and implant on the uterine wall. A potential drawback to laser assisted ICSI is that the thinning of the zona may result in duplicate hatching sites. This allows the embryo to escape via two openings, resulting in either degeneration or twinning. The theoretical concern is that the embryo would hatch through the site created during assisted hatching but also through the ICSI site as well (Abdelmassih *et al.*, 2002). Moser *et al.* (Moser *et al.*, 2004) discovered thinning the zona pellucida instead of completely opening it eliminated the concern for a second opening and incidentally improved blastulation rates through that site as well.

4.2.2 Sperm immobilization & selection

Sperm immobilization is critical when performing ICSI. The beating of the sperm tail in the oocyte after injection can cause damage. Typically during ICSI, the sperm tail is positioned under the glass microcapillary injection needle. The needle is brought down and across the tail causing it to break and immobilizing the sperm (Palermo *et al*, 1992; Nijs *et al*, 1996; Vanderzwalmen *et al.*, 1996; Yanagida *et al.*, *2001*). Fertilization rates are also closely linked to sperm immobilization, increasing from 54% to 68% (Vanderzwalmen *et al.*, 1996). Disruption of the sperm membrane aids the release of sperm factors important in oocyte activation (Dozortsev *et al.*, *1997).* Low level laser pulse can also be used to immobilize sperm, without affecting viability (Montag *et al.*, 1998, 2000d, 2009; Rienzi *et al*, 2004; Tadir *et al.*, 1990).
A rather unique application of laser is to identify and select viable sperm for ICSI. Usually motility is used as an indicator of living sperm. However in severe male factor cases such as asthenozoospermia, no motile sperm may be evident. This makes it very difficult to identify and select viable sperm for ICSI. A single laser pulse applied at the tip of a sperm's tail can aid in distinguishing living non-motile sperm from dead sperm. The tail of a viable sperm will curl, whereas the nonviable sperm will not respond to the laser pulse. Fertilization rates would be expected to be correspondingly higher if better sperm are selected for the injection (Montag *et al.*, 2000d, 2009). An alternative method for manipulating sperm includes optical trapping. Optical trapping uses a single beam non contact laser to move sperm during after immobilization or during ICSI (Clement-Sengewald *et al.*, 1996, 2002; Tadir *et al.*, 1991). The optical tweezers can hold actively moving sperm and determine their velocity (Clement-Sengewald *et al.*, 2002; Tadir *et al.*, 1991). Lasers used in optical trapping may be either infrared or ultraviolet (Clement-Sengewald *et al.*, 2002; Tadir 1989. Advantages of this technique include ease, no requirement for sophisticated micromanipulation skills or additional expensive disposable equipment. The capacity for the optical tweezers to determine velocity permits studies of medications on motility (Tadir *et al.*, 1989). It may also be used for polar body extraction or chromosomal manipulation (Tadir *et al.*, *1991).* Disadvantages include increased exposure time of the embryo to lasers, possible ultraviolet exposure depending on wavelength utilized and a potential adverse effect on the sperm (Tadir *et al.*, 1989).

4.2.3 Assisted hatching

To establish a successful pregnancy, the developing embryo must break out of its shell (zona pellucida) on day 5 or 6 by a process known as hatching. Once the embryo is hatched, it may implant on the endometrium and begin to grow but if it is unable to hatch, the pregnancy will not continue. Various factors contribute to failed hatching and implantation - increased maternal age, decreased egg quality, poor embryo and zona morphology to name a few, and the exact cause of failed hatching is unknown (Balaban *et al.*, 2002). An increase in zona hardness has also been implicated during in vitro fertilization (Inoue & Wolf, 1975; Montag *et al., 2000;* Balaban *et al.*, 2002). The physiologic mechanism leading to hatching is likely different in vivo than in vitro, with in vitro embryos hatching when a critical cell number has been reached. This is compared with hatching independently of cell mass in vivo, likely related to lytic enzymes found in vivo (Montag *et al.*, 2000a). It has become relatively common practice to facilitate the hatching of blastocysts by creating an artificial opening in the zona pellucida either by mechanical, chemical or optical methods, although the exact population benefiting most from this procedure is yet to be determined (Hammadeh *et al.*, 2011). Assisted hatching has been proposed to be potentially more beneficial in patients over 40, with thicker zonae or poor prognosis patients (Balaban *et al.*, 2002; De Vos & Van Steirteghem, 2000; Hammadeh *et al.*, 2011; Sagoskin *et al.*, 2007; Lanzendorf *et al.*, 1998).

In the late 1980s, Cohen *et al.* mechanically opened the zona pellucida, achieving higher implantation rates. Since that time, multiple methods have been proposed to facilitate hatching (De Vos & Van Steirteghem, 2000; Cohen *et al.*, 1990). Zona drilling uses Tyrode's acid solutions to create a defect in the zona (Malter & Cohen 1989; Ebner *et al.*, 2005; Neev *et al.*, 1992; Balaban *et al.*, 2002; De Vos & Van Steirteghem, 2000), whereas mechanical hatching utilizes a microneedle to slice off a thin piece of the zona (Malter & Cohen 1989; Ebner *et al.*, 2005; Balaban *et al.*, 2002; De Vos & Van Steirteghem, 2000). Enzymatic hatching using pronase to generally thin the zona pellucida is also an accepted method of assisted hatching (Balaban *et al.*, 2002; Fong *et al.*, 1998). Direct comparison of hatching methods is challenging due to inter-operator variability, differing depths of zona penetration and heterogeneous patient populations.

Laser provides an alternate means to facilitate hatching, and is faster and easier than other methods (Balaban *et al.*, 2002). The 2.94 um Er:YAG laser has been used for assisted hatching with a significant increase in pregnancy rates (Antinori *et al.*, 1996). The laser was deemed safe for clinical use after trials in animal models (Obruca *et al.*, 1994, as cited in Obruca *et al.*, 1997). The 1.48 micron infrared diode laser beam has been more widely used in clinical IVF labs as an efficient and simple method for embryo hatching. Multiple studies have demonstrated its safety (Sagoskin *et al.*, 2007; Lanzendorf *et al.*, 2007; Wong *et al.*, 2003) as well as efficacy when compared to acid hatching (Lanzendorf *et al.*, 2007; Balaban *et al.*, 2002; Jones *et al.*, 2006).

The optimal technique for laser assisted hatching is still being debated. The laser can be used to thin a large area of the zona, partially hatch by creating an incomplete hole or completely hatch by drilling completely through the zona (Figure 3). The number of shots and duration of pulse exposure is also subject to discussion with investigators varying parameters to achieve an appropriate tunnel size. Optimal hole size is as yet unclear, although >10 um leads to improved results (Ebner *et al.*, 2005). A study by Montag *et al.*, found no evidence of impaired growth or adverse effects as a result of laser hatching (Montag *et al.*, 2000a). Advocates of partial hatching argue increased safety using this method because the laser does not come in to direct contact with the embryo. Finally, proponents of the zona thinning technique contend that overall thinning will avoid inadequate hatching and be more likely to correspond with the natural hatch site due to a larger area being ablated (Moser *et al.,* 2004). Studies comparing multiple methods of hatching yield inconclusive results and no definitive recommendations can be made. A study comparing pulse intensity and number of pulses determined 50% intensity with 2 pulses was the optimal setting to increase blastocyst formation (Tinney *et al.,* 2005) by creating a complete hole rather than the less effective zona thinning. Specific settings to achieve those results would be expected to vary based on the power of different lasers. Mantoudis *et al.,* 2001, compared the three methods of laser hatching and determined partial hatching or thinning the zona is more effective. Implantation rates were 2.8%, 9.1% and 8.1% in the complete hatching, partial hatching and zona thinning groups. Clinical pregnancy rates were also significantly improved with 5.2%, 18.3% and 22.1%, respectively. Thinning in this study ablated the zona around 25% of the embryo, leaving only the inner membrane of the zona pellucida intact in that section. It is unclear what the diameter of the complete hatch site was in this study. Another concerning trend in this study was 22% of pregnancies were multiple pregnancies, more than typically seen (Mantoudis *et al.,* 2001), which is not unique to this trial (Hammadeh *et al.,* 2011). In contrast to the findings of Mantoudis *et al.,* Wong *et al.* found improved hatching rates with complete hatching compared to partial hatching, 38% vs. 25%, respectively (Wong *et al.,* 2003). Laser-assisted zona pellucida thinning prior to ICSI resulted in decreased oocyte degeneration rates, better blastocyst hatching rates and improved pregnancy rates after day 3 embryo transfer (Moser *et al.,* 2004). In this study embryos had their zona pellucida thinned by 50% via 5-6 laser pulses, covering at most 70 um of zona. A trial by Balaban *et al.* compared assisted hatching by laser, acid Tyrodes, pronase treatment and mechanical technique. These investigators concluded that all methods were comparable based on the outcome parameters studied, including implantation and pregnancy rates, multiple pregnancy rates and abortion rates (Balaban *et al.,* 2002). Additional studies comparing laser assisted hatching with acid drilling showed no significant differences with respect to pregnancy rates (Lanzendorf *et al.,* 2007; 1999; Jones *et al.,* 2006).

Laser assisted hatching is generally well-accepted in IVF labs, allowing improved standardization between operators (Lanzendorf *et al.,* 2007; Jones *et al.,* 2006). Children followed to one year of age after an assisted pregnancy using laser assisted hatching were

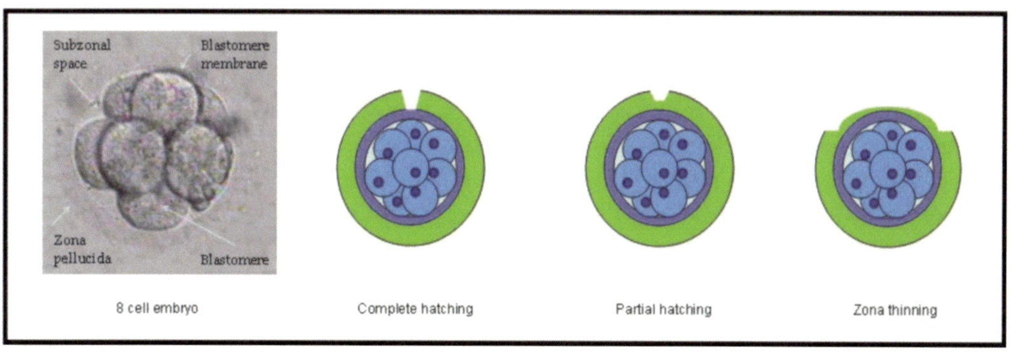

Fig. 3. Assisted hatching

found to have no increase in congenital malformations (Kanyo & Konc, 2003). Other pregnancies have also yielded healthy babies following laser assisted hatching (Lanzendorf *et al.*, 1998). The first 1.48 um laser to receive US FDA approval for clinical use in assisted hatching was the ZILOS-tk in 2004. This was followed by the Octax laser in 2006 and the Saturn Active Laser System in 2008.

4.2.4 Laser pulse blastocyst collapse

As the efficiency of embryo culture increases, supernumerary embryos are produced and cryopreserved for transfer in a future cycle (Iwayama *et al.*, 2010; Gardner *et al.*, 1998). One method of cryopreservation known as "vitrification" involves high molar concentrations of cryoprotectants and rapid cooling of the embryo at rates of -20,000 C°/min (Desai *et al.*, 2011). This cooling technique is extremely effective for embryos at all stages. The high cooling rate prevents ice crystal formation in cellular cytoplasm. Post-warming survival rates have been high with this technique. Yet it was observed that well-developed and expanded blastocysts had lower survival rates than the less mature blastocyst or the morula stage embryo (Vanderzwalmen *et al.*, 2002). The primary structural difference between the early stage blastocyst or morula and the later stage blastocyst is the presence of a fluid filled cavity in the expanded blastocyst, called a blastocoele.

Artificial shrinkage of the blastocyst to reduce fluid volume in the blastocoelic cavity before freezing was investigated as a technique to increase survival and ultimately increase clinical pregnancy and implantation rates (Vanderwalzmen *et al., 2002*). This has been carried out by either mechanical puncture of the blastocyst cavity with a needle and withdrawal of fluid (Vanderwalzmen *et al.*, 2002), use of osmotic shock to draw out fluid (Iwayama *et al.*, 2010) or by using laser pulses to collapse the blastocyst (Mukaida *et al.*, 2006) (Figure 4). In mechanical collapse, the inner cell mass of the blastocyst is positioned at 12 o'clock or 6 o'clock position. A glass micro needle is introduced into the cavity of the blastocoel and then withdrawn, which results in collapse of the cavity over 30 seconds to 2 minutes (Vanderwalzmen *et al.*, 2002; Mukaida *et al.*, 2006). During osmotic shock, the blastocyst is passed through media with high concentrations of sucrose to essentially "dehydrate" the embryo (Iwayama *et al.*, 2010). For laser collapse, a short duration laser pulse directed at the trophectoderm in a region away from the inner cell mass is delivered, shrinking the cavity immediately without additional manipulation of the embryo (Mukaida *et al.*, 2006). No statistical difference was seen on comparison of mechanical versus laser shrinkage (Mukaida *et al.*, 2006), or with osmotic versus laser shrinkage (Iwayama *et al.*, 2010), although results were improved in both cases as compared to controls (Mukaida *et al.*, 2006; Iwayama *et al.*, 2010). Human and mouse blastocsyts vitrified after mechanical or laser collapse have fewer damaged cells than untreated controls and total blastomere counts are higher after 24 hours of culture (Desai *et al.*, 2008). The rate of re-expansion after warming was also found to be higher (Desai *et al.*, 2008). In this study, an OCTAX 1.48 uM laser was used to deliver a single shot 10 ms pulse to the junction of cells located in the trophectoderm. The complete collapse of the blastocysts was seen within 2-4 minutes.

The major safety concern for use of laser is that the inner cell mass which ultimately becomes the fetus will inadvertently be exposed to the laser pulse. At this time the FDA has not approved this particular application of the laser in the U.S.

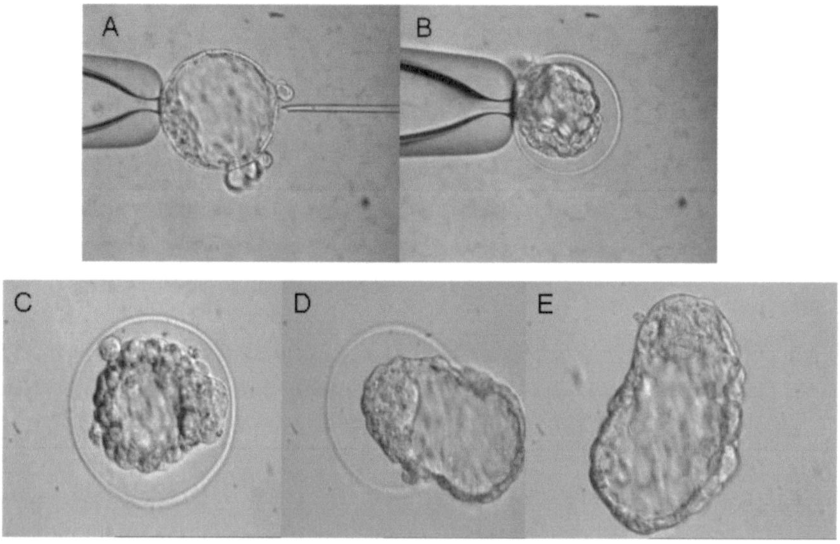

Fig. 4. (A) Blastocyst during mechanical collapse with ICSI needle. (B) Blastocyst immediately after collapse, (C) Blastocyst rewarmed after laser collapse, (D) 3 hours after rewarming, (E) after culture for 24 hours

4.2.5 Cellular microsurgery

Lasers may be used to remove material within the blastocyst that may prove detrimental to its development. This detrimental material includes cellular fragments or necrotic blastomeres. During embryo development it is possible to see cellular fragments appear. This is a process that may lead to impaired development as cells are dividing and natural planes are obstructed with fragments (Ebner *et al.*, 2005; Alikani *et al.*, 1999). Embryos with higher levels of fragmentation were found to have decreased implantation and pregnancy rates (Alikani *et al.*, 1999; Sathananthan *et al.*, 1990). Necrotic blastomeres are frequently observed in cryopreserved embryos upon warming. Release of toxic metabolites from dying cells may interfere with subsequent implantation. (Rienzi *et al.*, 2002). The laser can be used to create a small opening enabling extraction of fragments as well as dead cells. When necrotic blastomeres are removed, cleavage and implantation rates improve, and pregnancy rates increase from 17% to 45% (Rienzi *et al.*, 2002).

Another type of microsurgery that is well suited to the laser technology is preparation of zonae for the hemizona assay. The hemizona assay is used as a diagnostic tool to assess the binding capacity of sperm to the oocyte zona and also as a research model to study the effects of the environment or administered medications on the zona pellucida (Schopper *et al.*, 1999; Montag *et al.*, 2000c, 2009). For this procedure, the test oocyte is sliced into two sections, one to be used as the control and the other for the test treatment. A critical aspect of maintaining the accuracy of the test is that the oocyte is evenly divided so comparisons can be made. This bi-section can be accomplished using a mechanical technique or with the laser. Consecutive adjacent laser shots can be used to drill a series of holes through an oocyte immobilized using a micropipette (Montag *et al.*, 2000c). A study comparing laser to mechanical hemizona creation showed no difference in sperm binding between the two methods, and the laser drilling produced very even, flat hemizonae (Montag 2000c). The hemizona assay is performed more easily using lasers via mechanical techniques with a microscalpel (Schopper *et al.*, 1999).

The laser is particularly well suited for cellular microsurgery. The introduction of a laser with a femtosecond pulse to be used as a laser scalpel may further increase the accuracy of diagnostic and interventional procedures performed on the embryo (Rakityansky *et al.*, 2011). Biopsy of the trophectodermal cells of the blastocyst for pre-implantation genetic screening is one possibility. Currently the ability to accurately deliver the laser pulse to a very fine area and minimize heat transfer to adjacent cells has been a concern, limiting this use of lasers to research. Lasers may also be further developed to aid in elucidating a proteomic profile for embryos to help predict their success (Vela *et al.*, 2009).

4.2.6 Stem cell derivation

Embryonic stem (ES) cell lines are derived from the inner cell mass of blastocysts. Once isolated the inner cell mass can be used to establish pluripotent stem cell lines for use in transplants and to study cellular differentiation (Turetsky *et al.*, 2008). The ICMs from embryos that have been diagnosed with genetic disorders after PGD screening are potential source material for developing cell lines containing specific genetic conditions (i.e. cystic fibrosis, hemophilia) for use in research. Mechanical dissection of the inner cell mass from the trophectoderm is highly operator dependant, and chemical dissolution of the trophectoderm with Tyrode's acid subjects the inner cell mass to possible damage from the corrosive fluid (Turetsky *et al.*, 2008). Removal of the inner cell mass using several laser pulses has been shown to be an effective and an easy method to extract this stem cell source material from the blastocysts to establish ES cell lines (Turetsky *et al.*, 2008; Tanaka *et al.*, 2006). Laser also facilitates ICM isolation for cryopreservation of the stem cell source material (Desai et al., 2011).

4.2.7 Oocyte enucleation

Oocyte enucleation is similar to the process of dissecting the inner cell mass away from the outer layer of cells in an embryo. It is more challenging in the sense that only one cell, the oocyte, exists rather than the many cells in a blastocyst. Enucleation separates the nucleus of the oocyte from the remaining cellular material, effectively removing all genetic potential from the oocyte (Hirata *et al.*, 2011). This is done to establish cell lines for research purposes and to explore the genetic reprogramming potential of the oocyte cytoplasm (Hirata *et al.*, 2011; Malenko *et al.*, 2009; Raabe *et al.*, 2009). Once the nucleus with chromosomes is removed using a micropipette, new genomic material from somatic cells is introduced in to the enucleated oocyte (Malenko *et al.*, 2009; Hirata *et al.*, 2011). This may be done to develop embryonic cell lines for future therapeutic use (Hirata *et al.*, 2011). Using this procedure the cytoplasm of the oocyte reprograms differentiated somatic chromosomes into embryonic cells (Hirata *et al.*, 2011). Women who may feel uncomfortable donating eggs for research or therapeutic uses because the oocyte contains their genetic material may be more willing to donate knowing their genome will be removed (Hirata *et al.*, 2011).

The 1.48 uM diode laser may be used in conjunction with oocyte enucleation procedures in a similar manner as with ICSI and assisted hatching. A small hole is drilled in the zona pellucida, through which the nucleus is removed while leaving most of the cytoplasm (Li *et al.*, 2009). A picosecond pulsed 405 nm diode laser also effectively aids in enucleation with extremely short pulse duration of 1-2 seconds. This laser has not been approved for use in humans, however it is of interest due to its effect on intracellular structures (Raabe *et al.*, 2009). Although intracellular organelles were not found to be directly harmed following irradiation, function during the cell division process was prolonged when compared to non-irradiated cells. This indicates non-specific damage may have occurred (Raabe *et al.*, 2009) and cautions for judicious study of non-specific effects of irradiation in human embryology.

5. Safety & regulations

Lasers are currently considered by the Food and Drug Administration as a Class II device, special controls. As a Class II device, lasers must go through more than the general control measures regarding marketing and safety standards. They do not, however, have the stringent requirements and prolonged approval process prior to marketing required of the more highly regulated Class III devices. Class III devices are considered to be high risk, to the level of supporting life or presenting an unreasonable risk of harm. Three lasers have been approved by the FDA for use in reproductive technology: Saturn Active Laser System, Octax Laser Shot System and Hamilton Thorne Zona Infrared Laser Optical System. These lasers have only been approved for ablation of a small hole in the zone pellucida or thinning of the zona pellucida in approved patients.

The use of lasers in reproductive technology, particularly with respect to embryology, has stirred numerous concerns since its initial application. Areas of concern focus on the safety of the procedure as related to embryos at the time of development and for the children those embryos ultimately become. Primary aspects of laser function related to this issue are wavelength, heat generation and direct injury to blastomeres or oocytes through additional manipulation or imprecise beams.

The wavelength of lasers in reproductive technology falls into either the ultraviolet or infrared spectrum. Those lasers that have ultraviolet wavelengths provoke concern for possible mutagenic damage to embryonic DNA. The peak absorption rate of DNA is at 260 nm. Any laser with a wavelength in the UV range of the spectrum, 10-380 nm, increases likelihood of genetic damage or cytotoxicity. This includes excimer lasers with wavelengths at 193 nm, 308 nm and nitrogen 337 nm (Clement-Sengewald *et al.*, 2002). Data collected after zona drilling on mouse embryos with a 1.48 um laser found no significant differences in DNA methylation or early gene expression (Peters *et al.*, 2009; Kochevar, 1989).

Thermal damage occurs with absorption of heat by media surrounding the cells of interest. This is particularly true of the Er:YAG laser, which has a wavelength in the infrared spectrum but may pose a threat to cells by elevating the temperature of the culture media while in use (Clement-Sengewald *et al.*, 2002). Cells subjected to elevated temperatures may produce heat shock proteins as a protective mechanism, particularly HSP70i. When produced, these heat shock proteins help to stabilize other proteins and prevent apoptosis (Al-Katanani & Hansen, 2002). In a study examining the production of heat shock protein after 1.48 um laser drilling, no increase in levels of HSP70i were noted. Of note, the embryos were exposed to larger doses of laser energy during experiments than during routine zona drilling (Hartshorn *et al.*, 2005). This lends credence to the belief that the 1.48 uM laser has no immediate adverse effects on the embryo as a result of heat generation. Additionally, embryos exposed to laser drilling continue to develop at the same, if not better, rates than control embryos, and thus do not exhibit the retardation of growth seen if a cell is heat shocked (Hartshorn *et al.*, 2005). An associated problem lies within optimal laser settings for

a given procedure and the differing damage sustained by two routes to the same objective. For example, although visible results and initial growth may be unchanged, the amount of thermal spread anticipated to emerge from a lower power but longer duration pulse is greater than a higher power but much shorter pulse (Taylor *et al.,* 2010; Tucker *et al.,* 2009). This could lead to abnormal development later due to thermal spread (Tucker *et al.,* 2009). Although the peak temperature is much lower when a low powered laser is used, the prolonged pulse time leads to more extensive heating of the media and cells within that media (Tucker & Ball, 2009; Taylor *et al.,* 2010). It is currently uncertain how this type of thermal spread affects outermost blastomeres. A study examining oocyte lysis, cytogenic development and oocyte development following polar body biopsy via laser determined no deleterious effects were seen after the procedure (Hammoud *et al.,* 2010).
Long term data on childrens' health after use of the 1.48 um diode laser for zona opening is still limited. A study by Kanyo and Konc (2003) found no increase in congenital malformations after this procedure which is quite reassuring. As the use of laser technology in reproductive medicine becomes more widespread, more long term studies will be needed to evaluate both congenital defects and DNA abnormalities that may not manifest until later in life.

6. Conclusions

Lasers are useful in IVF as an additional tool with which to perform delicate procedures. The most commonly used laser in clinical IVF labs is the 1.48 um diode laser. This laser appears to be relatively safe for polar body or blastomere biopsy, sperm manipulation, drilling through the zona pellucida, stem cell derivation and cellular microsurgery. Laser technology may make performance of these tasks faster and easier. Definitive recommendations regarding whether or not to use lasers in reproductive technology are lacking. No conclusive data exists regarding long term safety of laser assistance in reproductive techniques and should be investigated more closely in the future.

7. References

Abdelmassih S., Cardoso, J., Abdelmassih V., Dias, J., Abdelmassih, R., & Z. Nagy. (2002) Laser-assisted ICSI: a novel approach to obtain higher oocyte survival and embryo quality rates. *Human Reproduction,* Vol. 17, No. 10, pp. 2694-99.

Al-Katanani, Y. & P. Hansen. (2002) Induced thermotolerance in bovine two-cell embryos and the role of heat shock protein 70 in embryonic development. *Molecular Reproduction and Development,* Vol. 62, pp. 174-80.

Alikani, M., Cohen, J., Tomkin, G., Garrisi, J., Mack, C., & R. Scott. (1999) Human embryo fragmentation in vitro and its implications for pregnancy and implantation. *Fertility and Sterility,* Vol. 7, No. 5, pp. 836-42.

Antinori, S., Panci, C., Selman, H., Caffa, B., Dani, G., & C. Versaci. (1996) Zona thinning with the use of laser: a new approach to assisted hatching in humans. *Human Reproduction,* Vol. 11, No. 3, pp. 590-94.

Balaban, B., Urman, B., Alatas, C., Mercan, R., Mumcu, A., & A. Isiklar. (2002) A comparison of four different techniques of assisted hatching. *Human Reproduction,* Vol. 17, No. 5, pp. 1239-43.

Clement-Sengewald, A., Schutze, K., Ashkin A., Palma, G., Kerlen, G., & G. Brem. (1996) Fertilization of bovine oocytes induced solely with combined laser microbeam and optical tweezers. *Journal of Assisted Reproduction and Genetics,* Vol. 13, No. 3, pp. 259-65.

Clement-Sengewald, A., Buchholz, T., Schutze, K., Berg, U. & F. Berg. (2002) Noncontact, laser-mediated extraction of polar bodies for prefertilization genetic diagnosis. *Journal of Assisted Reproduction and Genetics,* Vol. 19, No. 4, pp. 183-94.

Cohen, J., Elsner, C., Kort, H., Malter, H., Massey, J., Mayer, M. & K. Wiemer. (1990) Impairment of the hatching process following IVF in the human and improvement of implantation by assisting hatching using micromanipulation. *Human Reproduction,* Vol. 5, No. 1, pp. 7-13.

Cohen, J., Garrisi, G., Congedo-Ferrara, T., Kieck, K., Schimmel, T., & R. Scott. (1997) Cryopreservation of a single human spermatozoa. *Human Reproduction,* Vol. 12, No. 5, pp. 994-1001.

Dawson, A., Griesinger, G., & K. Diedrich. (2005) Screening oocytes by polar body biopsy. *Reproductive BioMedicine Online,* Vol. 13, No 1, pp 104-9.

De Vos, A., & A. Van Steirteghem. (2000) Zona hardening, zona drilling and assisted hatching: new achievements in assisted reproduction. *Cells Tissues Organs,* Vol. 166, pp. 220-27.

Desai, N., Szeptycki, J., Scott, M., AbdelHafez, A., & J. Goldfarb. (2008) Artificial collapse of blastocysts before vitrification: mechanical vs. laser technique and effect on survival, cell number, and cell death in early and expanded blastocysts. *Cell Preservation Technology,* Vol. 6, pp. 181-90.

Desai, N., Xu., J., Tsulaia, T., Szeptycki-Lawson, J., AbdelHafez, F., Goldfarb, J., & T. Falcone. (2011) Vitrification of mouse embryo-derived ICM cells: a tool for preserving embryonic stem cell potential? *Journal of Assisted Reproductive Genetics,* Vol. 28, pp. 93-99.

Dozortsev, D., Qian, C., Ermilov, A., Rybouchkin, A., Sutter, P., & M. Dhont. (1997) Sperm-associated oocyte-activating factor is released from the spermatozoa within 30 minutes after injection as a result of the sperm-oocyte interaction. *Human Reproduction,* Vol. 12, No. 12, pp. 2792-96.

Dumoulin J., Coonen, E., Bras, M., Bergers-Janssen, J., Ignoul-Vanvulchelen, R., van Wissen, L., Geraedts, J., & J. Evers. (2001) Embryo development and chromosomal anomalies after ICSI: effect of the injection procedure. *Human Reproduction,* Vol. 16, No. 2, pp. 306-12.

Ebner T., Moser M., & G. Tews. (2005) Possible applications of a non-contact 1.48 um wavelength diode in assisted reproduction technologies. *Human Reproduction Update,* Vol. 11, No. 4, pp 425-35.

Feichtinger, W., Strohmer, H., Fuhrberg, P,. Radivojevic, K., Antinori, S., Pepe, G., & C. Versaci. (1992) Photoablation of oocyte zona pellucida by erbium-yag laser for in vitro fertilization in severe male infertility. *Lancet,* Vol. 339, p. 811.

Fong, C., Bongso, A., Ng, S., Kumar, J., Trounson, A., & S. Ratnam. (1998) Blastocyst transfer after enzymatic treatment of the zona pellucida: improving in-vitro fertilization and understanding implantation. *Human Reproduction,* Vol. 13, No. 10, pp. 2926-32.

Gardner, DK., Lane, M., Stevens, J, Schlenker, T., & W. Schoolcraft. (2000) Blastocyst score affects implantation and pregnancy outcome: towards a single blastocyst transfer. *Fertility and Sterility*, Vol. 73, No. 6, pp. 1155-58.

Hammadeh, M., Fischer-Hammadeh, C., & K. Ali. (2011) Assisted hatching in assisted reproduction: a state of the art. *Journal of Assisted Reproductive Genetics*, Vol. 28, pp. 199-228.

Hammoud, I., Molina-Gomes, D., Albert, M., Bergere, M., Bailley, M., Wainer, R., Selva, J. & F. Vailard. (2010) Are zona pellucida laser drilling and polar body biopsy safe for in vitro matured oocytes? *Journal of Assisted Reproductive Genetics*, Vol. 27, pp. 423-27.

Hardarson, T., Lundin, K., & L. Hamberger. (2000) The position of the metaphase II spindle cannot be predicated by the location of the first polar body in the human oocyte. *Human Reproduction*, Vol. 15, No. 6, pp. 1372-76.

Hartshorn, C., Anshelevich, A., & L. Wangh. (2005) Laser zona drilling does not induce hsp70i transcription in blastomeres of eight-cell mouse embryos. *Fertility and Sterility*, Vol. 84, No. 5, pp. 1547-50.

Hirata, S., Fukasawa, H., Wakayama, S., Wakayama, T., & K. Hoshi. (2011) Generation of healthy cloned mice using enucleated cryopreserved oocytes. *Cellular Reprogramming*, Vol. 13, No. 1, pp. 7-11.

Inoue, M., & D. Wolf. (1975) Comparative solubility properties of rat and hamster zonae pellucidae. *Biology of Reproduction*, Vol. 12, pp 535-40.

Inoue, M., & D. Wolf. (1975) Fertilization-associated changes in the murine zona pellucida: a time sequence study. *Biology of Reproduction*, Vol. 13, pp. 546-51.

Iwayama, H., Hochi, S., & M. Yamashita. (2010) In vitro and in vivo viability of human blastocysts collapsed by laser pulse or osmotic shock prior to vitrification. *Journal of Assisted Reproductive Genetics*, published online ahead of print December 2010.

Jones, A., Wright, G., Kort, H., Straub, R., & Z. Nagy. (2006) Comparison of laser-assisted hatching and acidified Tyrode's hatching by evaluation of blastocyst development rates in sibling embryos: a prospective randomized trial. *Fertility and Sterility*, Vol. 85, No. 2, pp. 487-91.

Joris, H., De Vos, A., Janssens, R., Devroey, P., Liebaers, I., & I. Van Steirteghem. (2003) Comparison of the results of human embryo biopsy and outcome of PGD after zona drilling using acid Tyrode medium or a laser. *Human Reproduction*, Vol. 18, No. 9, pp. 1896-1902.

Kanyo K. & J. Konc. (2003) A follow up study of children born after diode laser assisting hatching. *European Journal of Obstetrics and Gynecology*, Vol. 110, pp. 176-80.

Kochevar, I., (1989) Cytotoxicity and mutagenicity of excimer laser radiation. *Lasers in Surgery and Medicine*, Vol. 9, pp. 440-45.

Lanzendorf, S., Ratts, V., Moley, K., Goldstein, J., Dahan, M., & R. Odem. (2007) A randomized, prospective study comparing laser-assisted hatching and assisted hatching using acidified medium. *Fertility and Sterility*, Vol. 87, No. 6, pp. 1450-57.

Li, J., Liu, X., Wang, H., Zhang, S., Liu, F., Wang, X, & Y. Wang. (2009) Human embryos derived by somatic cell nuclear transfer using an alternative enculeation approach. *Cloning and Stem Cells*, Vol. 11, No. 1, pp. 39-50.

Malenko, G., Stepanov, O., Komissarov, A., Antipova, T., Pinyugina, M., & M. Prokofiev. (2009) Efficiency of asynchronously in vitro-matured oocytes as recipients for nuclear transfer and of blind enucleation in zona-free bovine cloning. *Cloning and Stem Cells*, Vol .11, No. 2, pp. 287-92.

Malter, H., & J. Cohen. (1989) Partial zona dissection of the human oocyte: a non traumatic method using micromanipulation to assist zona pellucida penetration. *Fertility and Sterility*, Vol. 51, No. 1, pp. 139-48.

Mantoudis E., Podsiadly BT., Gorgy A., Venkat, G., & I. Craft. (2001) A comparison between quarter partial and total laser assisted hatching in selected infertility patients. *Human Reproduction*, Vol. 16, No. 10, pp. 2182-86.

Montag, M., van der Ven, K., Delacretax, G., Rink, K., & H. van der Ven. (1998) Laser-assisted microdissection of the zona pellucid facilitates polar body biopsy. *Fertility and Sterility*, Vol. 69, No. 3, pp. 539-42.

Montag, M., Rink, K., Dieckmann, U., Delacretax, G., & H. van der Ven. (1999) Laser-assisted cryopreservation of single human spermatozoa in cell-free zona pellucida. *Andrologia*, Vol. 31, pp. 49-53.

Montag, M., Koll, B., Holmes, P., et al. (2000a) Significance of the number of embryonic cells and the state of the zona pellucida for hatching of mouse blastocysts in vitro versus in vivo. *Biology of Reproduction*, Vol. 62, pp. 1738-44.

Montag, M., Koll, B., & H. van der Ven. (2000b) Use of a laser to evaluate zona pellucida hardness at different stages of mouse embryonic development in vitro and in vivo. *Journal of Assisted Reproduction and Genetics*, Vol. 17, No. 3, pp. 78-81.

Montag, M., Lemola, R., & H. van der Ven. (2000c) A new method to produce equally sized hemizonae pellucidae for the hemizona assay. *Andrologia*, Vol. 32, pp. 179-80.

Montag, M., Rink, K., Delcretaz, G., & H. van der Ven. (2000d) Laser-induced immobilization and plasma membrane permeabilization in human spermatozoa. *Human Reproduction*, Vol. 15, No. 4, pp. 846-52.

Montag, M., van der Ven, K., Dorn C., & H. van der Ven. (2004) Outcome of laser-assisted polar body biopsy and aneuploidy testing. *Reproductive BioMedicine Online*, Vol. 9, No. 4, pp. 425-29.

Montag M., Klose R., Koster M., Rosing, B., van der Ven, K., Rink, K., & H. van der Ven. (2009) Application of non-contact laser technology in assisted reproduction. *Medical Laser Application*, Vol. 24, pp. 57-64.

Moser M., Ebner T., Sommergruber M., Gaisswinkler, U., Jesacher, K., Puchner, M., Wiesinger, R., & G. Tews. (2004) Laser-assisted zona pellucida thinning prior to routine ICSI. *Human Reproduction*, Vol. 19, No. 3, pp. 573-78.

Mukaida, T., Oka, C., Goto, T., & K. Takahashi. (2006) Artificial shrinkage of blastocoeles using either a micro-needle or a laser pulse prior to the cooling steps of vitrification improves survival rate and pregnancy outcome of vitrified human blastocysts. *Human Reproduction*, Vol. 21, No. 12, pp. 3246-52.

Nagy, Z., Oliveira, A., Abdelmassih, & R. Abdelmassih. (2001) Novel use of laser to assist ICSI for patients with fragile oocytes: a case report. *Reproductive BioMedicine Online*, Vol. 4, No. 1, pp. 27-31.

Neev J., Tadir Y., Ho P., Berns, M., Asch, R., & T. Ord. (1992) Microscope-delivered ultraviolet laser zona dissection: principles and practices. *Journal of Assisted Reproduction and Genetics*, Vol. 9, No. 6, pp. 513-23.

Nijs, M., Vanderzwalmen, P., Vandamme, B., Segal-Bertin, G., Lejeune, B., Segal, L., van Roosendaal, E., & R. Schoysman. (1996) Fertilizing ability of immotile spermatozoa after intracytoplasmic sperm injection. *Human Reproduction,* Vol. 11, No. 10, pp. 2180-85.

Obruca, A., Strohmer, H., Blaschitz, A., Schonickle, E., Dohr, G., & W. Feichtinger. (1997) Ultrastructural observations in human oocytes and preimplantation embryos after zona opening using an erbium-yttrium aluminium-garnet (Er:YAG) laser. *Human Reproduction*, Vol. 12, No. 10, pp. 2242-45.

Palanker, D., Ohad, O., Lewis, A., Simon, A., Shenkar, J., Penchas, S., & N. Laufer. (1991) Technique for cellular microsurgery using the 193 nm excimer laser. *Lasers in Surgery and Medicine*, Vol. 11, pp. 580-86.

Palermo, G., Joris, H., Devroey, P., & A. van Streirteghem. (1992) Pregnancies after intracytoplasmic injection of a single spermatozoon into an oocyte. *Lancet*, Vol. 340, p. 17.

Palermo, G., Alikani, M., Bertoli, M., Colombero, T., Moy, F., Cohen, J., & Z. Rosenwaks. (1996) Oolemma characteristics in relation to survival and fertilization patterns of oocytes treated by intracytoplasmic sperm injection. *Human Reproduction*, Vol. 11, No. 1, pp. 172-76.

Peters, D., Kepikhov, K., Rodenacker, K., Marschall, S., Boersma, A., Hutzler, P., Scherb, H., Walter, J., & M. Harabe de Angelis. (2009) Effect of IVF and laser zona dissection on DNA methylation pattern of mouse zygotes. *Mammalian Genome*, Vol. 20, pp. 664-73.

Raabe, I., Vogel S., Peychl, J., I. Tolic-Norrelykke. (2009) Intracellular nanosurgery and cell enucleation using a picoseconds laser. *Journal of Microscopy*, Vol. 234, pp. 1-8.

Rakityansky, M., Agranat, M., Ashitkov, S., Ovchinnikov, A., Semenova, M., Sergeec, S., Sitnikov, D., & I. Shevelev. (2011) Cell technology employing femtosecond laser pulses. *Cell Technologies in Biology and Medicine*, Vol. 1, pp. 54-56.

Rienzi, L., Greco, E., Filippo, U., Iacobelli, M., Martinez, F., & J. Tesarik. (2001) Laser-assisted intracytoplasmic sperm injection. *Fertility and Sterility*, Vol. 76, No. 5, pp. 1045-47.

Rienzi, L., Nagy, ZP., Ubaldi, F., Iacobelli, M., Anniballo, R., Tesarik, J., & E. Greco. (2002) Laser-assisted removal of necrotic blastomeres from cryopreserved embryos that were partially damaged. *Fertility and Sterility*, Vol. 77, No. 6, pp. 1196-1201.

Rienzi, L., Ubaldi, F., Martinez, F., Minasi, M., Iacobelli, M., Ferrero, S., Tesarik, J., & E. Grego. (2004) Clinical application of laser-assisted ICSI: a pilot study. *European Journal of Obstetrics and Gynecology*, Vol. 115S, pp. S77-79.

Rink, K., Delacretaz, G., Salathe R., Senn, A., Nocera, D., Germond, M., de Grandi, P, & S. Fakan. (1996) Non-contact microdrilling of mouse zona pellucida with an objective-delivered 1.48 um diode laser. *Lasers in Surgery and Medicine,* Vol. 18, pp. 52-62.

Sagoskin, A., Levy, M., Tucker, M., Richter, K., & E. Widra. (2007) Laser assisted hatching in good prognosis patients undergoing in vitro fertilization-embryo transfer: a randomized controlled trial. *Fertility and Sterility*, Vol. 87, No. 2, pp. 283-87.

Sathananthan, H., Menezes, J., & S. Gunasheela. (2003) Mechanics of human blastocyst hatching in vitro. *Reproductive BioMedicine Online.* Vol 7, No. 2, pp. 228-34.

Schopper, B., Ludwig, M., Edenfeld, J., Al-Hasani, S., & K. Diedrich. (1999) Possible applications of lasers in assisted reproductive technologies. *Human Reproduction,* Vol. 14, pp. 186-93.

Schutze, K., Clement-Sengewald, A., & A. Ashkin. (1994) Zona drilling and sperm insertion with combined laser microbeam and optical tweezers. *Fertility and Sterility*, Vol. 61, No. 4, pp. 783-86.

Schutze, K., & G. Lahr. (1998) Identification of expressed genes by laser-mediated manipulation of single cells. *Nature Biotechnology*, Vol. 16, pp. 737-42.

Tadir, Y., Wright, W., Vafa, O., Ord, T., Asch, R., & M. Berns. (1989) Micromanipulation of sperm by a laser generated optical trap. *Fertility and Sterility*,Vol. 52, No. 5, pp. 870-73.

Tadir, Y., Wright, W., Vafa, O., Ord, T., Asch, R., & M. Berns. (1990) Force generated by human sperm correlated to velocity and determined using a laser generated optical trap. *Fertility and Sterility*, Vol. 53, No. 5, pp. 944-47.

Tadir, Y., Wright, W., Vafa, O., Liaw, L., Asch, R., & M. Berns. (1991) Micromanipulation of gametes using laser microbeams. *Human Reproduction*, Vol. 6, No. 7, pp. 1011-16.

Tanaka, N., Takeuchi, T., Neri, Q., Sills, E., & G. Palermo. (2006) Laser-assisted blastocyst dissection and subsequent cultivation of embryonic stem cells in a serum/cell free culture system: applications and preliminary results in a murine model. *Journal of Translational Medicine*, Vol. 4, pp. 20-32.

Taylor, T., Gilchrist, J., Hallowell, S., Hanshew, K., Orris, J., Glassner, M. & J. Wininger. (2010) The effects of different laser pulse lengths on the embryo biopsy procedure and embryo development to the blastocyst stage. *Journal of Assisted Reproductive Genetics*, Vol. 27, pp. 663-67.

Tinney, G., Windt, M., Kruger T., & C. Lombard. (2005) Use of a zona laser treatment system in assisted hatching: optimal laser utilization parameters. *Fertility and Sterility*, Vol. 84, No. 6, pp. 1737-41.

Tsai, M., Huang, F., Kung, F., Lin, Y., Chang, S., Wu, J., & H. Chang. (2000) Influence of polyvinylpyrrolidone on the outcome of intracytoplasmic sperm injection. *Journal of Reproductive Medicine*, Vol. 45, No. 2, pp.115-20.

Tucker, M., & G. Ball. (2009) Assisted hatching as a technique for use in human in vitro fertilization and embryo transfer is long overdue for careful and appropriate study. *Journal of Clinical Embryology*, Vol. 12, No. 1, pp. 10-14.

Turketsky, T., Aizenman, E., Gil, Y., Weinberg, N., Shufaro, Y., Revel, A., Laufer, N., Simon, A., Abeliovich, D., & B. Reubinoff. (2008) Laser-assisted derivation of human embryonic stem cell lines from IVF embryos after preimplantation genetic diagnosis. *Human Reproduction*, Vol. 23, No. 1, pp. 46-53.

Vanderzwalmen, P., Bertin, G., Lejeune, B., Nijs, M., Vandamme, B., & R. Schoyman. (1996) Two essential steps for a successful intracytoplasmic sperm injection: injection of immobilized spermatozoa after rupture of the oolemma. *Human Reproduction*, Vol. 11, No. 3, pp. 540-47.

Vanderzwalmen, P., Bertin, G., Debauche, C., Standaert, V., van Roosendaal, E., Vandervorst, M., Bollen, N., Zech, H., Mukaida, T., Takahashi, D., & R. Schoysman. (2002) Births after vitrification at morula and blastocyst stages: effect of artificial reduction of the blastocoelic cavity before vitrification. *Human Reproduction*, Vol. 17, No. 3, pp. 744-51.

Vela, G., Luna, M., Sandler, B., & A Copperman. (2009) Advances and controversies in assisted reproductive technology. *Mount Sinai Journal of Medicine*, Vol. 76, pp. 506-20.

Verlinsky, Y., Ginsberg, N., Lifchez, A., Valle, J., Moise, J., & C. Strom. (1990) Analysis of the first polar body: preconception genetic diagnosis. *Human Reproduction,* Vol. 5, No. 7, pp. 826-29.

Wong, B., Boyd CA., Lanzendorf SE. (2003) Randomized controlled study of human zona pellucida dissection using the Zona Infrared Laser Optical System: evaluation of blastomere damage, embryo development, and subsequent hatching. *Fertility and Sterility,* Vol. 80, No. 5, pp. 1249-54.

Yanagida, K., Katayose, H., Hirata, S., Hayashi, S., & A. Sato. (2001) Influence of sperm immobilization on onset of Ca^{+2} oscillations after ICSI. *Human Reproduction,* Vol. 16, No. 1, pp. 148-52.

11

Dynamic Analysis of Laser Ablation of Biological Tissue by Optical Coherence Tomography

Masato Ohmi and Masamitsu Haruna
Course of Health Science, Graduate School of Medicine
Osaka University
Japan

1. Introduction

Laser ablation is widely used in optical material engineering but also in clinical medicine. Actually, it has been used for evaporation and cutting of biological tissue in surgical operations; for example, the refractive surgery of cornea (Trokel *et al.* 1983; Puliafito *et al.* 1985) and the surgery of vascular (Isner *et al.* 1987). In particular, various types of CW and pulsed lasers have been considered for removal of hard dental tissues. Laser ablation may potentially provide an effective method for removal of caries and hard dental tissues with minimal thermal and mechanical damage to surrounding tissue. An important issue is quantitatively determining the dependence of tooth ablation efficiency or the ablation rate on the laser parameters such as repetition rate and energy of laser pulses. Up to now, the measurement has been made by observation of the cross section of the tissue surface, using a microscope or SEM, after cutting and polishing of a tissue sample (Esenaliev *et al.* 1996). This sort of process is cumbersome and destructive. On the other hand, shape of the tissue surface may change gradually with time after irradiation of laser pulses. The deformation of tissue surface is due to dehydration. The surrounding tissue may also suffer serious damage from laser ablation if the laser fluence is too high. Therefore, in-situ observation of the cross section of tissue surface is strongly required.

A very promising candidate for such an in-situ observation is the so-called optical coherence tomography (OCT) (Huang *et al.* 1991). The OCT is a medical diagnostic imaging technology that permits in-situ, micron-scale, tomographic cross-sectional imaging of microstructures in biological tissues (Hee *et al.* 1995; Izatt *et al.* 1996; Brezinski *et al.* 1996). At present, in the practical OCT, a super luminescent diode (SLD) is used as the light source for the low-coherence interferometer, providing the spatial resolution of 10 to 20 μm along the depth. Therefore, the OCT is potential for monitoring of the surface change during tissue ablation with micrometer resolution. Boppart et al have first demonstrated OCT imaging for observation of *ex vivo* rat organ tissue (Boppart *et al.* 1999). Alfrado et al have demonstrated thermal and mechanical damage to dentin by sub-microsecond pulsed IR lasers using OCT imaging (Alfano *et al.* 2004). We have also demonstrated an effective method for the *in situ* observation of laser ablation of biological tissues based on OCT (Haruna *et al.* 2001; Ohmi *et al.* 2005; Ohmi *et al.* 2007). In the traditional OCT system using a super-luminescent diode as a light source, imaging speed is limited. In fact, our first reported laser-ablation system, a time-domain OCT (TD-OCT) at the center wavelength of 0.8-μm is combined with a laser ablation system, where the optical axis of OCT is aligned with the 1.06-μm Q-switched YAG laser beam using a dichroic mirror. In this system, the data acquisition of each OCT image takes four seconds. The tissue laser ablation and the OCT imaging are repeated in turn. In this system, with this time delay for data acquisition, it is impossible to observe deformation of a crater and damage to the surrounding tissue due to thermal accumulation effects.

On the other hand, the recent application of Fourier-domain techniques with high-repetition rate swept laser source to OCT has led to an improvement in sensitivity of several orders of magnitude, toward high-speed OCT imaging (Yun *et al.* 2003; de Bore *et al.* 2003). Recently, we demonstrated true real-time OCT imaging of tissue laser ablation. A swept source OCT (SS-OCT) with 25 frames / s is used for the *in situ* observation, while tissue laser ablation is made continuously by 10-Hz YAG laser pulses (Ohmi *et al.* 2010). With this system, dynamic analysis of laser ablation can be achieved, taking thermal accumulation effects into account.

In this chapter, we summarize overview of *in situ* observation of biological tissue in laser ablation using OCT imaging technique. At first, laser ablation system with the time-domain OCT (TD-OCT) including the experimental data is described. Next, real-time *in situ* imaging of tissue ablation using swept source OCT (SS-OCT) is described. Laser ablation of hard and soft tissues including the ablation rate are demonstrated. Furthermore, the 3-D OCT image of the crater of biological tissue can be constructed by volume rendering of several hundred B-mode OCT images.

2. In-situ observation of laser ablation of biological tissue by time-domain OCT

2.1 System configuration

In order to achieve in-situ tomographic observation of the crater surface just after laser ablation of biological tissue, the laser-ablation optics and OCT imaging optics are combined. The system configuration is shown in **Fig. 1**. In laser ablation of tissue, the Q-switched Nd:YAG laser is used as the light source, which supplies laser pulses of 10 ns at the wavelength of 1.06 μm with the repetition rate of 10 Hz. The laser pulse is focused on a tissue sample via an x 10 objective with a 20-mm focal length lens. The focused beam spot size of 20 μm in the focal plane with the length of the beam waist is calculated of 630 μm. The laser pulse energy is typically 6.4 mJ with the energy per unit area of 5.1×10^3 J / cm^2 on the tissue surface.

On the other hand, the OCT system is a time-domain OCT (TD-OCT) which consists of the optical-fiber interferometer with the fiber-optic PZT phase modulators (Bouma *et al.* 2002). The light source is a 1.3-μm SLD whose output light of 13mW is coupled into a single-mode fiber directional coupler. For optical delay scanning, two identical fiber-optic PZT modulators are places on both reference and signal arms. In each PZT modulators, a nearly 20-m long single-mode fiber was wrapped around a cylindrical piezoelectric transducer. Two PZT modulators were driven in push-pull operation. The scanning depth along the optical axis becomes 1.0 mm when a 250-V triangular voltage is applied to two PZTs. In the sample arm of the interferometer, the collimated light beam of 6 mm diameter is focused on a sample via a microscope. Fortunately, it is a common knowledge that zero dispersion of a silica fiber lies near 1.3 μm. A great advantage of the all-optical-fiber OCT of Fig. 1, therefore, is that the coherence length does not increase significantly even if there is a remarkable optical path difference between reference and signal arms. In fact, we measured the coherence length of 19.1 μm. This value was very close to the expected value of 18.2 μm from the spectral bandwidth of the SLD itself. This value determines the resolution of OCT image along the optical axis. On the other hand, the lateral resolution is 5.6 μm determined by the focusing spot size of the x 10 objective used in the experiment. This value determines the resolution of OCT image along the optical axis.

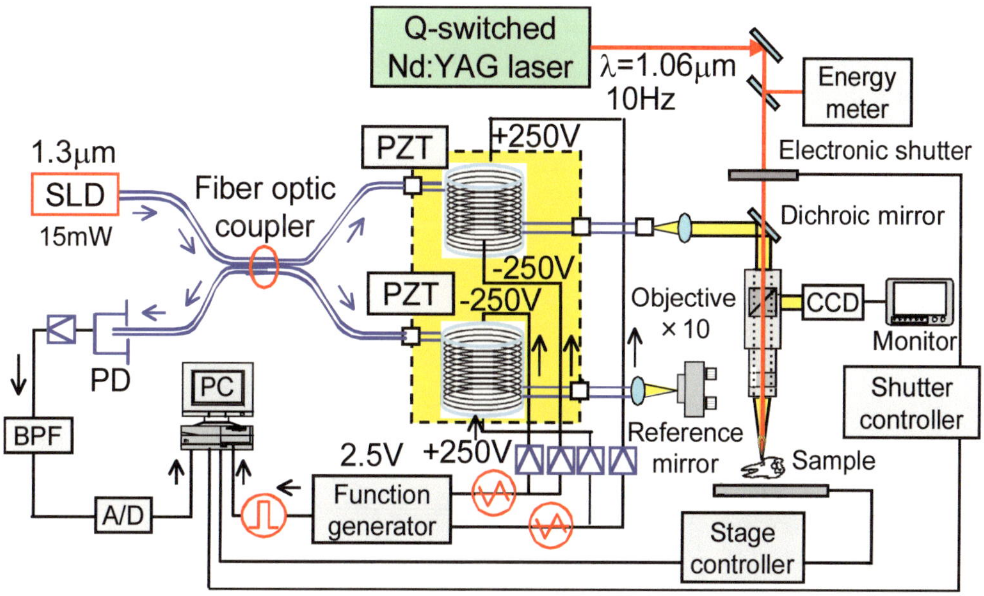

Fig. 1. System configuration of laser ablation with the time-domaion OCT (TD-OCT).

A key point for in-situ observation of the crater surface is that the YAG laser beam is aligned with the SLD light beam on the sample arm of the interferometer. These two light beams are combined or divided by a dichroic mirror, and an electronic shutter is placed in front of the YAG laser. Therefore, both the YAG laser and SLD light illuminate the same point on the tissue sample. In the experiment, at first, a certain number of YAG laser pulses are irradiated on the tissue sample, and a crater is formed on the sample surface. The YAG laser beam is then cut off with the electronic shutter, followed by obtaining an OCT image of the crater. The OCT imaging takes one second in the case where the image size is 1.0 x 1.0 mm^2 with a pixel size of 2.5 x 2.5 μm^2. After the OCT imaging, the laser ablation is again started with irradiation of a certain number of laser pulses. The laser ablation and OCT imaging are repeated by turn. This process is automatically controlled in our system. The characteristic of the system performance is summarized in Table 1, where the repletion rate of PZT phase modulator is 200 Hz at the OCT imaging area of 1 x 1 mm^2.

2.2 In-situ observation of ablation crater and the evaluation of ablation rate

In the experiment, human tooth enamel was used for the sample of laser ablation. A human tooth is a suitable representative for a hard tissue sample, because the tooth consists of two layers, enamel and dentine, and there is a remarkable difference in refractive index and hardness between these two materials. The interface between enamel and dentine is therefore recognized clearly in the OCT image. The ablation rate is quite different for enamel and dentine, as will be discussed later. The crater shape is also different between enamel and dentine because of the abrupt change in hardness at the interface.The Nd:YAG laser pulses were focused on the surface of human tooth enamel to make the ablation crater depending upon the laser-pulse shot number. **Figure 2** shows a series of OCT images of craters of human tooth enamel, where N is the laser-pulse shot number. From these OCT images, surface change of the ablation crater of the human tooth enamel is clearly observed. Moreover, showing all of OCT images continuously, time-serial tomographic observation of the crater in laser ablation is carried out.

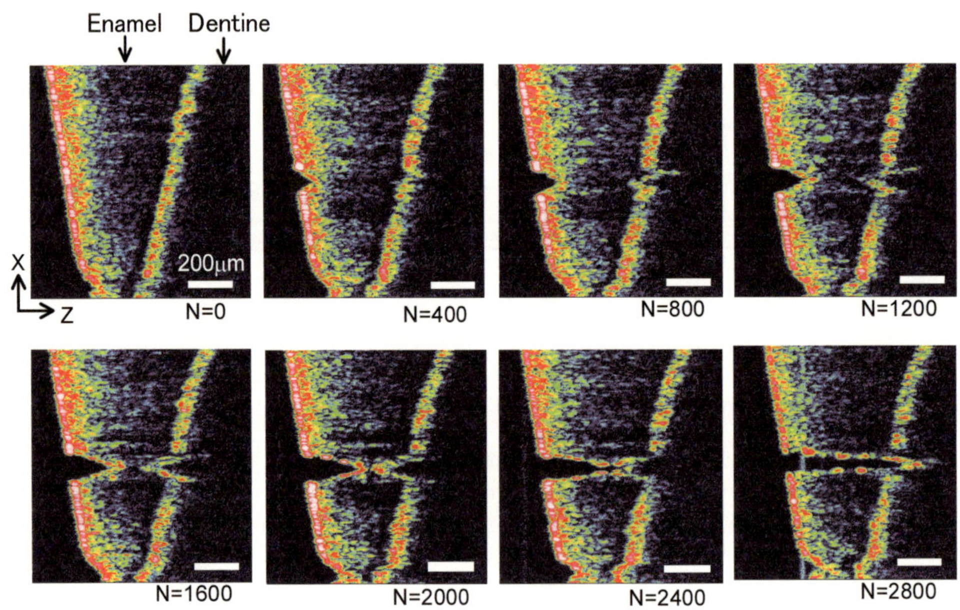

Fig. 2. A series of TD-OCT images of craters in laser ablation of human tooth.

The crater depth is also measured by the raster-scan signal of each OCT image. The measurement accuracy of the crater depth is 2.5 μm, which is determined by a pixel size of the OCT image. This value is smaller than the coherence length of 19μm of the SLD light source. The measured crater depths are plotted with respect to the laser-pulse shot number N, as shown in **Fig. 3**. From the data of N = 0 to 2000, a straight line was determined by the least squares method. The slope of the straight line yields the ablation rate of 0.11 μm / pulse with a standard deviation σ of 0.008 μm / pulse when the laser pulse energy is 16.0 mJ. Furthermore, from the data of N = 2200 to 2800, a straight line was determined by the least squares method. The slope of the straight line yields the ablation rate of 0.46 μm / pulse with a standard deviation σ of 0.015 μm / pulse in the human tooth dentine. The ablation rate of human tooth dentine is almost four times larger than human tooth enamel. Dentine is somewhat soft tissue rather than human tooth enamel. From the experimental results described above, one can find that OCT is really useful for monitor of the crater shape and the ablation rate with the damage of the surrounding tissues.

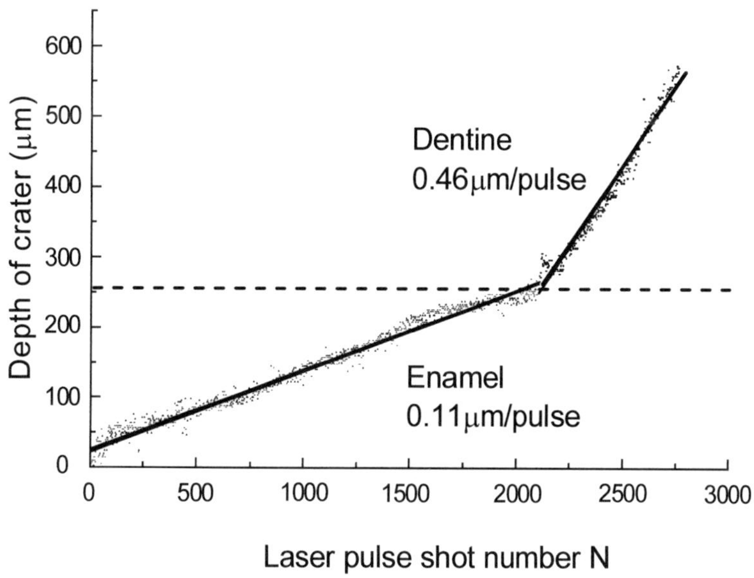

Fig. 3. Measurement of ablation rate of human tooth.

3. Real-time imaging of laser ablation of biological tissue by swept-source OCT

3.1 System configuration

In the former system, with this time delay for data acquisition, it is impossible to observe deformation of a crater and damage to the surrounding tissue due to thermal accumulation effects. In order to perform dynamic analysis of laser ablation of biological tissue, a swept-source OCT (SS-OCT) is combined with a YAG-laser ablation system, as shown in **Fig. 4**. In the SS-OCT, the optical source is an extended-cavity semiconductor wavelength-swept laser

employing an intracavity polygon scanner filter (HSL-2000, santec corporation). The lasing frequency is swept linearly with time, to obtain the reflected light distribution along the depth of the tissue sample. Fourier transformation of the interference signals results in reflected light distribution along the tissue depth. The SS-OCT consists of fiber-optic components, and the illuminating laser beam on the signal arm of the OCT interferometer is aligned with the YAG laser beam using a dichroic mirror. The light reflected from the reference mirror and the sample were recieved through magneto-optic circulators and combined by a 50/50 coupler. A fiber-optic polarization controller in the reference arm and the sample arm were used to align the polarization states of the two arms. The laser beam is then scanned with a Galvano mirror, resulting in a clear image of the ablation crater of the tissue sample. The center wavelength of the swept laser is 1.33 μm, with a wavelength scanning range of 110 nm. The sweep frequency of the laser source is 20 kHz at 25 frames / s, while the imaging area is 1 x 1 mm^2 with a pixel size of 8 x 5 μm. The real-time imaging of tissue laser ablation is thus realized in a fusion system of YAG-laser ablation and the fiber-optic SS-OCT. The measured coherence length of the SS-OCT system is 13 μm.
An electronic shutter is placed in front of the dichroic mirror to exactly adjust the ablation time. Both the YAG laser beam and the OCT probing laser beam are focused with the x 10 objective. The focused spot size is adequately adjusted by the laser beam width. In the experiment, the focused beam spot size is nearly 20 μm on the tissue surface. On the other hand, the focused spot size of the OCT probing beam is 5.6 μm, with a focal depth of only 40 μm. The out-of-focusing is unavoidable in the resulting OCT images, because there is no focus tracking mechanism in the present system.

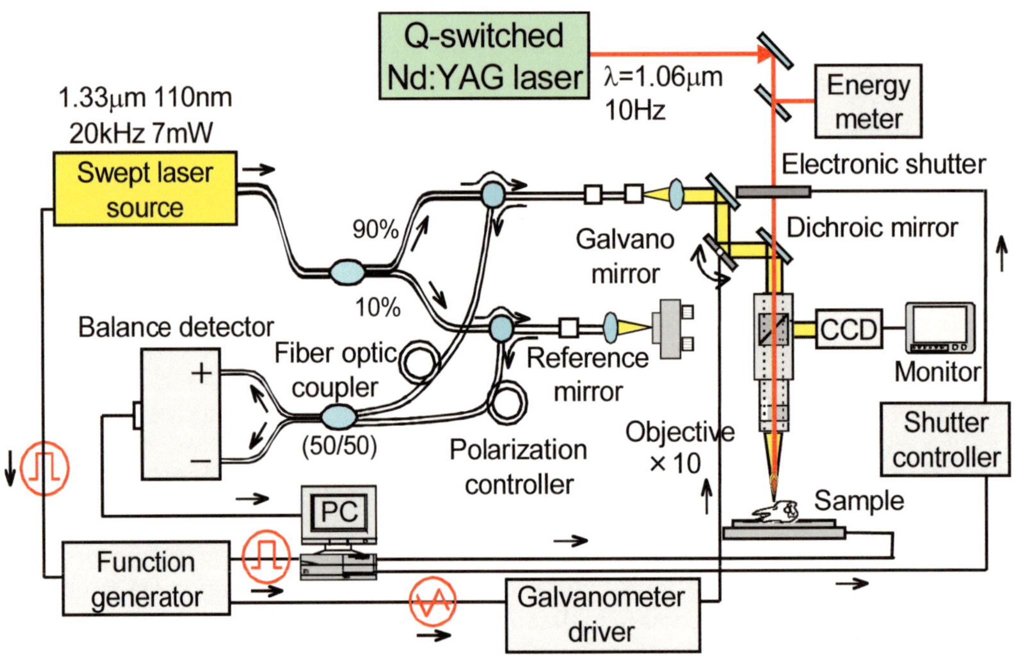

Fig. 4. System configuration of laser ablation with the swept-source OCT (SS-OCT).

3.2 Real-time imaging of tissue laser ablation and the evaluation of ablation rate
3.2.1 Hard tissue ablation

In the experiment, the laser pulse energy is typically 15.7 mJ with the energy per unit area of 5.0×10^3 J / cm^2 with a focused beam spot size of 20 μm on the tissue surface. The laser ablation of a human tooth made is continuously by the Q-switched YAG laser. Time-sequential OCT images of the crater of a human tooth are shown, where N is the shot number of the illuminating laser pulses, as shown in **Fig. 5**. The interface between enamel and dentine is clearly recognized in each OCT image because of the large refractive index difference between enamel (n = 1.652) and dentine (n = 1.546) (Ohmi *et al.* 2000). The crater depth increases gradually in the enamel, and it appears as if the interface between enamel and dentine juts out into the enamel. Near N = 2400, the YAG laser beam penetrates into the dentine through the enamel. The crater width becomes abruptly narrower in the dentine, reflecting the large difference in hardness between enamel and dentine. In addition, in the real-time imaging shown in **Fig. 5**, a small flying particle (debris), is observed in the crater, as indicated by a white circle, although the ablation plume is not imaged by OCT. The crater depth is measured in each OCT image, obtained by real-time imaging at 25 frames / s, where d is determined by the raster scan signal along the center of the crater. All measured values of d are plotted with respect to the shot number N of laser pulses, as shown in **Fig. 6**.

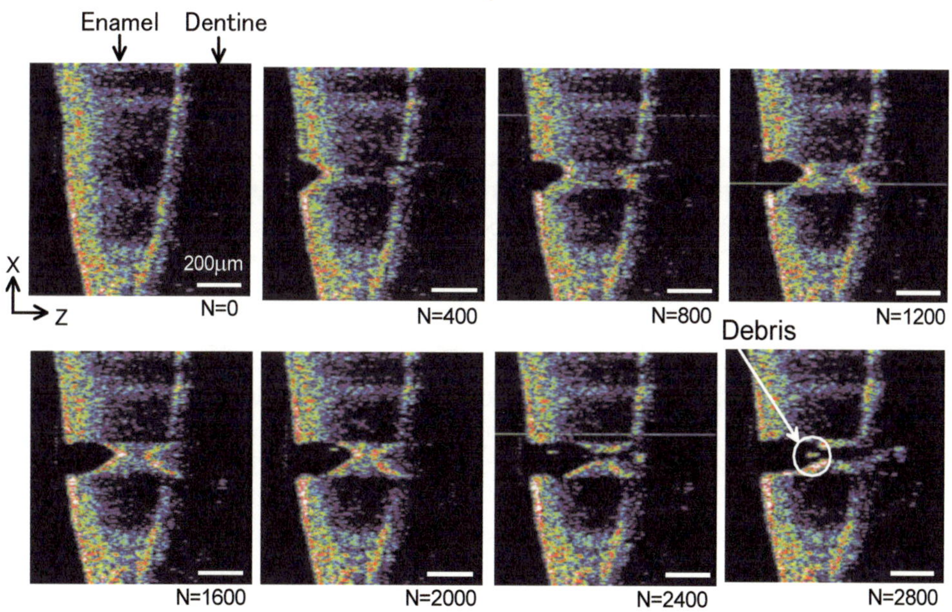

Fig. 5. A series of SS-OCT images of craters in laser ablation of human tooth.

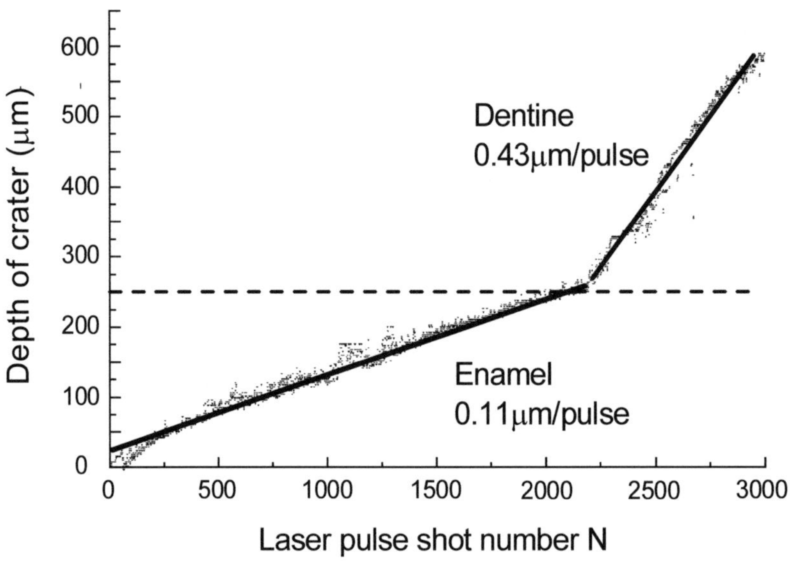

Fig. 6. Measurement of ablation rate of human tooth.

Furthermore, OCT images of craters formed after illuminating laser pulses in enamel and dentine are shown in **Fig. 7 (a)**, where the input laser fluence was 1.42×10^3 J / cm^2 to 6.87×10^3 J / cm^2. The ablation rate versus the input laser fluence for enamel and for dentine is also shown in **Fig. 7 (b)**. The ablation rate does not increase in linear proportion to the laser fluence, due to thermal accumulation effects, and it tends to saturate as the fluence increases. From the OCT image of the crater, the ablation volume of the crater increases according to the input laser fluence.

It is important to pay attention to the ablation rate and the volume of the crater. The ablation volume of a crater is evaluated in the following manner. In each frame of time-sequential OCT images, the crater is cut into 5-μm thick disks along the depth, under the assumption that the crater has a circular cross section. This assumption is consistent with the actual crater shape found in the 3-Dimensional OCT (3-D OCT) image of a human tooth, as will be shown later. The diameter is easily measured for each disk in the OCT image, and the crater volume is then counted by piling up 5-μm thick disks along the depth. All measured values of the ablation volume are plotted with respect to the shot number N of laser pulses, as shown in **Fig. 8**. From the slope of the straight line, the volume ablation rate of enamel and of dentine are obtained to be 1.31×10^4 μm^3 / pulse and 4.90×10^4 μm^3 / pulse, respectively. The volume ablation rate versus input laser fluence for enamel and for dentine is shown in **Fig. 9**. The volume ablation rate increases in linear proportion to the input laser fluence.

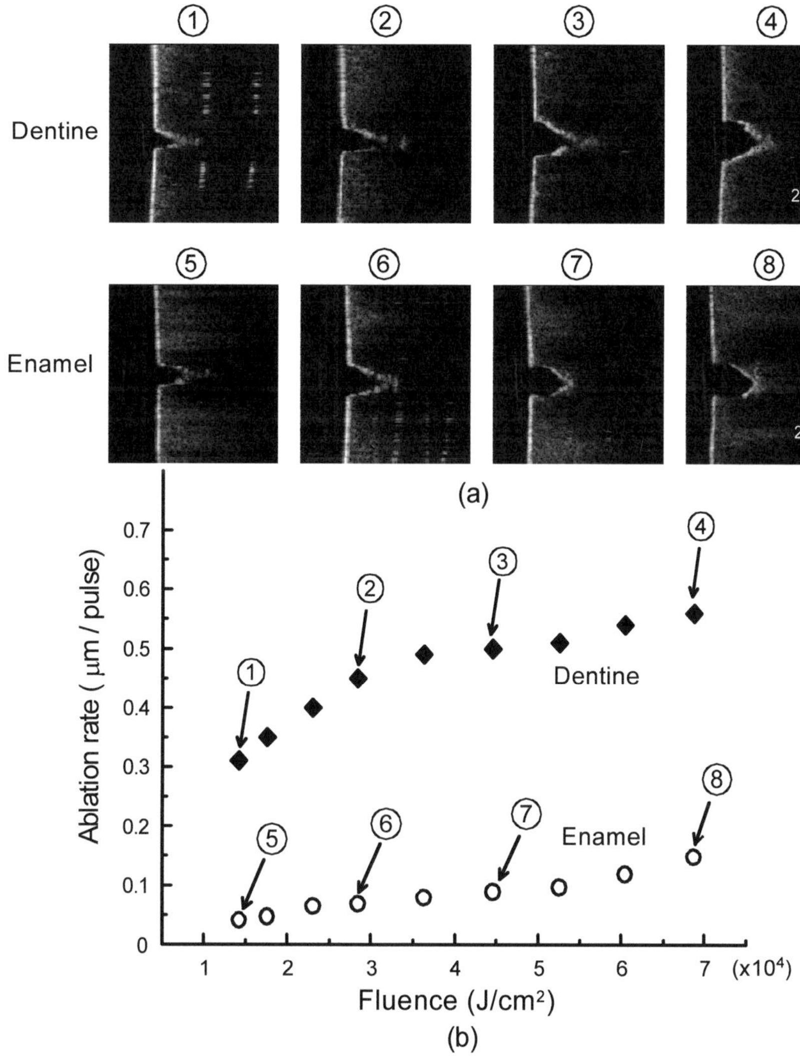

Fig. 7. Ablation rate versus laser fluence. (a) OCT images of the crater of enamel and dentine. (b) Ablation rate versus laser fluence.

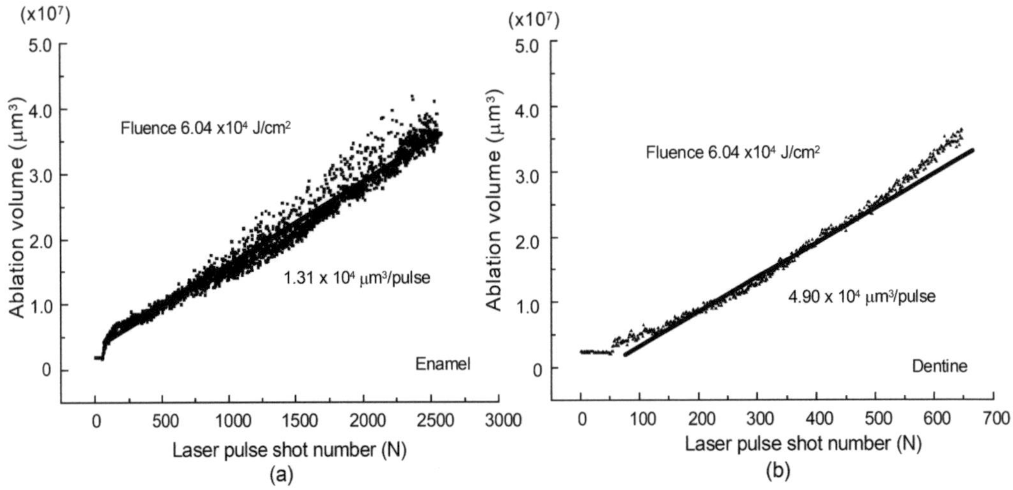

Fig. 8. Measured ablation volume versus laser shot number. (a) Enamel, (b) Dentine.

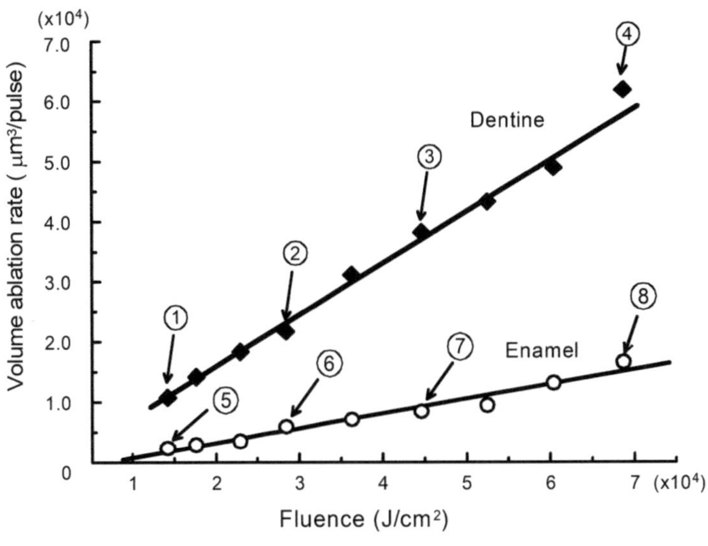

Fig. 9. Volume ablation rate versus laser fluence.

3.2.2 Soft tissue ablation

The aorta of a dog was used as an example of a soft tissue. An aorta has a three-layer wall that consists of the tunica intima, tunica media, and tunica adventitia. In the experiment, the YAG laser beam is focused on the inner surface of the aorta. In addition, the input laser fluence is reduced to 6.0×10^2 J / cm² to avoid the pronounced thermal accumulation effect that occurs in soft tissue ablation. Time-sequential OCT images of the dog aorta are shown, where the shot number N of laser pulses is 0 to 50, as shown in **Fig. 10**. In particular, thermal deformation of the crater is found where upheaval and removal of tissues are observed, when N is larger than 10. The crater diameter on the tissue surface increases with N, and is widened to nearly 260 μm due to the thermal deformation. The ablation rate of the aorta is 16.2 μm / pulse with a standard deviation σ of 0.41 μm / pulse. In comparison to the hard tissue ablation described in the **Section 3.2.1**, the ablation rate of the aorta is nearly 150 times larger than that for the human tooth, even though the input laser fluence is only one tenth. On the other hand, in the case where the input laser fluence is reduced to 6.0×10^2 J / cm^2, the thermal deformation of the crater is suppressed, resulting in a narrower crater, 600 μm deep with a diameter of 35 μm, without any damage to the surrounding tissues.

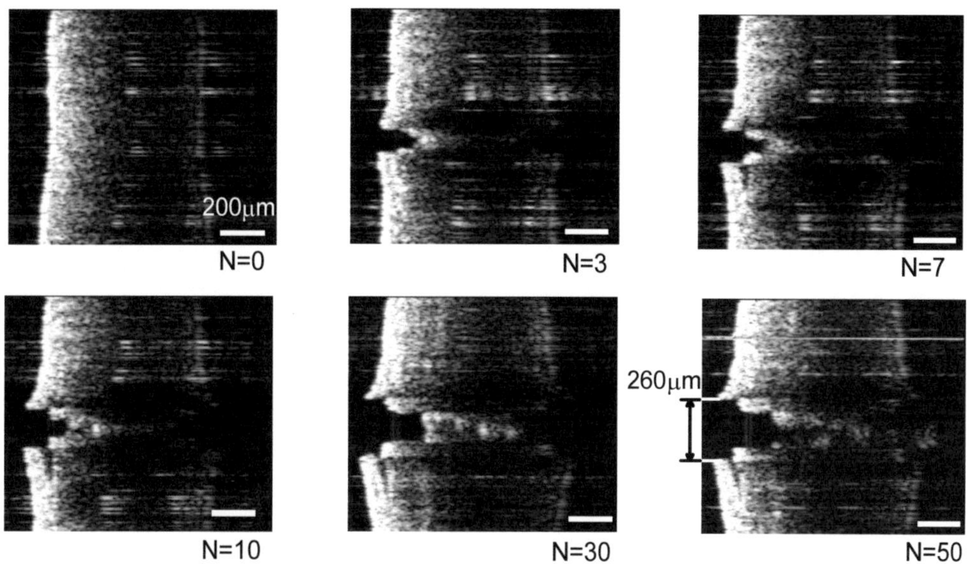

Fig. 10. Time-sequential OCT images of craters in laser ablation of dog aorta.

The 3-D OCT image of the crater of the aorta can be constructed by volume rendering of two hundred B-mode OCT images, obtained with a step of 5 μm over the distance of 0.5 mm, as shown in **Fig. 11 (a)**. The crater shape can be precisely observed in the 3-D OCT image. Under the condition where the input laser fluence is as large as 6.0×10^2 J / cm^2, smooth muscle fibers of the aorta surrounding the crater are coagulated and shrunken due to the thermal accumulation effect. As a result, the crater is expanded along the direction of the muscle fibers. The real-time OCT imaging is thus very useful for monitoring the thermal damage caused during soft tissue ablation. The 3-D OCT image of the crater of the human tooth is shown in **Fig. 11 (b)**. One can see that the crater of the human tooth has a circular cross section. This result is consistent with an assumption of the calculation of the ablation volume, as shown in **Figs. 8** and **Fig. 9**.

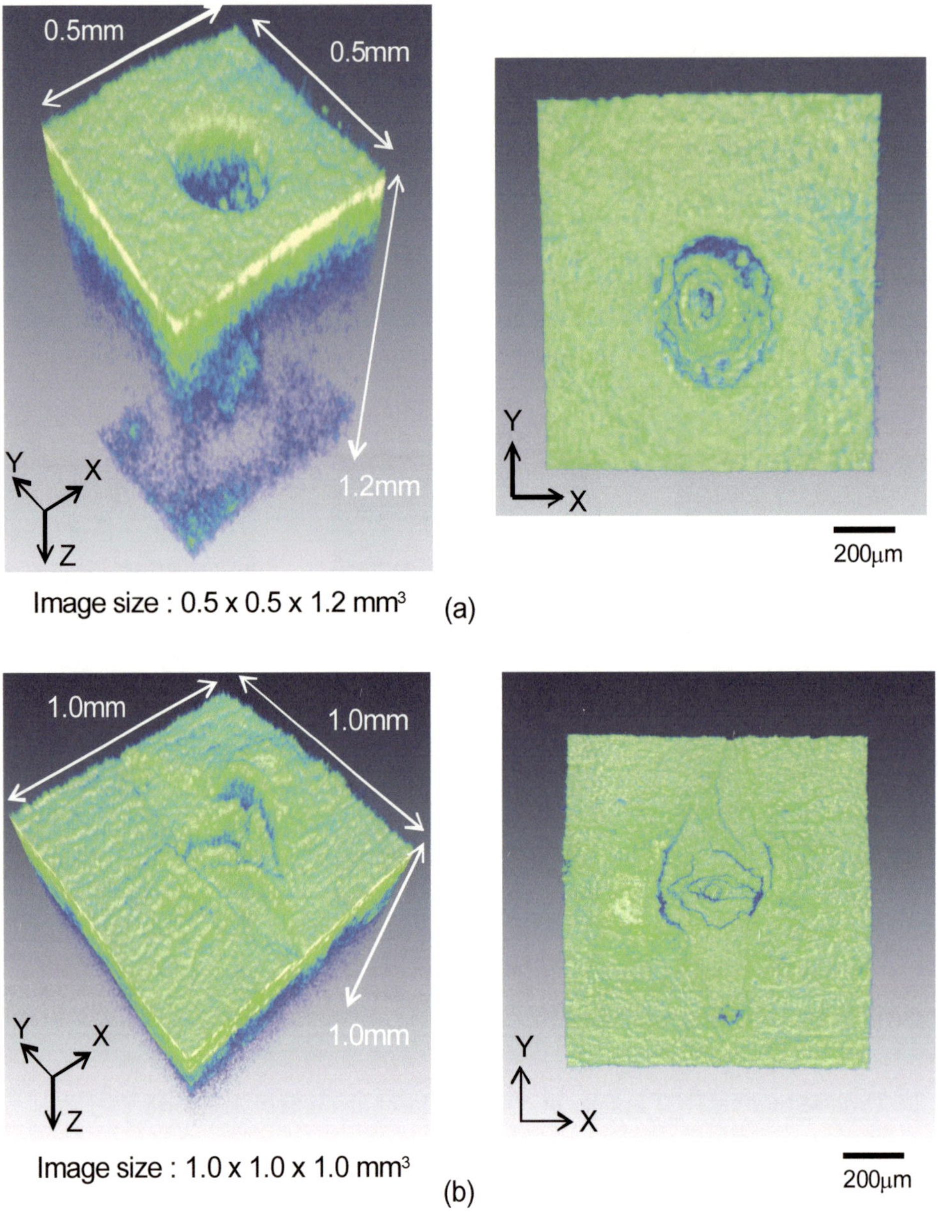

Fig. 11. 3D-OCT images of ablation crater. (a) dog aorta, (b) human tooth.

4. Discussion and conclusion

We have demonstrated the laser ablation system with a function of in-situ OCT observation of biological-tissue surface. In the experiment, time-serial OCT images of craters were carried out, and then the depth of the crater of tissue and the ablation rate were determined. Furthermore, dynamic analysis of tissue laser ablation has been demonstrated based on real-time OCT imaging of craters for both hard and soft tissues. In a human tooth, time variation of the crater depth can be measured very precisely with a standard deviation comparable to the coherence length of the SS-OCT. This results in a determination of the ablation rate with an accuracy below 0.01 μm / pulse.

At the interface between the enamel and the dentine, the ablation rate changes drastically, as does the crater shape, because of the difference in hardness between these two media. The higher ablation rate causes a narrower crater, and vice versa. The volume ablation rate increase can be evaluated from the OCT images of the crater and is in linear proportion to the input laser fluence. On the other hand, during laser ablation of soft tissue, such as the aorta of a dog, thermal deformation of the crater is found, including upheaval and removal of tissues. Thus, real-time OCT imaging is thus very useful for dynamic analysis of tissue laser ablation.

In the present fusion system of laser ablation and OCT, the image resolution is not yet sufficiently low for dynamic analysis of tissue ablation. The image resolution should be a few microns or less to allow monitoring of tissue treatment at the size of a cell. In this case, the focal depth of an objective becomes a few tens of microns, and proper focus tracking is then required for clear OCT imaging during tissue laser ablation.

5. References

Trokel, S. L.; Srinivasan, R. & Braren, B. (1983). Excimer laser surgery of the cornea. *American Journal of Ophthalmology* 96, 710-715.

Puliafito, C.A.; Steinnert, R. F.; Deeutsch, T. F.; Hillenkamp,F.; Dehm, E. J. & and Adler, C. M. (1985). Excimer laser ablation of the cornea and lens. *Ophthalmology* 92, 741-748.

Isner, J.M.; Steg, P. G. & Clarke, R. H.(1987). Current status of car- diovascular laser therapy. *IEEE Journal of Quantum Electronics* 23, 1756-1771.

Esenaliev, R. A.; Oraevsky, S.; Rastergar, C. Frederickson and M. Motamedi. Mechanism of dye-enhanced pulsed laser ablation of hard tissues: implications for dentistry. *IEEE Journal of Selected Topics in Quantum Electronics* 2, 836-846.

Huang D.; Swanson E. A., Lin C. P.; Schuman J. S.; Stinson W. G.; Chang W.; Hee M. R.; Flotte T.; Gregory K.; Puliafito C. A. & Fujimoto J. G. (1991). Optical coherence tomography. Science 254, 1178-1181.

Hee M. R.; Izatt J. A.; Swanson E. A.; Huang D.; Schuman J. S.; Lin C. P.; Puliafito C. A. & Fujimoto J. G. (1995). Optical coherence tomography of the human retina. (1995). Arch. *Ophthalmology* 113, 325-332.

Izatt, J.A.; Kulkarni, M. D.; Wang, H. D.; Kobayashi, K. & Sivak, M. V. (1996). Optical coherence tomography and microscopy in gastrointestinal tissues. *IEEE Journal of Selected Topics in Quantum Electronics* 2, 1017-1028, 1996.

Brezinski, M.E.; Tearney, G. J.; Bouma, B. E.; Izatt, J. A.; Hee, M. R.; Swanson, E. A.; Southern, J. F. & Fujimoto, J. G. (1996). Optical coherence tomography for optical biopsy. Properties and demonstration of vascular pathology.*Circulation* 93, 1206-1213.

Boppart, S. A.; Herrmann, J.; Pitris, C.; Stamper, D. L.; Brezinski, M. E. & Fujimoto, J. G. (1999). High-resolution optical coherence tomography-guided laser ablation of surgical tissue. *J Surg. Res.* 82, 275-284.

Alfredo, D. R.; Anupama, V. S.; Charles, Q. L.; Robert, S. J. & Daniel, F. (2004). Peripheral thermal and mechanical damage to dentin with microsecond and sub-microsecond 9.6 μm, 2.79 μm, and 0.355 μm laser pulses. *Lasers Surg. Med.* 35, 214-228.

Haruna ,M.; Konoshita, R.; Ohmi, M.; Kunizawa , N. & Miyachi, M. (2001). In-situ tomographic observation of tissue surface during laser ablation, *Proc. SPIE* 4257, 329-333.

Ohmi, M.; Tanizawa, M., Fukunaga, A. & Haruna, M. (2005). In-situ observation of tissue laser ablation using optical coherence tomography. *Opt. Quantum. Electron.* 37, 1175-1183.

Ohmi, M.; Nishino, M.; Ohnishi, M.; Hashishin, Y. & Haruna ,M. (2007). An approach to high-resolution OCT analyzer for laser ablation of biological tissue. *Proc. 3rd Asian and Pacific Rim Symp. Biophotonics (APBP2007) (Cairns)* 99-100.

Yun, S.; Tearney, G.; de Bore, J. F.; Iftimia, N. & Bouma, B. E. (2003). High speed optical frequency-domain imaging. *Opt . Express* 11, 2953-2963.

de Bore, J. F.; Cense,B.; Park, B. H.; Pierce, M. C., Tearney, G. J. & Bouma, B. E. (2003). Improved signal-to-noise ratio in spectral-domain compared with time-domain optical coherence tomography. *Opt.Lett.* 28, 2067-2069.

Ohmi, M.; Ohnishi, M.; Takada, D. & Haruna, M. (2010). Dynamic analysis of laser ablation of biological tissue using real-time optical coherence tomography. *Meas. Sci. Tecnol.* 21, 094030.

12

Polarization Detection of Molecular Alignment Using Femtosecond Laser Pulse

Nan Xu, Jianwei Li, Jian Li, Zhixin Zhang and Qiming Fan
National Institute of Metrology
China

1. Introduction

Femtosecond laser is becoming a powerful tool to manipulate the behaviors of molecules. When molecules are irradiated by strong laser field with intensity below the ionization threshold of molecules, the interaction between molecules and the laser electric field tends to align the molecules with the most polarizable axis along the laser polarization vector. If the laser pulse duration is larger than the rotational period of the molecule, the free rotor transforms into a pendular state that liberates about the polarization vector. Upon turning off the laser, the librator adiabatically returns to the isotropic free rotator from which it originates. If the laser pulse duration is less than the molecular rotational period, the laser-molecule interaction gives the molecules a rapid "kick" to make the molecular axis toward the laser field vector. After extinction of the laser, the transient alignment can periodically revive as long as the coherence of the rotational wave packet is preserved. Therefore, the former is also called adiabatic alignment and the latter field-free alignment. Even though both adiabatic alignment and field-free alignment can produce macroscopic ensembles of highly aligned molecules, field-free alignment has the obvious advantages that will not interfere with subsequent measurements. A variety of new and exciting applications of field-free aligned molecules are currently emerging. For example, Litvinyuk *et al.* measured strong field laser ionization of aligned molecules and obtained directly the angle dependent ionization rate of molecules by intense femtosecond laser field. Another novel application is to accomplish a tomographic reconstruction of the highest occupied molecular orbital of nitrogen by using high harmonic generation from intense femtosecond laser pulses and aligned molecules. Recently, Kanai *et al.* observed the quantum interference during high-order harmonic generation from aligned molecules and demonstrated that aligned molecules could be served as an ideal quantum system to investigate the quantum phenomena associated with molecular symmetries. The weak field polarization technique has homodyne and heterodyne detection modes. The alignment signal is proportional to $(\langle\cos^2\theta\rangle - 1/3)^2$ for homodyne detection and $(\langle\cos^2\theta\rangle - 1/3 + C)^2$ for heterodyne detection, where C describes the constant external birefringence contribution. Because the magnitude and the polarity of the external birefringence are hard to precisely control, homodyne detection is commonly used up to now. However, the homodyne signal cannot indicate whether the $\langle\cos^2\theta\rangle$ is larger or smaller than $1/3$. In other words, the homodyne signal cannot demonstrate whether the aligned molecule is parallel or perpendicular to the laser polarization direction. Using the heterodyne method, the alignment signals directly reproduce the alignment parameter<$\cos^2\theta$>.

1.1 Angle-dependent AC stark shift

Any non-spherical polarizable particle placed in an electric field will experience a torque due to the angular-dependent interaction (potential) energy U between the induced dipole moment $\vec{p}=\vec{\alpha}\cdot\vec{\varepsilon}$ and the field $\vec{\varepsilon}$. Consider, for simplicity, a linear particle having one dominant axis of polarizability $\alpha_{\|} > \alpha_{\perp}$ as shown in Figure 1. When placed in the field $\vec{\varepsilon}$ the potential energy is given by $U=-\vec{p}\cdot\vec{\varepsilon}$. The change in the potential energy for a small change of the field strength d$\vec{\varepsilon}$ would be

$$dU=-\vec{p}\cdot d\vec{\varepsilon}=-p_{//}d\varepsilon_{//}-p_{\perp}d\varepsilon_{\perp} \tag{1}$$

where the directions || and $\perp$ are parallel and perpendicular to the dominant axis of the particle. After substitution of the components of the induced dipole moment $p_i = \alpha_i\varepsilon_i$, dU becomes

$$dU=-\alpha_{//}\varepsilon_{//}d\varepsilon_{//}-\alpha_{\perp}\varepsilon_{\perp}d\varepsilon_{\perp} \tag{2}$$

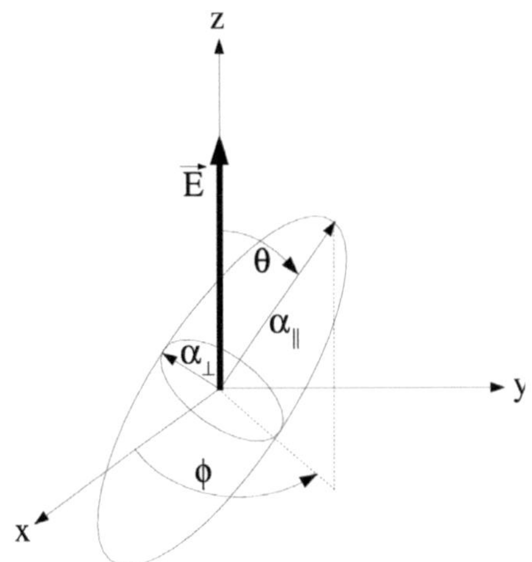

Fig. 1. Geometry of an anisotropic particle in an electric field $\vec{\varepsilon}$.

which can be integrated to give

$$U=-\frac{1}{2}[\alpha_{//}{}^{2}\varepsilon_{//}{}^{2}+\alpha_{\perp}{}^{2}\varepsilon_{\perp}{}^{2}] \tag{3}$$

By using the angle θ between the dominant axis of the particle and the electric field $\vec{\varepsilon}$ this can be written as

$$\begin{aligned} U(\theta)&=-\frac{1}{2}[\alpha_{//}{}^{2}\varepsilon^{2}\cos^{2}\theta+\alpha_{\perp}{}^{2}\varepsilon^{2}\sin^{2}\theta] \\ &=-\frac{1}{2}\alpha_{\perp}\varepsilon^{2}-\frac{1}{2}\Delta\alpha\varepsilon^{2}\cos^{2}\theta \end{aligned} \tag{4}$$

with $\Delta\alpha=(\alpha_{\parallel} - \alpha_{\perp})$.
This potential contains a constant term and an angular-dependent term. The constant term, however, is just a coordinate-independent shift which does not introduce any torques and can hence be dropped for convenience. Furthermore, when dealing with the particular case of diatomic molecules placed in infrared or near-infrared laser fields $\varepsilon(t) \sim \varepsilon_0 \sin\omega t$ which are far off-resonant with rotational frequencies, as is typical in experiments of strong field control of molecular rotations, the oscillating electric field switches direction too fast for the nuclei to follow directly. These oscillations can be removed from the potential energy by considering instead the time-average of the energy $U(\theta)$ over one cycle

$$U(\theta,t) = \int_0^{2\pi} \frac{1}{2} \Delta\alpha\varepsilon_0{}^2 f^2(t) \sin^2(t^{/}) \cos^2\theta dt^{/} \\ = -\frac{1}{4} \Delta\alpha\varepsilon_0{}^2 f^2(t) \cos^2\theta \tag{5}$$

where ε_0 is the maximum field strength of the laser and f(t) represents the envelope of the laser pulse which varies much slower than the field oscillations. This laser induced potential energy is known as the angular AC Stark shift. Note that any permanent dipole of the molecule would give a zero contribution to the potential energy upon time-averaging over one cycle of the laser field.

1.2 Quantum evolution

When the laser pulse interacts with the molecular gas, rotational wave packets are created in each molecule. The particular wave packet created in a given molecule will depend on its initial angular momentum state. Hence, to calculate the response of the molecular medium, the induced wave packet starting from each initial state in the thermal distribution must be calculated.
Consider a laser pulse with the electric field linearly polarized along the z-axis as in Figure 1. The interaction of laser pulse with the molecule is described by the Schrödinger equation

$$i\hbar\frac{\partial\psi(t)}{\partial t} = (BJ^2 - U_0(t)\cos^2\theta)\psi(t) \tag{6}$$

where θ is the angle between the laser polarization and the molecular axis, BJ^2 is the rotational energy operator, and

$$U_0(t) = \frac{\Delta\alpha\varepsilon_0^2}{4}\sin^2(\frac{\pi t}{2\tau_{on}}) \tag{7}$$

where τ_{on} gives the time for the pulse to rise from zero to peak amplitude and is also the full width at half maximum (FWHM) of the $\sin^2$ pulse.
The evolution of the wave function for the duration of the aligning pulse was calculated numerically in the angular momentum basis |J, M> .The time-dependent wave function is first expanded in the |J, M> basis

$$\psi(t) = \sum_{J,M} A_{J,M}(t)|J,M\rangle \tag{8}$$

Where |J, M> is the spherical harmonics function, and $A_{J,M}$ (t) is the expansion coefficient. In this basis, the Hamiltonian H(t) = [BJ^2 -U_0(t) $\cos^2\theta$] becomes

$$\begin{aligned}&\langle J,M|H(t)|\Psi(t)\rangle \\ &= B_0 J(J+1)A_{J,M} - U_0(t)C_{J,J+2,M}A_{J+2,M} - U_0(t)C_{J,J,M}A_{J,M} - \\ &U_0(t)C_{J,J-2,M}A_{J-2,M}\end{aligned} \tag{9}$$

Where

$$\begin{aligned}C_{J,J,M} &= \langle J,M|\cos^2\theta|J,M\rangle \\ C_{J,J+2,M} &= \langle J,M|\cos^2\theta|J+2,M\rangle \\ C_{J,J-2,M} &= \langle J,M|\cos^2\theta|J-2,M\rangle\end{aligned} \tag{10}$$

The Hamiltonian (9) does not couple even and odd J. All transitions occur between J ~J + 2 and J ~ J-2. This is a consequence of the symmetry of the angular potential $\cos^2\theta$ with respect to the point $\theta=\pi/2$. Furthermore, different M states do not couple. This is a consequence of the cylindrical symmetry of the angular potential.
With the rotational superposition at the end of the pulse expanded in angular momentum states

$$\psi(t) = \sum_{J,M} A_{J,M}|J,M\rangle \tag{11}$$

the field-free evolution of the wave packet becomes

$$\psi(t) = \sum_{J,M} A_{J,M}e^{-i(E_J/\hbar)t}|J,M\rangle \tag{12}$$

where E_J is the eigenenergy，E_J=BhcJ(J+1).
Using these energies, the field-free evolution given by Equation (12) is

$$\psi(t) = \sum_{J,M} A_{J,M}e^{-iBhcJ(J+1)t/\hbar}|J,M\rangle \tag{13}$$

Setting t =π/B_0 gives

$$\begin{aligned}\psi(t=1/2Bc) &= \sum_J A_J e^{-iBhcJ(J+1)(1/2Bc)/\hbar}|J,M\rangle \\ &= \sum_J A_J e^{-iJ(J+1)\pi}|J,M\rangle \\ &= \sum_J A_J|J,M\rangle = \psi(t=0)\end{aligned} \tag{14}$$

where the fact that J(J+1) is always an even integer and hence exp[-iJ(J+1)π] =1 was used. This shows that after a field-free evolution of t = π/B_0 the wave function will exactly reproduce the wave function at t = 0. Such behavior is called a wave-packet revival.

1.3 Measurement of alignment

The standard measure of alignment is defined in a slightly different way and is given by the average value of $\cos^2\theta$, where θ is the angle between the laser polarization direction and the molecular axis.

$$<\cos^2\theta>=\langle\Psi|\cos^2\theta|\Psi\rangle \tag{15}$$

This measure would give a value of $<\cos^2\theta> = 1$ for an angular distribution perfectly peaked along the 'poles' $\theta = 0$ and π, $<\cos^2\theta> = 0$ for a distribution peak along the 'equator' $\theta = \pi/2$, and $<\cos^2\theta> = 1/3$ for an isotropic distribution evenly distributed across all θ. If $<\cos^2\theta> > 1/3$, the molecule is predominantly aligned along the laser polarization direction. If $<\cos^2\theta> < 1/3$, the probability distribution for the axis of the molecule is concentrated around a plane orthogonal to the laser polarization direction and labeled as an antialignment molecule.

During the interaction with the laser pulse, this measure is simply obtained by numerical integration over the computed wave function. For field-free propagation, the time-dependent measure of alignment is given by

$$\begin{aligned}<\cos^2\theta>(t)_{J_0,M_0} &= \langle\Psi|\cos^2\theta|\Psi\rangle \\ &= \sum_J |A_{J,M}|^2 C_{J,J,M} + |A_{J,M}||A_{J+2,M}|C_{J,J+2,M}\cos(\omega_J t+\varphi)\end{aligned} \tag{16}$$

where $\omega_J=(E_{J+2}-E_J)$.ϕ denotes the relative phase between the states $|J, M>$ and $|J+2, M>$ at the start of the field-free evolution. Note that during the field-free evolution the $<\cos^2\theta>(t)$ signal is composed of the discrete frequencies ω_J.

The alignment signal is further averaged over an initial Boltzmann distribution of angular momentum states for a given initial temperature T. This is accomplished by calculating the rotational wave-packet dynamics for each initial rotational state in the Boltzmann distribution, and then incoherently averaging the $<\cos^2\theta>(t)_{J,M}$ signal from each initial state $|J, M>$ weighted by the Boltzmann probability

$$<\cos^2\theta>(t)=\frac{\sum_{J_0,M_0}\exp(-E_{J_0}/kT)<\cos^2\theta>(t)_{J_0,M_0}}{\sum_{J_0}(2J_0+1)\exp(-E_{J_0}/kT)} \tag{17}$$

2. Measurement of molecular alignment

Now, the experimentalists have developed two typical methods to evaluate experimentally the alignment degree of molecules. The first one is realized by breaking the aligned molecule through multielectron dissociative ionization or dissociation followed by ionization of the fragments. The alignment degree $<\cos^2\theta>$ was thus deduced from the angular distribution of the ionized fragments. The disadvantage for this method is that the probe laser is so strong that destroys the aligned molecules. The second one is the weak field polarization spectroscopy technique based on the birefringence caused by aligned molecules. The advantage for this method is that the probe laser is so weak that it neither affects the alignment degree nor destroys the aligned molecules.

The first section outlines the homodyne detection method to measure alignment of different gas molecules. The enhanced field-free alignment is also demonstrated here. The second section outlines the heterodyne detection method and the numerical calculation of molecular alignment. In this section, field-free alignment signals and the population of rotational states of diatomic molecules are present. The last section is the detection of gas component using molecular alignment, in which a feasibility of rapid detection of gas component is shown.

2.1 Measurement of molecular alignment

We report our results about field-free alignment of diatomic molecules (N_2, O_2, CO) and polyatomic molecules (CO_2, CS_2, C_2H_4) at room temperature under the same laser properties. We also demonstrated experimentally that the alignment degree could be strongly enhanced by using double pulses at a separated time delay. These researches provide a feasible approach to prepare field-free highly aligned molecules in the laboratory for practical applications.

2.1.1 Experimental setup

Figure2 shows the experimental setup of the molecular alignment measurement. The laser system consists of a chirped pulse amplified Ti:sapphire system operating at 800nm and a repetition rate of 10Hz. The laser pulse of 110fs was split into two parts to provide a strong energy pump beam and a weak energy probe beam both linearly polarized at 45° with respect to each other. For double pulses alignment of molecules, the strong pump laser was split into another two aligning pulses with equal intensity. The relative separated times between the two pulses is precisely adjusted using an optical translational stage controlled by a stepping motor. Both the pump beam and the probe beam are focused with a 30cm focal length lens into a 20cm long gas cell at a small angle. The gas cell was filled with different gases at room temperature under one atmosphere pressure. The field-free aligned molecules induced by the short pump laser will cause birefringence and depolarize the probe laser. After the cell, the depolarization of the probe, which represents the alignment degree, is analyzed with a polarizer set at 90°with respect to its initial polarization detection. In order to eliminate the laser fluctuation, a reference laser was introduced. The alignment signals and the reference laser signals were detected by two homotypical photoelectric cells and transformed into a computer via a four-channel A/D converter for analysis.

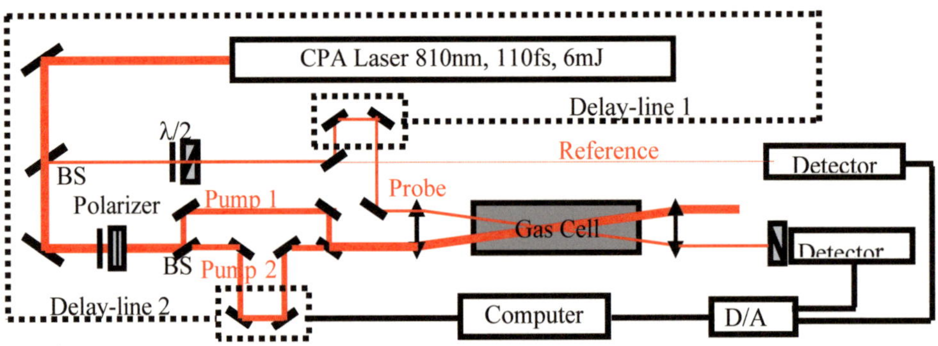

Fig. 2. Experimental setup for measuring field-free alignment of molecules induced by femtosecond laser pulse. BS: beam splitter.

2.1.2 Results and discussion

Figure 3 shows the alignment signal for diatomic molecules (a) N_2, (b) O_2, and (c) CO irradiated by 800nm, 110fs at an intensity of 6×10^{13} W/cm^2. The classical rotational period T_r of molecules is determined by the equation $T_r = 1/2\, B_0 c$ where B_0 is rotational constant in the ground vibronic state and c is the speed of the light. For N_2, O_2 and CO, B_0 is 2.010, 1.4456, 1.9772 cm^{-1}, respectively. The corresponding rotational period T_r is therefore 8.3 ps for N_2, 11.6 ps for O_2 and 8.5 ps for CO. It is clearly noted from figure 3 that the alignment signal fully revives every molecular rotational period. However, there are also moments of strong alignment that occur at smaller intervals. The difference at quarter full revival for N_2, O_2 and CO can be well explained by the different nuclear spin weights of the even and odd J states in the initial distribution. At 1/4, 3/4, 5/4, ... full revivals, the odd wave packet has maxima (minima) whereas the even wave packet has minima (maxima). For homonuclear diatomic molecules, the nuclear spin statistics controls the relative weights between even and odd J states. In the case of N_2, the relative weights of the even and odd J are 2:1. As a result, the temporary localization of the even wave packet at $T_r/4$ is only partially cancelled by its odd counterpart. Thus, some net N_2 alignment and antialignment is observed near t = $n\, T_r/4$, where n is an odd number. In the case of O_2, only odd J states are populated. Since only a single localized wave packet exists, strong net alignment and antialignment is observed near the time of a quarter revivals. For heteronuclear diatomic molecule CO, the even and odd J states are equally populated, the opposite localizations would cancel and therefore no net alignment would be observed at the time of the quarter revival.

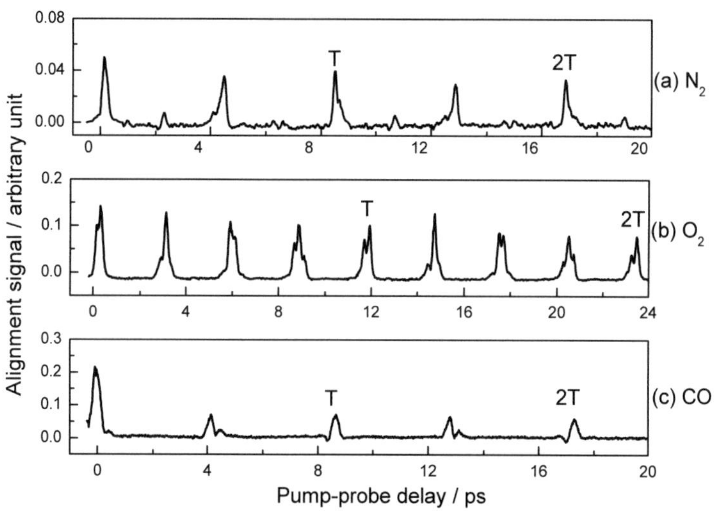

Fig. 3. Field-free alignment signal for diatomic molecules (a) N_2, (b) O_2, and (c) CO irradiated by 800nm, 110fs at an intensity of 6×10^{13} W/cm^2.

Figure 4 shows the alignment signal for polyatomic molecules (a) CO_2, (b) CS_2, and (c) C_2H_4 irradiated by 800nm, 110fs at an intensity of 6×10^{13} W/cm^2. The classical rotational period T_r is 42.7 ps for CO_2, 152.6 ps for CS_2 and 9.3 ps for C_2H_4. It can clearly be seen that the alignment signal repeats every molecular rotational period. Note that although CO_2 is not actually a homonuclear diatomic, the two O atoms are indistinguishable. Hence symmetrization of the wave function with respect to these two particles require that only even *J* states are populated. Since only a single localized wave packet exists, strong net alignment and antialignment is observed near the time of a quarter revivals. For the same reason, the net alignment and antialignment is also observed near the time of a quarter revival for CS_2. In a recent theoretical paper, Torres *et al.* explicitly calculated the angular distribution of CS_2 ensemble as they evolve through a rotational revival. They found the ensemble deploys a rich variety of butterfly-shaped distribution, presenting always some degree of order between the aligned and antialigned distributions. Unlike the linear molecules, complicated revival signals were observed for C_2H_4 because of its asymmetric planar structure. Our experimental observation of C_2H_4 well agreed with theoretical calculation carried out by Underwood *et al.* Those authors also proposed a theoretical scheme to realize three-dimensional field-free alignment of C_2H_4 by using two orthogonally polarized, time-separated laser pulses.

In Figure 4, it can also be seen that the alignment signal does not return to background signal with probe laser preceding the aligned laser, especially for CS_2. The increased background signal results from the permanent alignment of the molecules, in which the laser-molecule interaction spreads each initial angular momentum state to higher *J* but does not change *M*. Thus, rather than being uniformly distributed, the angular momentum vectors of each *J* state in the wavepacket are preferentially oriented perpendicular to the aligning pulse polarization. Due to the relaxation of the rotational population, the permanent alignment will decay monotonically under field-free conditions towards its thermal equilibrium.

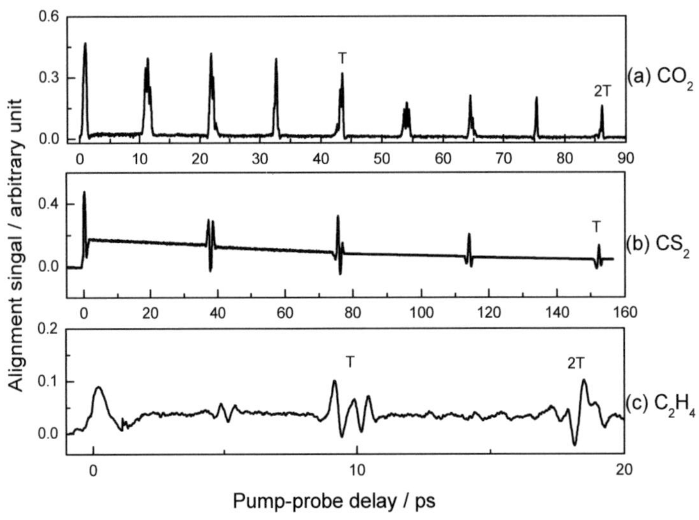

Fig. 4. Field-free alignment signal for diatomic molecules (a) CO_2, (b) CS_2, and (c) C_2H_4 irradiated by 800nm, 110fs at an intensity of 6×10^{13} W/cm^2.

For real applications, it is important to ensure the higher degree of alignment obtained under field-free condition. Theoretical investigation indicated that the degree of alignment could be improved by minimizing the rotational temperature of the molecules or by increasing the laser intensity. For practical application, minimizing the rotational temperature is not a good approach. Therefore, we studied the field-free alignment of molecules by varying laser intensity.

However, the maximum degree of alignment thus obtained is limited by ionization of the molecule in the laser. In order to obtain highly aligned molecules without destroying the molecule, theorists proposed multiple pulse method, in which alignment is created with a first pulse, and then the distribution is squeezed to a higher degree of alignment with subsequent pulses. Thus multiple-pulse method gets around the maximum intensity limit for single laser pulse and highly aligned molecules can be obtained without destroying the molecule.

The enhanced field-free alignment of CS_2 by means of two-pulse laser was also experimentally performed, in which the aligning laser was divided into two beams with equal intensity of 2×10^{13} W/cm^2. Figure 5 clearly shows the timing for the two aligning laser pulses and the probe laser pulse. The first aligning laser pulse prepares a rotational wave packet at time zero and the second aligning laser pulse modifies this rotational wave packet at $T_r/4$. The probe laser pulse measures the alignment degree of molecules at $3T_r/4$. Thus the probe laser measured the alignment signal at $3T_r/4$ when the first aligning laser worked alone, which is shown in red line in the inset of Figure 5. The probe laser measured the alignment signal at $T_r/2$ when the second aligning laser worked alone, which is shown in blue line in the inset of Figure 5. Depending on the delay time between the first aligning and the second aligning laser pulses, the field-free alignment can be instructive or destructive. With a proper adjustment of the delay between the two aligning laser pulses, an obvious enhanced alignment signal is observed in the probe region, as well as the permanent alignment, which is shown in black line in the inset in Figure 5. The optimal delay of the second aligning laser pulses is typically located before the maximum alignment during a strong revival after the first aligning laser pulse. With such a timing, the second aligning laser pulse catches the molecules as they are approaching the alignment peak and pushes them a bit more toward an even stronger degree of alignment. The region of increased alignment will appear in subsequent full revivals from this point. Therefore, it is very promising that field-free highly aligned molecules can be obtained using multiple pulses.

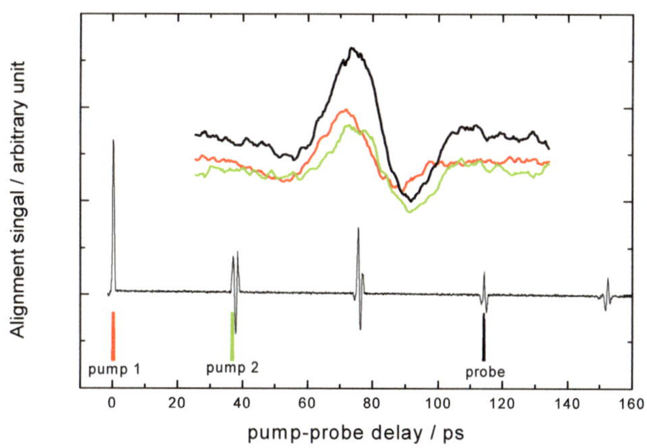

Fig. 5. (Lower) Single-pulse alignment signal illustrates pulse timing for double-pulse experiment. (Upper) Red line represents the alignment signal at $3T_r/4$ induced by the first aligning laser pulse alone, blue line represents the alignment signal at $T_r/2$ induced by the second aligning laser pulse alone, black line represents the enhanced alignment signal induced by the two aligning laser pulses with appropriate separated times.

2.2 Heterodyne detection of molecular alignment

The weak field polarization technique has homodyne and heterodyne detection modes. The alignment signal is proportional to $(<\cos^2\theta>-1/3)^2$ for homodyne detection and $(<\cos^2\theta>-1/3)$ for pure heterodyne detection. Comparing with the homodyne signal, pure heterodyne signal had the merit of directly reproducing the alignment parameter $<\cos^2\theta>$ except a 1/3 baseline shift. Unfortunately, the pure heterodyne signal is hardly obtained in the experimental measurement; homodyne detection is still commonly used till now. However, the homodyne signal does not indicate whether the aligned molecule is parallel or perpendicular to the laser polarization direction.

We modified the typical weak field polarization technique. Both homodyne and pure heterodyne detection were realized in this experimental apparatus. They were employed to quantify the post-pulse alignment of the diatomic molecules irradiated by a strong femtosecond laser pulse. The alignment signal and its Fourier transform spectrum were analyzed and compared with the numerical calculation of the time-dependent Schrödinger equation.

2.2.1 Theory

The state vector of the free molecule denoted by $\Phi(t)$ was probed by a non-resonant weak laser pulse

$$\overrightarrow{\varepsilon_d} = \vec{e}E_{probe}\exp(-i\omega(t-\tau)) \tag{18}$$

where E_{probe} denotes the electric field envelope of the incident probe laser and τ is the time delay between the pump and the probe laser pulses. After traveling in the aligned molecules, the linearly polarized probe laser depolarized and became elliptical. The ellipticity was determined by the average of the field-induced dipole moment under the state vector $\Phi(t)$. Using a polarizer orthogonal to the probe field, the depolarization of the probe laser was measured. With the approximation of slowly varying envelope and small amplitude, the signal field was described by the wave equations [18]. After the integral over the state vector $\Phi(t)$, the signal field was:

$$E_s(\tau) = \frac{3\,l\Delta\alpha N\omega}{8c}(<\cos^2\theta>_\tau - \frac{1}{3})E_{probe}\exp(i\pi/2) \tag{19}$$

where l is the distance that the probe laser traveled in the aligned molecules, ω is the laser frequency, $\Delta\alpha=(\alpha_{\|} - \alpha_{\perp})$is the anisotropy of the molecular dynamical polarizability, N is the molecular number density, C is the speed of the light. It should be mentioned that there was a $\pi/2$ phase shift between the signal field $E_S(\tau)$ and the probe laser electric field E_{probe}. The aforementioned alignment signal is commonly measured homodyne signal. The field-induced birefringence is accessed by measuring the ellipticity of an initially linearly polarized laser field traveling through the aligned molecules.

When the probe laser polarization was a little off from the optic axis of the quarter wave plate ($\delta \sim 5°$), the linearly polarized probe laser became elliptical after the quarter wave plate.

$$\vec{E}_{probe}(\tau) = \vec{e}_X E_X \exp[-i\omega(t-\tau)] + \vec{e}_Y E_Y \exp[-i\omega(t-\tau) \pm i\pi/2] \tag{20}$$

There is a π/2 phase shift between E_X and E_Y. The sign of the phase shift is determined by the polarization direction of the linearly polarized laser relative to the main optical axis of the quarter wave plate. In addition to $E_S(\tau)$, a constant external electric field E_Y is also collected by the detector. The detection becomes heterodyne. The signal intensity is determined by:

$$\begin{aligned} I_{sig}(\tau) &= \eta\int_{\tau-T_d/2}^{\tau+T_d/2}\left|E_s(\tau)+E_Y\exp[-i\omega(t-\tau)\pm i\pi/2]\right|^2 dt \\ &= \eta\int_{\tau-T_d/2}^{\tau+T_d/2}\left|\frac{3\,l\,\omega\,\Delta\alpha\,N}{8c}(<\cos^2\theta>_\tau-\frac{1}{3})E_X\exp(-i\omega(t-\tau)+i\frac{\pi}{2})+E_Y\exp[-i\omega(t-\tau)\pm i\frac{\pi}{2}]\right|^2 dt \\ &\propto [(<\cos^2\theta>_\tau-\frac{1}{3})+C]^2 \end{aligned} \quad (21)$$

where T_d is the response time of the detector and much longer than the pulse width of the probe laser, η is the detection efficiency. The magnitude of the parameter

$$C=\frac{8\,c\,tg\delta}{3\,l\,\omega\,\Delta\alpha\,N}, \quad (22)$$

which denotes the contribution of the external electric field, is determined by the ellipticity

$$\varepsilon=\frac{E_Y}{E_X}=\left|tg\delta\right| \quad (23)$$

The sign of the parameter C, which denotes the polarity of the external electric field, is determined by the rotation direction of the elliptical polarized probe laser after the quarter wave plate.

The pure heterodyne signals are derived from the difference between the two heterodyne signals under the existence of an external electric field with opposite polarity and equal magnitude.

$$\begin{aligned} & I_{sig}^{positive}(\tau)-I_{sig}^{negative}(\tau) \\ & \propto\left\{[(<\cos^2\theta>_\tau-\frac{1}{3})+\left|C\right|]^2-[(<\cos^2\theta>_\tau-\frac{1}{3})-\left|C\right|]^2\right\} \\ & =4\left|C\right|(<\cos^2\theta>_\tau-\frac{1}{3}) \end{aligned} \quad (24)$$

The above equation clearly demonstrates that the alignment signal is proportional to ($<\cos^2\theta>-1/3$) for pure heterodyne detection.

2.2.2 Experimental setup

An 800 nm, 110 fs laser pulse was divided into two parts to provide a strong energy pump beam and a weak energy probe beam, both linearly polarized at 45° with respect to each other. An optical translational stage controlled by a stepping motor was placed on the pump beam path in order to precisely adjust the relative separation times between the two pulses. Both the pump beam and the probe beam were focused with a 30 cm focal length lens into a

20 cm long gas cell at a small angle. The gas cell was filled with different gases at room temperature under one atmospheric pressure. The field-free aligned molecules induced by the strong pump laser caused birefringence and depolarized the probe laser. The depolarization of the probe laser, which represents the alignment degree, was analyzed with a polarizer set at 90° with respect to its initial polarization direction. The alignment signals were detected by a photoelectric cell and transformed into a computer via a four-channel A/D converter for analysis.

The main modification was that a $\lambda/4$ wave plate was inserted on the probe laser path before the gas cell. Figure 1 also shows the relative directions of the laser polarizations, the optic axis of the quarter wave plate and the signal field. The optic axis of the quarter wave plate was along X direction, 45° with respect to the pump laser polarization. The signal electric field in Y direction was collected by a detector. When the probe laser polarization was along the optic axis of the quarter wave plate, this was the common used homodyne detection. When the probe laser polarization was a little off from the optic axis of the quarter wave plate ($\delta \sim 5°$), the linearly polarized probe laser became elliptical after the quarter wave plate. In addition to the transient birefringence caused by the aligned molecules, a constant external electric field is also collected by the detector. The detection becomes heterodyne. The pure heterodyne signals are derived from the difference between the two heterodyne signals under the existence of an external electric field with opposite polarity and equal magnitude.

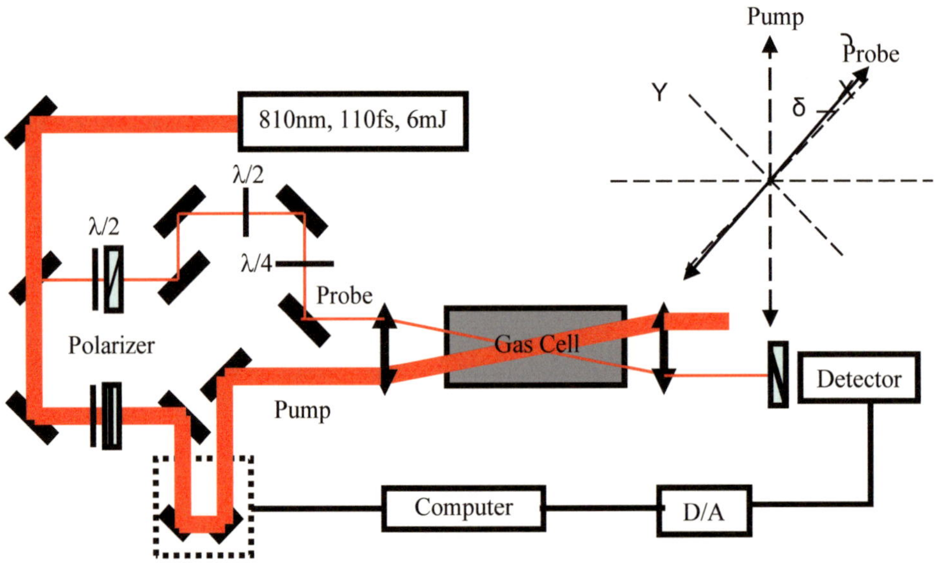

Fig. 6. Experimental setup for measuring field-free alignment of molecules induced by strong femtosecond laser pulses. The optic axis of the quarter wave plate was along X direction, 45° with respect to the pump laser polarization. The signal electric field in Y direction was collected by a detector.

2.2.3 Results and discussion

1. Field-free alignment

The calculated revival structures of N_2, O_2 and CO irradiated by 800 nm, 110 fs laser pulses at an intensity of 2×10^{13} W/cm^2 are shown in Figures 7a, 8a and 9a, respectively. The baseline value of $<\cos^2 \theta>$ is about 0.334, approximating an isotropic distribution of 1/3. The classical rotational period Tr of molecules is determined by the equation $Tr = 1/(2B_0\, c)$, where B0 is the rotational constant of the diatomic molecule in the vibrational ground state and c is the speed of the light. For N_2, O_2 and CO, B0 is 2.010, 1.4456 and 1.9772 cm^{-1}, respectively. The corresponding rotational period Tr is therefore 8.3 ps for N_2, 11.6 ps for O_2 and 8.5 ps for CO.

The alignment signal fully revives every molecular rotational period. There are also moments of strong alignment that occur at shorter intervals. However, the three molecules exhibit different behaviors at the quarter full revivals. The ratios of the alignment signal at quarter revivals to that at full revivals were nearly 1/3 for N_2, 1 for O_2 and 0 for CO. The large difference at quarter full revivals for N_2, O_2 and CO results from the different nuclear spin weights of the even and odd J states in the initial distribution. At quarter full revivals, the odd wave packet has maxima (minima), whereas the even wave packet has minima (maxima). For homonuclear diatomic molecules, the nuclear spin statistics control the relative weights between even and odd J states. In the case of N_2, the relative weights of the even and odd J are 2:1. As a result, the temporary localization of the even wave packet at quarter full revival was partially cancelled by its odd counterpart. Thus, the alignment signal at the quarter full revival was about 1/3 of that at the full revival for N_2. In the case of O_2, only odd J states were populated. Since only a single localized wave packet existed, the alignment signal at the quarter full revival was almost equal to that at the full revival for O_2. For the heteronuclear diatomic molecule CO, the opposite localizations of the even and the odd wave packets would cancel each other. Therefore, no net alignment would be observed at the time of the quarter full revival.

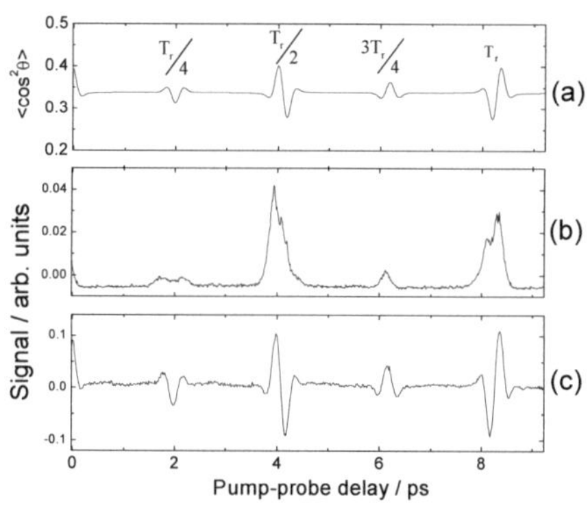

Fig. 7. Revival structure of N_2 irradiated by 800 nm, 110 fs at an intensity of 2×10^{13} W/cm^2 (a) numerical calculation, (b) homodyne signal, (c) pure heterodyne signal.

Figures 7b, 8b and 9b display the homodyne signal versus the pump-probe delay for N_2, O_2 and CO irradiated by 800 nm, 110 fs laser pulses at an intensity of 2×10^{13} W/cm^2. The signal was proportional to $(<\cos^2 \theta> -1/3)^2$. Each peak denotes the alignment moment with the molecular axis parallel to the pump laser polarization direction ($<\cos^2 \theta> > 1/3$) or perpendicular to the pump laser polarization direction ($<\cos^2 \theta> < 1/3$). For the intervals between the alignments, the angular distribution of the molecules was isotropic relative to the laser polarization direction ($<\cos^2 \theta> = 1/3$). Although the homodyne signal clearly determined the moment that the alignment occurred, it could not indicate whether the aligned molecules were parallel or perpendicular to the laser polarization direction.

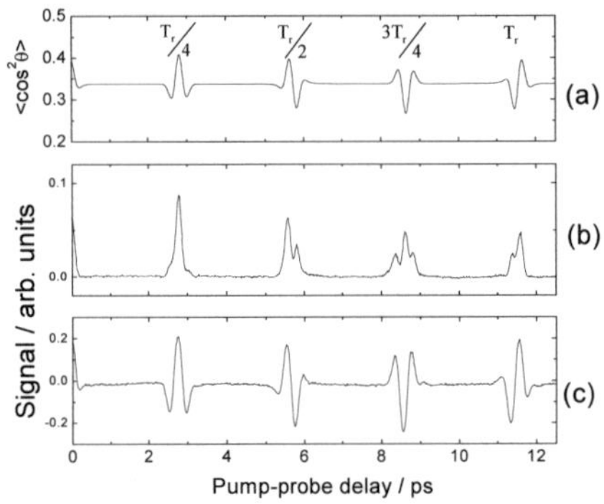

Fig. 8. Revival structure of O_2 irradiated by 800 nm, 110 fs at an intensity of 2×10^{13} W/cm^2 (a) Numerical calculation, (b) homodyne signal, (c) pure heterodyne signal.

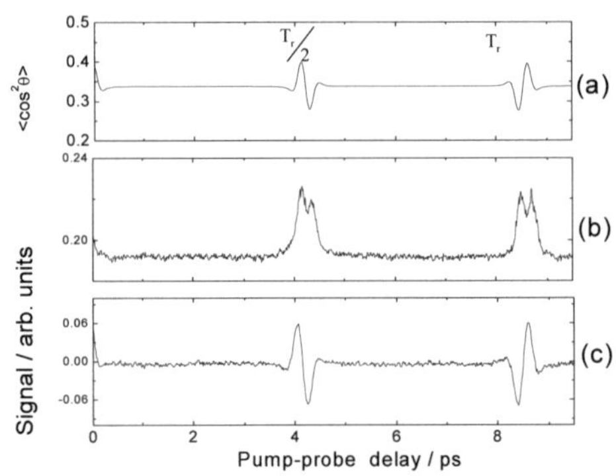

Fig. 9. Revival structure of CO irradiated by 800 nm, 110 fs at an intensity of 2×10^{13} W/cm^2 (a) Numerical calculation, (b) homodyne signal, (c) pure heterodyne signal.

Figures 7c, 8c and 9c display the pure heterodyne signal versus the pump-probe delay for N_2, O_2 and CO irradiated by 800 nm, 110 fs laser pulses at an intensity of 2×10^{13} W/cm^2. The signal was proportional to ($\langle\cos^2\theta\rangle$-1/3). Comparing with the numerical calculated alignment parameter $\langle\cos^2\theta\rangle$, there is only a baseline (~1/3) shift. Thus, the heterodyne signal directly reproduced the revival structure of molecules under the field-free condition.

2. Fourier transforms of the time-dependent alignment signals

The Fourier transform spectrum of the time-dependent alignment parameter $\langle\cos^2\theta\rangle$ signal contains a serial of beat frequencies $\Delta\omega$ between adjacent J states, which are given by:

$$\Delta\omega_{J,J+2} = \frac{E_{J+2} - E_J}{\hbar} = (4J+6)\omega_0 \tag{25}$$

where $\omega_0=2\pi B_0 c$ is the fundamental phase frequency. The amplitudes of the beat frequencies are proportional to the products of the expanding coefficients. These coefficients denoted the populations of the different |J> states in the rotational wave packet.

Figs. 10a, 11a and 12a show the Fourier transform spectra of the calculated $<\cos^2\theta>$ in Figs. 7a, 8a and 9a. In the present study, the beat frequency $\Delta\omega$ is directly replaced by the rotational quantum number J and all the Fourier transforms of the alignment signals span three full periods of the molecules. Each spectrum describes the revival structure decomposing into different |J> states. There is a ~2:1 intensity alternation between even J and odd J states for N_2, but there are only odd J states for O_2. The difference of the relative weights between even and odd J states resulted in different alignment signals at quarter revivals for these molecules.

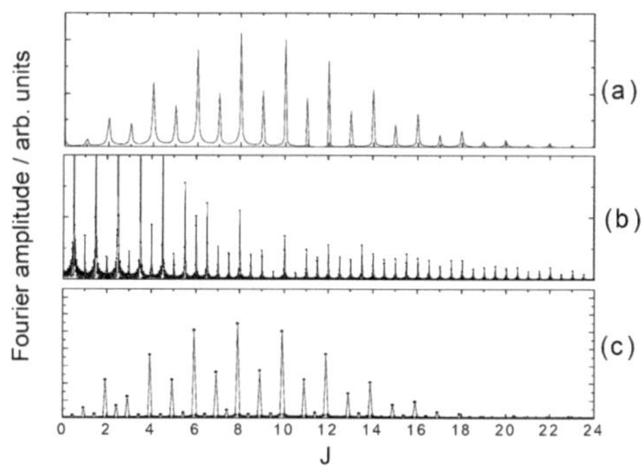

Fig. 10. Fourier transforms of the revival structure of N_2 shown in Figure 7.

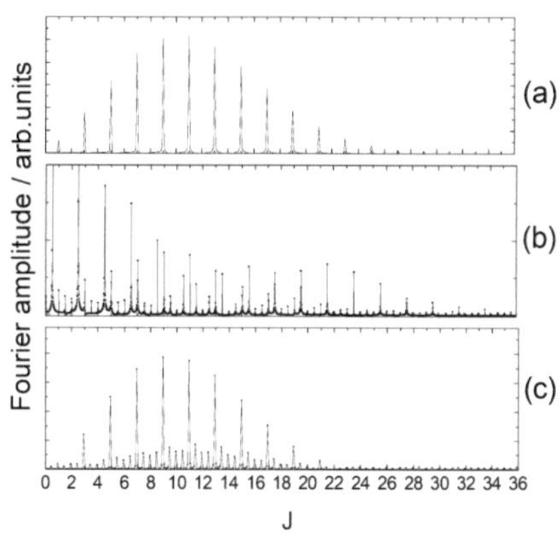

Fig. 11. Fourier transforms of the revival structure of O_2 shown in Figure 8.

Figs. 10b, 11b and 12b show the Fourier transform spectra of the homodyne signals in Figs. 7b, 8b and 9b. The beat frequencies were more than the fundamental frequencies. They also include the sum and the difference frequencies. The front progression is the difference frequencies, the middle progression is the fundamental frequencies and the end is the sum frequencies. However, the Fourier amplitudes of the fundamental frequencies were minor for the Fourier transform of the homodyne signal, even though they reflected the populations of different J states in the rotational wave packet.

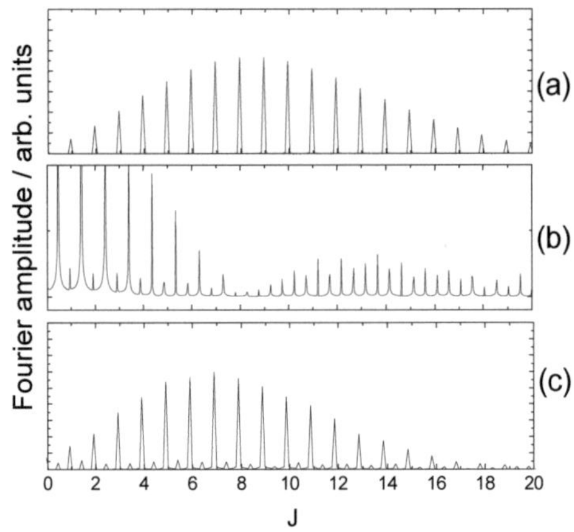

Fig. 12. Fourier transforms of the revival structure of CO shown in Figure 9.

Figs. 10c, 11c and 12c show the Fourier transform spectra of the pure heterodyne signals in Figs. 7c, 8c and 9c. In comparison with the contribution from the complicated beat frequencies in the homodyne signal, the contribution from the fundamental frequencies dominated in the Fourier transform spectrum of the pure heterodyne signal. The Fourier transform spectrum of the heterodyne signal was very similar to that of the calculated alignment parameter $<\cos^2 \theta>$, which reflected the actual populations of different J states in the rotational wave packet.

2.3 Detection of gas component using molecular alignment

Due to the lower molecular density, field-free alignment of gas sample is more obvious than liquid and solid, which could be used in rapid detection of gas component. The experimental results bellows also demonstrated that gas mixture of N_2 and O_2 present a mixed alignment structure in which N_2 and O_2 present their own alignment structure. This result shows a feasibility of rapid detection of gas component.

Figure 13 shows the alignment signal for (a) N_2, (b) O_2, and (c) air at room temperature under one atmosphere pressure. As we know, air mainly contains N_2 and O_2. It is clearly

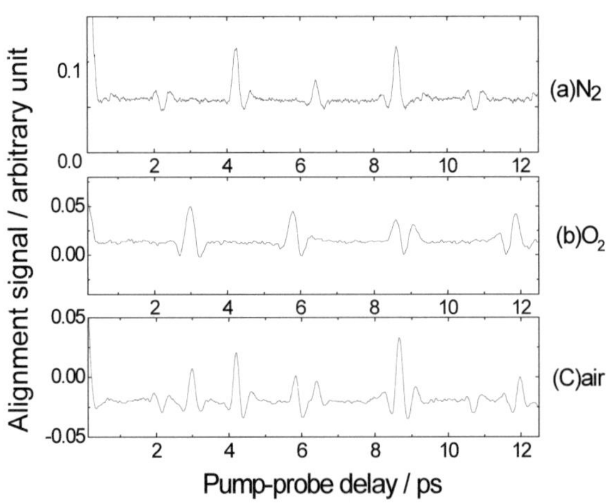

Fig. 13. Alignment signal for (a) N_2, (b) O_2, and (c) air.

seen that alignment structure of gas sample can be derived from alignment signal of pure N_2 and O_2. It is possible that one can identify the component from gas mixture rapidly if the alignment structure of pure component is obtained. Precision of this alignment detection method just depends on the value of polarization anisotropy for different molecules.This technique has two weaknesses. First, if the gas molecule is spherical, which means it has no polarization anisotropy; there will be no alignment signal. Second, if different molecules have same rotational periods, it is also hard to distinguish each molecule during the mixed alignment structure. Although this technique is not perfect, it can be used to detect different gas component easily and rapidly.

3. Conclusion

In summary, we have realized field-free alignment of N_2, O_2, CO, CO_2, CS_2, and C_2H_4 at room temperature using strong femtosecond laser pulses. We also demonstrated that the degree of alignment could be greatly improved by using two-pulse scheme with appropriate separated time. These researches indicate that multiple-pulse alignment is a feasible approach to obtain macroscopic ensembles of highly aligned molecules in the laboratory. We believe our results will promote the practical applications of field-free aligned molecules.

We modified the typical weak field polarization technique. The homodyne detection and the heterodyne detection were realized in an apparatus. They were utilized to quantify the field-free alignments of diatomic molecules N_2, O_2 and CO irradiated by strong femtosecond laser pulses. The alignment signal is proportional to $(\langle\cos^2\theta\rangle - 1/3)^2$ for homodyne detection and $(\langle\cos^2\theta\rangle - 1/3)$ for pure heterodyne detection. Fourier transform spectra of the homodyne signal and the pure heterodyne signal were also studied. Comparing with the homodyne detection, the pure heterodyne detection had the following advantages. First, the pure heterodyne signal directly reproduced the alignment parameter $\langle\cos^2\theta\rangle$ except a 1/3 baseline shift. Second, the Fourier transform spectrum of the pure heterodyne signal was very similar to that of the calculated alignment parameter $\langle\cos^2\theta\rangle$ and reflected the actual populations of different *J* states in the rotational wave packet.

Different gas samples present different alignment structure. We also demonstrated that N_2 and O_2 component can be identified by measuring alignment structure of an air sample. This result will promote the applications of femtosecond laser in gas component detection and other fields.

4. References

T. Seideman, *Time-resolved photoelectron angular distributions as a means of studying polyatomic nonadiabatic dynamics*, J. Chem. Phys., 2000. 113: p. 1677 .

T. Seideman, *Time-resolved photoelectron angular distributions as a probe of coupled polyatomic dynamics*, Phys. Rev. A, 2001. 64: p. 042504.

T. P. Rakitzis, A. J. van den Brom, and M. H. M. Janssen, *Directional dynamics in. photodissociation of oriented molecules*, Science, 2004. 303: p. 852.

V. Aquilanti, M. Bartolomei, F. Pirani, D. Cappelletti, Vecchiocattivi, Y. Shimizu, and T. Kasai, *Orienting and aligning molecules for stereochemistry and photodynamics*, Phys. Chem. Chem. Phys., 2005.7: p. 291.

N. Mankoc-Borstnik, L. Fonda, and B. Borstnik, *Coherent rotational states and their creation and time evolution in molecular and nuclear systems*, Phys. Rev. A, 1987. 35: p. 4132.

R. A. Bartels, T. C. Weinacht, N. Wagner, M. Baertschy, C. H. Greene, M. M. Murnane, and H. C. Kapteyn, *Phase Modulation of Ultrashort Light Pulses using Molecular Rotational Wave Packets*, Phys. Rev. Lett., 2002. 88: p. 013903.

S. Ramakrishna and T. Seideman, *Intense Laser Alignment in Dissipative Media as a Route to Solvent Dynamics*, Phys. Rev. Lett., 2005. 95: p. 113001.

K. F. Lee, D. M. Villeneuve, P. B. Corkum, and E. A. Shapiro, *Phase Control of Rotational Wave Packets and Quantum Information*, Phys. Rev. Lett., 2004. 93: p. 233601.

J. Itatani, J. Levesque, D. Zeidler, H. Niikura, H. Pepin, J. C. Kieffer, P. B. Corkum, and D. M. Villeneuve, *Tomographic imaging of molecular orbitals*, Nature (London) 2004. 432: p. 867.

J. G. Underwood, B. J. Sussman, A. Stolow, *Field-Free Three Dimensional Molecular Axis Alignment*, Phys. Rev. Lett., 2005. 94: p. 143002.

Hertz E, Daems D, Guerin S, et al., Field-free molecular alignment induced by elliptically polarized laser pulses: Noninvasive three-dimensional characterization, Phys. Rev. A, 2007. 76: p. 043423.

Lee KF, Villeneuve DM, Corkum PB, et al. *Field-free three-dimensional alignment of polyatomic molecules*, Phys. Rev. Lett., 2006. 97: p. 173001.

R. de Nalda, C. Horn, M. Wollenhaupt, M. Krug, L. Banares and T. Baumert, *Pulse shaping control of alignment dynamics in* N_2, J. Raman Spectrosc., 2006. 38 (5): p. 543.

Christer Z. Bisgaard, Mikael D. Poulsen, Emmanuel Pe´ronne, Simon S. Viftrup, and Henrik Stapelfeldt, *Observation of Enhanced Field-Free Molecular Alignment by Two Laser Pulses*, Phys. Rev. Lett., 2004. 92: p. 173004.

P.W. Dooley, I.V. Litvinyuk, K.F. Lee, D.M. Rayner, M. Spanner, D.M. Villeneuve, P.B. Corkum, *Direct imaging of rotational wave-packet dynamics of diatomic molecules*, Phys. Rev. A, 2003. 68: p. 023406.

V. Renard, M. Renard, S. Guerin, Y.T Pashayan, B. Lavorel, O. Faucher, H.R. Jauslin, *Postpulse Molecular Alignment Measured by a Weak Field Polarization Technique*, Phys. Rev. Lett., 2003. 90: p. 153601.

A. Rouzee, V. Renard, B. Lavorel, O. Faucher, *Laser spatial profile effects in measurements of impulsive molecular alignment*, J. Phys. B, 2005. 38 : p. 2329.

V. Renard, O.Faucher, B. Lavorel, *Measurement of laser-induced alignment of molecules by cross defocusing*, Opt. Lett., 2005. 30: p. 70.

S.Zamith, Z. Ansari, F. Lepine, M.J.J. Vrakking, *Measurement of laser-induced alignment of molecules by cross defocusing*, Opt. Lett., 2005. 30: p. 2326.

R.W. Boyd. *Nonlinear Optics*. Academic Press, California USA, 1992.

W.H. Press et al., *Numerical Recipes*, 2nd ed. Cambridge University Press, Cambridge, England, 1992.

C.Z. Bisgaard, M.D. Poulsen, E. Peronne, S.S. Viftrup, H. Stapelfeldt, *Observation of Enhanced Field-Free Molecular Alignment by Two Laser Pulses*, Phys. Rev. Lett., 2004. 92: p. 173004.

G. Herzberg, *Molecular spectra and molecular structure*, Van Nostrand Reinhold Company Ltd., 1966.

S.Ramakrishna,T. Seideman, *Dissipative dynamics of laser induced nonadiabatic molecular alignment*, J. Chem. Phys., 2006. 124: p. 034101.

Rouzée, A.; Guérin, S.; Boudon, V.; Lavorel, B.; Faucher, O. *Field-Free One-Dimensional Alignment of Ethylene Molecule*. Phys. Rev. A 2006, 73, 033418.

P.B. Corkum, C. Ellert, M. Mehendale, P. Dietrich, S. Hankin, S. Aseyev, D. Rayner, D. Villeneuve, Faraday Discuss., 113, 47 (1999)

Part 4

Other Applications

13

Deconvolution of Long-Pulse Lidar Profiles

Ljuan L. Gurdev, Tanja N. Dreischuh and Dimitar V. Stoyanov
Institute of Electronics, Bulgarian Academy of Sciences 72, Tzarigradsko shosse, Sofia Bulgaria

1. Introduction

Active remote-sensing methods and instruments such as microwave radars, optical radars (lidars), and acoustical radars (sodars, sonars) have widely been used for in-depth or surface probing of atmosphere, ocean and earth (Doviak & Zrnic, 1984; Measures, 1984; Kovalev & Eichinger, 2004; Van Trees, 2001; Marzano & Visconti, 2002). The recent active sensing methods are based mainly on the so-called lidar (LIght Detection And Ranging) or Time-Of-Flight (TOF) principle (Measures, 1984; Kovalev & Eichinger, 2004). This principle consists in the detection of backscattering-due radiative returns (at angle π) from the probed media after irradiating them by penetrating narrow-beam pulsed radiation. Then, the return signal profile detected in the time domain contains range-resolved information about the radiation-matter interaction (absorption and scattering) processes and the related material characteristics along the line of sight (LOS). The range-resolution scale (along the LOS) is determined by the (larger of the) characteristic pulse response length and the sampling interval $\Delta z_0 = c\Delta t_0/2$ of the lidar system, and by the noise level and bandwidth (Gurdev et al., 1998, 1993); Δt_0 is the sampling interval (the digitizing step) in the time domain. The value of Δz_0 is usually assumed to be less than the least variation scale of the investigated extinction and backscattering inhomogeneities.

Thus, a "hardware" way of improving the accuracy and resolution of lidars is to use as fast as possible analog-to-digital converters (ADC) and as short as possible sensing laser pulses. Consequently, the realization of the hardware approach depends on the development of the electronic and laser technologies and is connected with overcoming different technological difficulties. For instance, shortening the laser pulses is often connected with lowering the pulse energy or increasing the (peak) pulse power. Then, in the former case one should amplify the shortened pulses while in the latter case the pulse power should be restricted to avoid nonlinear disturbance of the investigated (sensed) medium. Let us also note that in coherent heterodyne lidars the sensing pulse length should be above a threshold determined by the required resolution of measuring the Doppler velocity and the wavelength of the sensing radiation (Hannon & Thomson, 1994). The only way of improving the range resolution in this case is to use shorter laser pulses of proportionally shorter radiation wavelength. As another example one may consider GRAYDAR (Gamma RAY Detection

And Ranging) (Gurdev et al., 2007a, 2007b; Dreischuh et al., 2007) where, because of the absence of short-pulse gamma ray lasers, the δ–pulse sensing procedure is based on the use of electron-positron annihilation-due gamma-photon pairs.
At the same time, there exist some "software" approaches to improving the resolution and accuracy of the lidars. One of them consists in the use of deconvolution techniques (algorithms) for recovering the short-pulse lidar profiles on the basis of the measured long-pulse lidar profiles and known sensing pulse shape (Gurdev et al., 1993, 1998; Dreischuh et al., 1995, 1996; Stoyanov et al., 1996; Park et al., 1997; Bahrampour & Askari, 2006). Specific approaches have also been developed to improving the resolution of coherent heterodyne pulsed Doppler lidars (Gurdev et al., 2001, 2002, 2003, 2008a). Mention as well an original and effective approach to achieving lidar-signal sampling intervals shorter than the data acquisition step based on the random delay of the sensing laser shots with respect to the ADC start pulses (Stoyanov et al., 2004, 2007, 2010).
The purpose of the present chapter is to give a brief review of the works and to generalize the results obtained there about the advantages and limitations of some above-mentioned software approaches to improving the resolution and the accuracy of different TOF-based (lidar type) sensing methods. The first circle of problems considered is devoted mainly to deconvolution techniques for improving the resolution of long-pulse elastic lidars for sensing the atmosphere. The features are marked of Fourier and Volterra deconvolution algorithms at different levels and types of the measurement noise, and different types of uncertainties of the sensing laser pulses. The well-defined pulses of special concrete shape obtained by pulse-shaping are also of interest because they allow the design of special effective deconvolution algorithms. Here we also briefly describe a double-sided linear-strategy variant of lidar-type optical tomography. The following topic of interest concerns a novel (center-of-mass wavelength) Thomson scattering lidar method for measuring electron temperature profiles in thermonuclear plasmas (Gurdev et al., 2008b; Dreischuh et al., 2009) as well as some recent results about the Fourier-deconvolution due improvement of the sensing accuracy and resolution in this case. The concluding part of the chapter contains a brief discussion of the investigations described and the results obtained as well as of the importance of the software approaches to improving the lidar sensing accuracy and resolution.

2. Lidar equations

Let us consider a material object irradiated by penetrating quasi-monochromatic narrow-beam pulsed radiation of wavelength λ_i (Fig.1). The direct detection of the backscattering-due radiative return transforms it into an electrical signal (return signal) F measured as a function of the time delay t after the instant of pulse emission. In this way a temporal return signal profile $F(t)$ is obtained. At practically constant speed of propagation c of the sensing radiation and single-scattering conditions, there exists one-to-one correspondence $t\equiv 2z/c$ ($z\equiv ct/2$) between the time t and the LOS distance z to the sensing-pulse front that is in fact the front of the scattering volume contributing to the signal at this time (Measures, 1984; Kovalev & Eichinger, 2004). Then, one can write that $F=F(t=2z/c)=F(z=ct/2)$, which is an expression of the basic feature of the lidar (or TOF) principle. That is, the return signal profile in the time domain contains range-resolved information about the radiation-matter interaction (absorption and scattering) processes and the related material characteristics along the LOS.

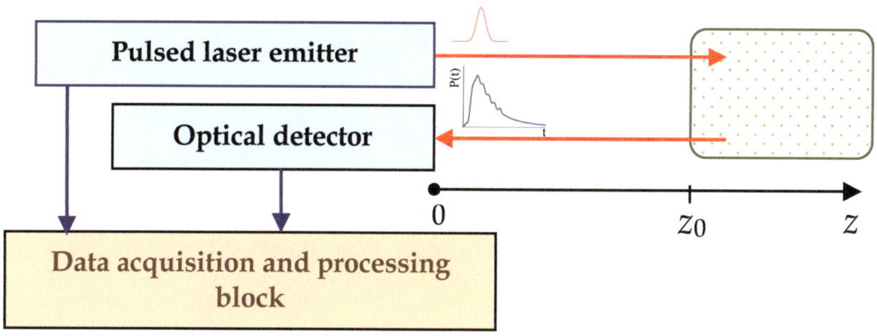

Fig. 1. Illustration of the lidar principle.

In the general case of inelastic scattering and presence of broadening effects, the lidar return will be frequency shifted and spectrally broadened. Then, the detected return power $P_l(\lambda_{s1},\lambda_{s2};z=ct/2)$ within a wavelength interval $[\lambda_{s1},\lambda_{s2}]$ is given by the following most general lidar equation (e.g. Measures, 1984; Gurdev et al., 2008b, 1998):

$$P_l(\lambda_{s1},\lambda_{s2};z) = AE_0(\lambda_i)\int_{\lambda_{s1}}^{\lambda_{s2}} d\lambda_s K(\lambda_i,\lambda_s)\int_0^z dz' f[2(z-z')/c]\Phi(\lambda_i,\lambda_s;z'), \quad (1)$$

where A is the lidar receiving aperture area, $E_0(\lambda_i)$ is the incident (sensing) pulse energy, $K(\lambda_i, \lambda_s)$ is a characteristic of the transceiving spectral transparency and sensitivity of the lidar, $f(\theta)$ is the effective pulse response function of the lidar system, θ is time variable,

$$\Phi(\lambda_i,\lambda_s;z) = \eta(\lambda_i,\lambda_s;z)\beta(\lambda_i;z)L(\lambda_i,\lambda_s;z)T(\lambda_i,\lambda_s;z)/z^2 , \quad (2)$$

η is receiving efficiency of the lidar, λ_i and λ_s are wavelengths of the incident and the backscattered radiation, respectively, β is the volume backscattering coefficient, $L(\lambda_i,\lambda_s;z)$ is the spectral contour of the scattered radiation,

$$T(\lambda_i,\lambda_s;z) = \exp\left\{-\int_0^z [\alpha_t(\lambda_i,z') + \alpha_t(\lambda_s,z')]dz'\right\} \quad (3)$$

is the two-way transparency of the investigated medium (from $z'=0$ to $z'=z$), and $\alpha_t(\lambda_i, z')$ and $\alpha_t(\lambda_s, z')$ are respectively the forward and backward extinction coefficients.

When the system response length [concerning $f(\theta)$] is less than the least variation scale of the properties of the medium, Eq.(1) is reduced to the following (short-pulse, δ-pulse, or maximum-resolved, Gurdev et al., 1993) lidar equation:

$$P_s(\lambda_{s1},\lambda_{s2};z) = \frac{cA}{2}E_0(\lambda_i)\int_{\lambda_{s1}}^{\lambda_{s2}} d\lambda_s K(\lambda_i,\lambda_s)\Phi(\lambda_i,\lambda_s;z). \quad (4)$$

At last, in the case of a single line shape $L(\lambda_s)$ that is essentially narrower than the dependence of K on λ_s, instead of the long-pulse and short-pulse Eqs.(1) and (4), respectively, we obtain

$$P_l(\lambda_{sc};z) = AE_0(\lambda_i)K(\lambda_i,\lambda_{sc})\int_0^z dz' f[2(z-z')/c]\Phi(\lambda_i,\lambda_{sc};z') \quad (5)$$

and

$$P_s(\lambda_{sc};z)=\frac{cA}{2}E_0(\lambda_i)K(\lambda_i,\lambda_{sc})\Phi(\lambda_i,\lambda_{sc};z)\,, \qquad (6)$$

where λ_{sc} is the central wavelength of $L(\lambda_s)$ and

$$\Phi(\lambda_i,\lambda_{sc};z)=\eta(\lambda_i,\lambda_{sc};z)\beta(\lambda_i;z)T(\lambda_i,\lambda_{sc};z)/z^2\,. \qquad (7)$$

In case of elastic scattering, $\lambda_{sc}=\lambda_i$. Let us also note that the effective pulse response function of the lidar, $f(\theta)$, is a convolution

$$f(\theta)=\int_{-\infty}^{\infty}d\theta' q(\theta-\theta')s(\theta') \qquad (8)$$

of the receiving-system (including the ADC unit) pulse response $q(\theta)$ ($\int_0^{\infty}q(\theta)d\theta=1$) and the sensing-pulse shape $s(\theta)=P_p(\theta)/E_0$, where $P_p(\theta)$ is the pulse power shape.
The above-described lidar equations are basic instruments for quantitative analysis of data obtained by direct-detection lidars. They are adaptable to photon-counting mode of detection by using the formal substitutions:

$$P_l\rightarrow N_l,\;\; P_s\rightarrow N_s,\;\; E_0\rightarrow N_0,\;\; L(\lambda_s)\rightarrow L(\lambda_s)\,\lambda_s/\lambda_i\;, \qquad (9)$$

where N_l and N_s are photon counting rates, and N_0 is the number of photons in the incident laser pulse.

3. Deconvolution techniques for improving the resolution of long-pulse direct-detection elastic lidars

In the case of elastic, e.g., aerosol or Rayleigh scattering in the atmosphere, the lidar return is characterized by too small spectral broadening and is described in general by Eq.(5) at $\lambda_{sc}=\lambda_i$. Instead of Eq.(5), it is convenient to write

$$P_l(z)=(2/c)\int_{\eta}^{\mu}dz' f[2(z-z')/c]P_s(z') \qquad (10)$$

For pulse response functions $f(\theta)$ with asymptotically decreasing tails, the integration limits in Eq.(10) may be retained the same as in Eq.(5), that is, η=0 and μ=z. At the same time, one may choose to write η=-∞ and μ=∞ because the functions $P_l(z)$, $P_s(z)$ and $f(\theta=2z/c)$ are supposed defined and integrable over the interval (-∞,∞). The finite integration limits η=0 and μ=z indicate only the points where the integrand becomes identical to zero. When the response function is restricted, say rectangular, with duration τ, the integration limits are $\eta=z-c\tau/2$ and μ=z. In any case, the software approach to improving the lidar resolution consists in solving the integral equation (10) with respect to the maximum-resolved lidar profile $P_s(z)$ at measured long-pulse profile $P_l(z)$ and measured or estimated system response shape $f(\theta)$.
With η = -∞ and μ =∞, Eq.(10) represents $P_l(z)$ as convolution of $P_s(z)$ and $f(\theta=2z/c)$. Then, the solution with respect to $P_s(z)$ is obtainable in principle by Fourier deconvolution, but attentive noise analysis should be performed and noise-suppressing techniques should be used to ensure satisfactory recovery accuracy. When the spectral density $I_f(\omega)$ of $f(\theta)$ has

zeros or is considerably narrower than the spectral density $I_n(\omega)$ of the noise (see below), the Fourier deconvolution becomes impracticable and Eq.(10), with η=0 and μ=z, could be considered and solved as the first kind of Volterra integral equation with respect to $P_s(z)$. The retrieval of $P_s(z)$ for some special, e.g., rectangular, rectangular-like or exponentially-shaped response functions can also be performed analytically at relatively low and controllable noise influence.

Eq.(10) can naturally be given in a discrete form based on sampling the signal and the lidar response function. Then, the solution with respect to $P_s(z)$ is obtainable by using matrix formulation of the problem (Park et al., 1997). Other deconvolution techniques such as Fourier-based regularized deconvolution, wavelet-vaguelette deconvolution and wavelet denoising, and Fourier-wavelet regularized deconvolution can also be effective in this case (Bahrampour & Askari, 2006; Johnstone et al., 2004). A retrieval of the maximum-resolved lidar profile with improved accuracy and resolution is achievable as well using iterative deconvolution procedures (Stoyanov et al., 2000; Refaat et al., 2008). Note by the way that the applied problems concerning deconvolution give rise to a powerful development of the mathematical theory of deconvolution (e.g., Pensky and Sapatinas, 2009, 2010).

Below we shall describe an extended, more complete analysis, in comparison with our former works, of the above-mentioned general (Fourier and Volterra) and special (for concrete response functions) deconvolution approaches. The fact will be taken into account that the signal-induced (say Poisson or shot) noise or the background-due noise is smoothed by the lidar response function. Let us first consider some features of the Fourier-deconvolution procedure. Suppose in general that the noise N accompanying the signal $P_s(z)$ consists of two components, N_1 and N_2, where N_1 is induced by the signal itself, and N_2 is a stationary background independent of the signal. Then the measured lidar profile to be processed is

$$P_{lm}(z) = P_l(z) + (2/c)\int_{-\infty}^{\infty} dz'\{f[2(z-z')/c]N_1(z') + q[2(z-z')/c]N_2(z')\}\ . \qquad (11)$$

The Fourier deconvolution based on Eq.(10), with $P_{lm}(z)$ [Eq.(11)] instead of $P_l(z)$, is straightforward and leads to the following expression of the restored profile $P_{sr}(z)$:

$$P_{sr}(z) = (2\pi)^{-1}\int_{-\infty}^{\infty} \tilde{P}_s(k)\exp(-jkz)dk + \varepsilon(z) = (2\pi)^{-1}\int_{-\infty}^{\infty} [\tilde{P}_l(k)/\tilde{f}(\omega)]\exp(-jkz)dk + \varepsilon(z)\,, \qquad (12)$$

where ω=$ck/2$, j is imaginery unity, t=$2z/c$,

$$\tilde{P}_l(k) = \int_{-\infty}^{\infty} P_l(z)\exp(jkz)dz\,,\ \tilde{f}(\omega) = \int_{-\infty}^{\infty} f(t)\exp(j\omega t)dt\,,\ \text{and}\ \tilde{P}_s(k) = \int_{-\infty}^{\infty} P_s(z)\exp(jkz)dz \qquad (13)$$

are respectively Fourier transforms of $P_l(z)$, $f(t)$, and $P_s(z)$, and

$$\varepsilon(z) = N_1(z) + (2\pi)^{-1}\int_{-\infty}^{\infty} [\tilde{N}_2(k)\tilde{s}(\omega)]\exp(-jkz)dk \qquad (14)$$

is a formally written realization of the random error due to the noise;

$$\tilde{N}_2(k) = \int_{-z_l}^{z_l} N_2(z)\exp(jkz)dz\ ,\quad \tilde{s}(\omega) = \int_{-\infty}^{\infty} s(t)\exp(j\omega t)dt\ , \qquad (15)$$

and $[-z_l,z_l]$ is the real integration interval instead of $[-\infty,\infty]$ supposed to be sufficiently large that $P_s(z)$ is fully restored to some characteristic distance $z_c<z_l$ for which $P_s(z_c)$ practically vanishes. Assuming that the correlation radius r_{c2} of $N_2(z)$ is much smaller than z_l and using Eqs.(14) and (15), we obtain (in the limit $z_l\to\infty$) the following expression for the error variance:

$$D\varepsilon(z) = \left\langle \varepsilon^2(z) \right\rangle = D_{N_1}(z) + (2\pi)^{-1}\int_{-\infty}^{\infty}[I_{N_2}(k) / I_s(\omega)]dk \,, \tag{16}$$

where, respectively, $I_s(\omega) = |s(\omega)|^2$ and $I_{N_2}(k) = \lim_{z_l\to\infty} D_{N_2}\int_{-2z_l}^{2z_l} K_{N_2}(\zeta)\exp(jk\zeta)d\zeta$ are spectral densities of $s(t)$ and $N_2(z)$, and $D_{N_1}(z) =< N_1^2(z) >$ and $D_{N_2} =< N_2^2(z) >$ are variances of $N_1(z)$ and $N_2(z)$; $K_{N_2}(\zeta) = \langle N_2(z)N_2(z+\zeta)\rangle / D_{N_2}$ is the correlation coefficient of $N_2(z)$, and <.> denotes an ensemble average. According to Eq.(16), when the noise spectrum $I_{N_2}(k)$ is wider than $I_s(\omega = ck/2)$, the variance $D\varepsilon$ would have infinite value. Consequently, some type of low-pass filtering is always necessary for decreasing the noise influence, retaining an improved retrieval resolution.

When the measured long-pulse lidar profile $P_{lm}(z)$ is smoothed by a low-pass filter $\varphi(z-z')$ with spectral characteristic $\tilde{\varphi}(k) = \int_{-\infty}^{\infty}\varphi(z)\exp(jkz)dz$, Eqs.(12), (14), and (16) retain their forms, where only the following substitutions should be introduced

$$\tilde{P}_l(k) \to \tilde{P}_l(k)\tilde{\varphi}(k)\,;\; N_1(z) \to (2\pi)^{-1}\int_{-\infty}^{\infty}\tilde{N}_1(k)\tilde{\varphi}(k)\exp(-jkz)dk\,;\; \tilde{N}_2(k) \to \tilde{N}_2(k)\tilde{\varphi}(k)\,;$$

$$I_{N_2}(k) \to I_{N_2}(k)|\tilde{\varphi}(k)|^2\,;\; D_{N_1}(z) \to (2\pi)^{-1}\int I_{N_1}(k,z)|\tilde{\varphi}(k)|^2 dk\,; \tag{17}$$

where

$$\tilde{N}_1(k) = \int_{-z_l}^{z_l} N_1(z)\exp(jkz)dz \,, \tag{18a}$$

and $N_1(z)$ is assumed to be statistically quasihomogeneous random function (Rytov, 1976) such that its local spectral density and covariance are, respectively,

$$I_{N_1}(k,z) = \lim_{z_l\to\infty}\int_{-2z_l}^{2z_l} Cov(\rho,z)\exp(jk\rho)d\rho \,, \tag{18b}$$

$$Cov(\rho,z) = \langle N_1(z+\rho/2)N_1(z-\rho/2)\rangle \,. \tag{18c}$$

An improved retrieval resolution may be achieved as well with increasing the computing step $\Delta z=c\Delta t/2$, whose least value $\Delta z_0=c\Delta t_0/2$ is the sampling interval. The finite-computing-step systematic (bias) error depends, in general, on the value of Δz and on the shape of $P_s(z)$ (Gurdev et al., 1993). Naturally, for a lower value of Δz and a smoother shape of $P_s(z)$, the bias error is smaller. In the absence of noise, at short-enough computing step a high accuracy in the restoration of $P_s(z)$ is achievable.

To estimate the effect of a finite computing step on the value of $D\varepsilon$, Eq.(16) should be rewritten as

$$D\varepsilon(z) = D_{N_1}(z) + (2\pi)^{-1}\int_{-\pi/\Delta z}^{\pi/\Delta z}[I_{N_2}(k) / I_s(\omega)]dk \,. \tag{19}$$

According to Eq.(19), when Δz increases above r_{c2}, the effect of the noise decreases because of narrowing its spectral band. When the spectrum $I_{N_2}(k)$ is narrow compared with $I_s(\omega = ck/2)$, i.e., when r_{c2} exceeds the pulse length, from Eq.(19) the lower limit is obtained, $D\varepsilon_{\min} = D_{N_1}(z) + D_{N_2}$, of the variance $D\varepsilon$.

The Fourier-deconvolution systematic retrieval error due to uncertainties in the pulse response function $f(\theta)$ is investigated in depth and detail in Dreischuh et al., 1995. It is shown that various, deterministic or random uncertainties give rise to two main effects on the retrieval accuracy. First, depending on the sign of the uncertainty, an elevation or lowering takes place of the smooth component of the lidar profile. This shift up or down is proportional to the smooth component and to the ratio of the uncertainty area to the true pulse area. The smooth uncertainties affect the whole lidar profile in the same way. The fast varying high-frequency uncertainties lead in addition to amplitude and phase distortions of the small-scale high-frequency structure of the lidar profile. Extremely sharp characteristic-spike cuts and fast-varying alternating-sign (deterministic or random) uncertainties lead to small retrieval errors because of their small areas. The results from investigating the influence of the pulse response uncertainties on the retrieval error allow one to estimate the order and the character of the possible recovery distortions and to choose ways to reduce or prevent them. For instance, in the case of a spike-cut uncertainty in the laser pulse shape, the use of a suitable approximation, instead of the unknown true spike spectrum, leads to effective error reduction (Stoyanov et al., 1996).

In the cases when the Fourier deconvolution becomes impracticable, when for instance the spectrum $I_{N_2}(k)$ is much wider than $I_s(\omega = ck/2)$ or $I_s(\omega)$ has zero spectral components, Eq.(10) can be considered in the form

$$P_l(z) = (2/c)\int_0^z dz' f[2(z-z')/c] P_s(z') \ , \tag{20}$$

which is the first kind of Volterra integral equation. By the substitution $t'=2z'/c$ ($t=2z/c$), and with double differentiation assuming that $f(0)=0$, we obtain

$$P_s(ct/2) = \Im(t) + \int_0^t K(t-t') P_s(ct'/2) dt' , \tag{21}$$

where $\Im(t) = P_l^{II}(t = 2z/c) / f^I(0)$, $K(t-t') = -f^{II}(t-t') / f^I(0)$, $f^I(0) = f^I(t-t')|_{t'=t}$, and the symbols such as $\varphi^J(y)$ (J = I,II,...) denote the J th derivative of the function φ with respect to y. Eq.(21) is the second kind of Volterra integral equation with respect to $P_s(ct/2=z)$, which has a unique continuous solution within the interval $[t_0, t]$ ($[z_0, z]$, respectively), when $\Im(t)$ is a continuous function within the same interval and the kernel $K(t - t')$ is a continuous or square-summable function of t and t' over some rectangle $\{t_0 \le t, t' \le \theta\}$. The solution of Eq.(21) is obtainable in the form

$$P_s(ct/2) = \Im(t) + \int_0^t R(\xi) \Im(t-\xi) d\xi , \tag{22}$$

where the substitution $t'=t-\xi$ is used meanwhile. Here $R(\xi) = \sum_{i=1}^{\infty} K_i(\xi)$ is the resolvent, $K_i(\xi) = \int_0^{\xi} K_{i-1}(\zeta) K_1(\xi - \zeta) d\zeta$, and $K_1(\xi) = K(\xi)$. The bias error $\delta(z=ct/2)=P_{sc}(z=ct/2)$-

$P_s(z=ct/2)$ caused by the finite calculation step Δt is obtainable by using Eq.(22), provided that the resolvent R is known almost without error as if it is calculated with a computing step much less than Δt. The result is that

$$\delta(z=ct/2)=-(2/30)\Delta t^4[P_s^{IV}(t)-\Im^{I}(t_0)R^{II}(t-t_0)-\Im^{II}(t_0)R^{I}(t-t_0)-\Im^{III}(t_0)R(t-t_0)]\,.(23)$$

$P_{sc}(z = ct/2)$ is the numerically restored profile in the absence of noise.
The noise influence on the retrieval accuracy can be estimated taking into account the fact that the noise N_1 is convolved with the overall lidar response function $f(\theta)$, while the noise N_2 is convolved with the receiving system response function $q(\theta)$. Assume that the durations of $f(\theta)$ and $q(\theta)$ are respectively τ_f and τ_q. They are in practice the correlation times of the effective additive noises obtained by the convolution of N_1 and N_2 [see Eq. (11)]. Following the approach employed in Gurdev et al., 1993, the variance $D\varepsilon(z)=<\varepsilon^2(z)>$ of the random error $\varepsilon(z)$ is estimated as

$$D\varepsilon(z)\sim[f^{I}(0)]^{-2}[D_{N_1}(z)\tau_{c1}/\tau_f^5+D_{N_2}\tau_{c2}/\tau_q^5], \tag{24a}$$

where $\tau_{c1,2}$ (assumed here $<<\tau_f,\tau_q$) are the correlation times of N_1 and N_2, respectively. Because of the real discrete calculation procedure the computing step Δt plays in fact the role of minimum correlation time with respect to N_1 and N_2 and their convolutions with the corresponding response functions [Eq. (11)]. In this case, when $\tau_{f,q} <\Delta t$

$$D\varepsilon(z)\sim[f^{I}(0)]^{-2}[D_{N_1}(z)+D_{N_2}](\Delta t)^{-4}\ . \tag{24b}$$

In the opposite case, when $\tau_{c1,2}>>\tau_f,\tau_q>\Delta t$, it is obtained that

$$D\varepsilon(z)\sim[f^{I}(0)]^{-2}[D_{N_1}(z)/\tau_{c1}^4+D_{N_2}/\tau_{c2}^4]\ . \tag{24c}$$

According to Eqs.24a-c, as in the case of Fourier deconvolution, a fast fluctuating broadband noise leads to higher statistical deconvolution error compared to a slowly fluctuating narrowband noise whose effect is lowered by the deconvolution.
The sensing laser pulse shape conditions entirely the processes of convolution and deconvolution when its duration $\tau_s>>\tau_q$. Such is for instance the case of atmospheric lidars, where the receiving system response time τ_q is substantially less than the laser pulse duration τ_s and practically $f(\theta)\equiv s(\theta)$. There are some types of laser pulse shapes in this case that lead to simple, accurate and fast deconvolution algorithms permitting one by suitable scanning to investigate in real time the fine spatial structure of atmosphere or other objects penetrated by the sensing radiation. Such pulses are the so-called rectangular, rectangular-like, and exponentially-shaped pulses to which it is impossible or difficult to apply Fourier or Volterra deconvolution techniques. The contemporary progress in the pulse shaping art would allow one to obtain various desirable laser pulse shapes.
In the case of rectangular laser pulses with duration τ, when $f(\theta)=\tau^{-1}$ for $\theta\in[0,\tau]$ and $f(\theta)=0$ for $\theta\notin[0,\tau]$, Eq.(10) acquires the form

$$P_l(z)=(2/c\tau)\int_{z-c\tau/2}^{z}dz'P_s(z')\ . \tag{25}$$

The differentiation of Eq.(25) leads to the relation

$$P_s(z) = (c\tau / 2)P_l^I(z) + P_s(z - c\tau / 2) \ , \tag{26}$$

that is,

$$P_s(z) = (c\tau / 2)\sum_{i=1}^{Q} P_l^I(z - ic\tau / 2) + P_s(z - (Q+1)c\tau / 2) \ , \tag{27}$$

where Q is the integer part of $t/\tau=2z/c\tau$. The distortion $\delta(z=ct/2)$ caused by a finite computing step $\Delta z=c\Delta t/2$ is estimated on the basis of Eq.(26) as

$$\delta(z) = -(1/30)(\Delta z)^4 P_s^{\mathrm{IV}}(z) \ . \tag{28}$$

On the basis of Eqs.(11) and (27), the variance $D\varepsilon(z)=<\varepsilon^2(z)>$ of the random rectangular-pulse deconvolution error $\varepsilon(z)$ is estimated as

$$D\varepsilon(z) \sim \tau^2(Q+1)[D_{N_1}(z)\tau_{c1} / \tau_f^3 + D_{N_2}\tau_{c2} / \tau_q^3] , \tag{29a}$$

when $\tau_{c1,2} << \tau_{f,q}$, and

$$D\varepsilon(z) \sim \tau^2(Q+1)[D_{N_1}(z)\tau_{c1}^{-2} + D_{N_2}\tau_{c2}^{-2}] , \tag{29b}$$

when $\tau_{c1,2} >> \tau_{f,q}$; $\tau_f \equiv \tau$. When $\tau_{f,q} < \Delta t$, instead of (29a) we have

$$D\varepsilon(z) \sim \tau^2(Q+1)[D_{N_1}(z) + D_{N_2}](\Delta t)^{-2} . \tag{29c}$$

So it is seen that the essential random errors are due in fact to the broadband noise such that $\tau_{c1,2}<<\tau_{f,q}<\Delta t$. Also, because of the recurrent character of the algorithm the statistical retrieval error is accumulated with z so that its variance $D\varepsilon(z)$ is proportional to the number of recurrence cycles Q.

A rectangular-like pulse shape $f(\theta)$ with rise and decay time τ_r and duration τ is given by the expression

$$f(\vartheta) = \begin{cases} 0 & \text{for } \vartheta < 0 \\ \tau^{-1}[1 - \exp(-\vartheta / \tau_r)] & \text{for } \vartheta \in [0, \tau] \\ \tau^{-1}[1\text{-}\exp(\text{-}\tau / \tau_r)]\exp[-(\vartheta - \tau) / \tau_r] & \text{for } \vartheta \geq \tau \end{cases} \ . \tag{30}$$

Such a shape has zero spectral components. Therefore, the Fourier deconvolution algorithm is not applicable in this case. The Volterra-deconvolution algorithm also leads to some problems. Nevertheless, the following recurrence deconvolution algorithm has been derived (Dreischuh et al., 1996; Gurdev et al., 1998):

$$P_s(z) = (c\tau / 2)[P_l^I(z) + (c\tau_r / 2)P_l^{II}(z)] + P_s(z - c\tau / 2) \ . \tag{31}$$

The deconvolution error $\delta(z)$ caused by the discrete data processing is obtained in the form

$$\delta(z) = -(1/30)\{(\Delta z)^4 P_s^{\mathrm{IV}}(z) + \sum_{i=0}^{Q}[2(c\tau / 2)(c\tau_r / 2)(\Delta z)^4]P_l^{\mathrm{VI}}(z - ic\tau / 2) \ . \tag{32}$$

In the case of broadband noise N with correlation times $\tau_{c1,2} < \tau_f, \tau_q$ ($\tau_f = \tau$), the random error variance $D\varepsilon$ is estimated to be

$$D\varepsilon(z) \sim (Q+1)[D_{N_1}(z)(\tau_{c1}/\tau_f)(1+\tau_r^2/\tau_f^2) + D_{N_2}(\tau_f^2\tau_{c2}/\tau_q^3)(1+\tau_r^2/\tau_q^2)]. \quad (33)$$

If in addition $\tau_{f,r,q}<\Delta t$, instead of the estimate (33) we obtain

$$D\varepsilon(z) \sim (Q+1)[D_{N_1}(z) + D_{N_2}][1+\tau_r^2/(\Delta t)^2]. \quad (34)$$

The simplest exponentially-shaped pulses have the following shape:

$$S(\vartheta) = \begin{cases} 0 & \text{for } \vartheta < 0 \\ (\vartheta/\tau^2)\exp(-\vartheta/\tau) & \text{for } \vartheta \geq 0 \end{cases}. \quad (35)$$

Although the Fourier and Volterra deconvolution algorithms are applicable in this case, we have obtained another simpler and faster algorithm (Gurdev et al., 1996), namely

$$P_s(z) = P_l(z) + c\tau P_l^{I}(z) + (c\tau/2)^2 P_l^{II}(z) \;. \quad (36)$$

The calculation error and the variance of the error due to the noise for $\tau_{c1,2} << \tau_{f,r,q}$ are evaluated as follows:

$$\delta(z) = -(c\tau/30)(\Delta z)^4[P_l^{\mathrm{V}}(z) + (c\tau/2)P_l^{\mathrm{VI}}(z)] \quad (37)$$

and

$$D\varepsilon(z) \sim (\tau_{c1}/\tau_f)(1+4\tau^2/\tau_f^2 + \tau^4/\tau_f^4)D_{N_1}(z) + (\tau_{c2}/\tau_q)(1+4\tau^2/\tau_q^2 + \tau^4/\tau_q^4)D_{N_2}. \quad (38)$$

For $\tau_{f,r,q}<\Delta t$, instead of (38) we have

$$D\varepsilon(z) \sim [D_{N_1}(z) + D_{N_2}][1+4\tau^2/(\Delta t)^2 + \tau^4/(\Delta t)^4] \;. \quad (39)$$

The restoration of the short-pulse lidar profile $P_s(z)$ allows one not only to improve the accuracy and the resolution of the lidar sensing but to develop methods as well for linear-strategy optical tomography of translucent scattering objects. For this purpose, one should measure, in combination with a lateral scan, the backscattering signal profile and the pulse energy passing through the object along each current line of sight at both the mutually opposite directions of sensing as it is shown in Fig.2.

In this way, the spatial distribution of the backscattering and extinction coefficients within the objects can be determined (Gurdev et al., 1998). Indeed, the forward illumination short-pulse lidar equation can be written in the form [see Eqs.(6) and (7)]

$$S(z) = S_1(z) = E_{01}\beta(z)\exp[-2\int_{z_1}^{z}\alpha_t(z')dz'] \;, \quad (40)$$

where E_{01} is the forward propagating sensing-pulse energy, $S(z)=S_1(z)=2P_{S1}(z)z^2/[cAK\eta(z)]$ is the so-called lidar S-function, $P_{S1}(z)$ is the lidar profile, and z_1 is the longitudinal coordinate (along the LOS) of the entrance of the sensing pulse/beam into the object. The final coordinate z_2 of the beam axis through the object is in fact the coordinate of the entrance into

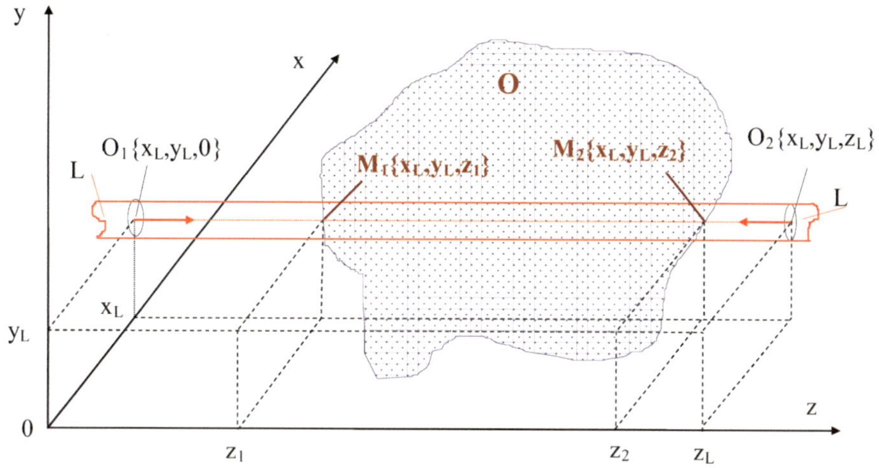

Fig. 2. Illustration of the backscattering and extinction coefficient reconstruction approach based on lidar principle. A right-handed rectangular coordinate system {0xyz} is used to determine uniquely the coordinates of the points within the investigated object O, the positions ($O_1\{x_L,y_L,0\}$ and $O_2\{x_L,y_L,z_L\}$) and orientations (**O_1O_2** and **O_2O_1**) of the lidar transceiver system L, the sensing-radiation path of propagation (the line of sight, $\overline{O_1O_2}$), and the coordinates $M_1\{x_L,y_L,z_1\}$ and $M_2\{x_L,y_L,z_2\}$ of the initial and the final scattering volumes, respectively, along the LOS. The object O is irradiated from two reciprocally opposite directions along each LOS chosen here to be parallel to axis 0z.

the object of the backward propagating (along **O_2O_1** direction) sensing pulse. The backward sensing *S*-function $S_2(z)=2P_{S2}(z_L-z)\,(z_L-z)^2/[cAK\eta(z_L-z)]$ is described by the equation

$$S_2(z) = E_{02}\beta(z)\exp[-2\int_z^{z_2}\alpha_t(z')dz'] \ , \tag{41}$$

where E_{02} and $P_{S2}(z)$ are the corresponding sensing-pulse energy and lidar profile, and z_L is the new longitudinal coordinate of the transceiver lidar system (Fig.2). On the basis of Eqs.(40) and (41) it is not difficult to obtain that

$$\beta(z) = [S_1(z)S_2(z)/(E_{t1}E_{t2})]^{1/2} \ , \tag{42}$$

and

$$\alpha_t(z) = 0.25\{\ln[S_2(z)/S_1(z)]\}' \ , \tag{43}$$

where the corresponding lidar profiles $P_{S1}(z)$ and $P_{S2}(z)$ (in S_1 and S_2) and transmitted pulse energies $E_{t1} = E_{01}\exp[-\int_{z_1}^{z_2}\alpha_t(z)dz]$ and $E_{t2} = E_{02}\exp[-\int_{z_1}^{z_2}\alpha_t(z)dz]$ are to be measured experimentally; the prime in Eq.(43) denotes first derivative with respect to *z*.
The noise-induced random errors $\delta_\beta(z)$ and $\delta_\alpha(z)$ in the determination of $\beta(z)$ and $\alpha_t(z)$, respectively, are estimated (Gurdev et al., 1998) as follows:

$$\delta_\beta(z) = <[\beta_m(z)-\beta(z)]^2>^{1/2}/\beta(z) \sim \{0.25[\delta^2 P_{s1}(z) + \delta^2 P_{s2}(z)] + \delta_E^2\}^{1/2} \tag{44}$$

and

$$\delta_\alpha(z) = \left\langle [\alpha_{tm}(z) - \alpha_t(z)]^2 \right\rangle^{1/2} < 0.25[(D\varepsilon)^{1/2}/\rho][P_{s1}^{-2}(z) + P_{s2}^{-2}(z)]^{1/2}\{1 + \rho[r_1^{-2}(z) + r_2^{-2}(z)]\}^{1/2}, \quad (45)$$

where $\beta_m(z)$ and $\alpha_{tm}(z)$ are the backscattering and extinction profiles, respectively, calculated on the basis of the experimental data, $\beta(z)$ and $\alpha_t(z)$ are the corresponding true profiles, $\delta^2 P_{s1,2}(z) = D\varepsilon_{1,2}(z)/P^2_{s1,2}(z)$ are the relative variances of the random errors ε_1 and ε_2 in the determination of P_{s1} and P_{s2}, $\delta_E^2 = <(E_{tm}-E_t)^2>/E_t^2$ is the relative variance of the transmitted pulse energy with measured value E_{tm} and true value E_t, $D\varepsilon(z)=\max\{D\varepsilon_{1,2}(z)\}$, ρ is an estimate of the correlation radius of the random functions $\varepsilon_{1,2}(z)$, and $r_{1,2}(z)=|P_{s1,2}(z)/P'_{s1,2}(z)|$. When ρ is smaller than the computing step Δz, one should replace it by Δz in Eq.(45). According to Eqs.(44) and (45), the higher the signal-to-noise ratio (the smaller $\delta P_{s1,2}$ and δ_E) the smaller the random errors δ_β and δ_α. In addition, δ_α depends on the spectral properties of the noise (ρ) in combination with the signal variability ($r_{1,2}$).

The efficiency of the deconvolution techniques discussed in this section and their performance are tested and confirmed by detailed computer simulations. Some of the models employed and results obtained are illustrated in Figs.3-5. The sampling interval Δt_0 is assumed to be equal to 0.1 μs corresponding to Δz_0= 15m. Models of a maximum-resolved lidar profile $P_s(z)$ and the corresponding detected lidar return $P_l(z)$ [see Eq.(10)] in the case of pulse response function $f(\theta)$ given in the inset are shown in Fig.3. As can be seen, $P_s(z)$ consists of some mean profile, a high-resolution component in the near field, and a double-peak structure introducing discontinuities at a further range. The system response function $f(\theta)$ is chosen to have a shape close to this of the typical TEA-CO_2 laser pulses. It consists of an initial spike followed by a long tail. As a result of the effect of convolution, important information about the small-scale variations of the backscattering within the long-resolution cell (about 200-300 m) is lost in the registered long-pulse profile $P_l(z)$. In the absence of noise the deconvolution procedures ensure accurate retrieval of the short-pulse profile $P_s(z)$. Then the restored profiles $P_{sc}(z)$ do not differ visibly from the original model $P_s(z)$. As it is shown in Gurdev et al., 1993, the systematic errors due to discrete data processing can be of the order of or smaller than 1% on the average. The random noise influence on the retrieval accuracy is simulated assuming that $\tau_{c1,2} << \tau_{f,q}$, $\tau_q << \tau_f$ and even $\tau_q < \Delta t_0$ as it is in the atmospheric lidars. In this case, at comparable noise levels N_1 and N_2, the influence of the stationary background component N_2 will be dominating [see Eqs.(11), (17), (24a), (29a), (33), and (38)]. Therefore, we have simulated a stationary effective additive noise n corresponding to the convolution of N_2 and the receiving system response q. The correlation time τ_c of the noise n is of the order of τ_q and may be both larger and smaller than Δt_0. In the latter case we have in practice a white noise with restricted frequency band ($\omega < \pi/\Delta t_0$) due to sampling. The effective correlation time of such a noise is equal to Δt_0. In the simulations we have generated white noise ($\tau_c \sim \Delta t_0$) and Gaussian-correlation noise ($\tau_c > \Delta t_0$). The noise level is specified by the (signal-to-noise, SNR) ratio of the minimum of the double-peak structure of $P_s(z)$ (see Fig.3) to the standard deviation of the noise n.

In Fig.4, the original short-pulse profile $P_s(z)$ is compared with the profiles $P_{sr}(z)$ restored by using Fourier deconvolution in the presence of white noise with SNR=50. As seen in Fig.4a, the deconvolution leads to an increase of the noise influence and the error magnitude considerably exceeds the oscillation amplitude of the retrieved profile. So, some type of controllable low-pass filtering is necessary, retaining at the same time an improved retrieval resolution. In

Fig.4b such a filtering is realized by increasing the computing step up to $\Delta t=4\Delta t_0$. The results from filtering the measured lidar profiles $P_{lm}(z)$ by a smooth monotonic low-pass filter with $4\Delta t_0$-wide window are shown in Fig.4c. As seen, both types of processing lead to similar restored profiles with considerable reduction of the noise effect [see Eq.(19)].

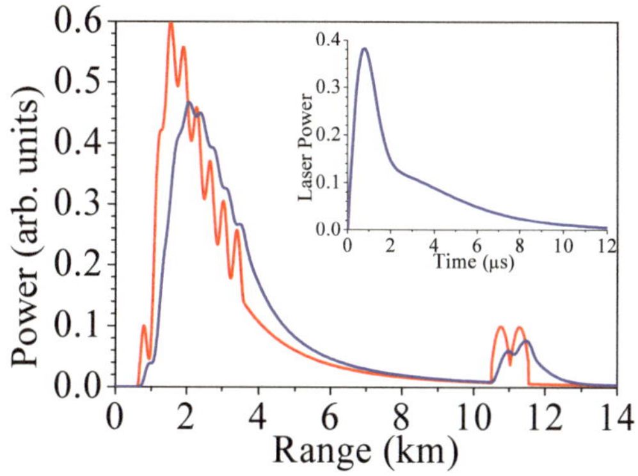

Fig. 3. Short-pulse lidar profile $P_s(z)$ (red) and the corresponding detected lidar return $P_l(z)$ (blue) obtained for the pulse response shape $f(\theta)$ (inset).

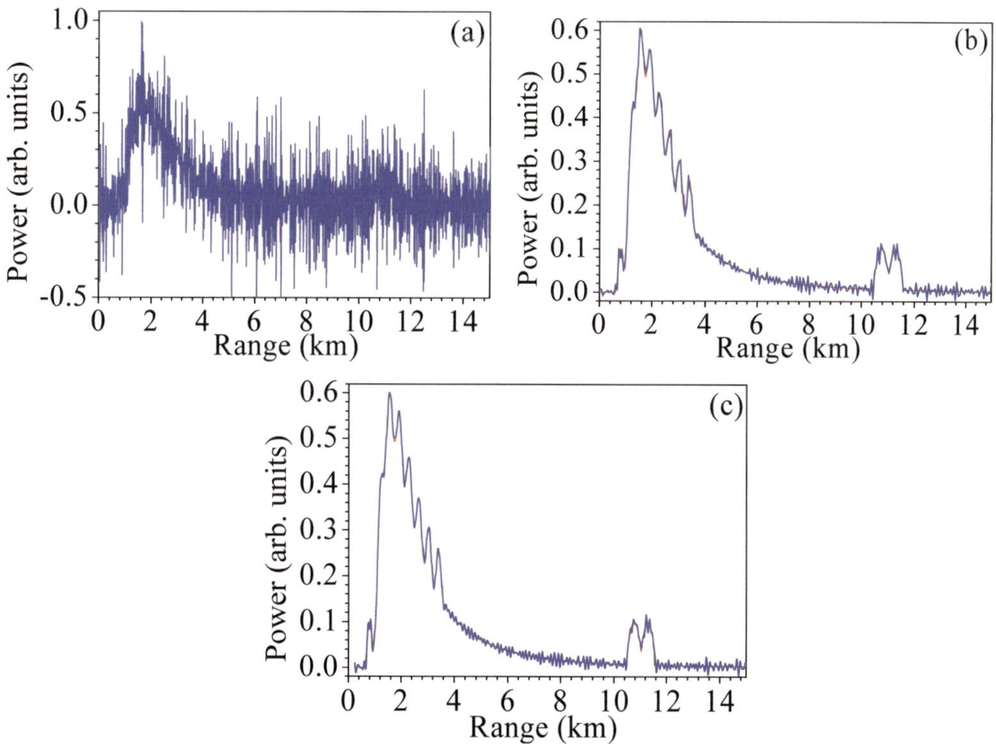

Fig. 4. Profile $P_s(z)$ (red) and the profile restored by use of Fourier deconvolution (blue), in the presence of white Gaussian-distributed noise with SNR=50, at $\Delta t=\Delta t_0$ (a), $\Delta t=4\Delta t_0$ (b), and when using a smooth monotonic filter with a $4\Delta t_0$-wide window applied to the measured lidar profile (c).

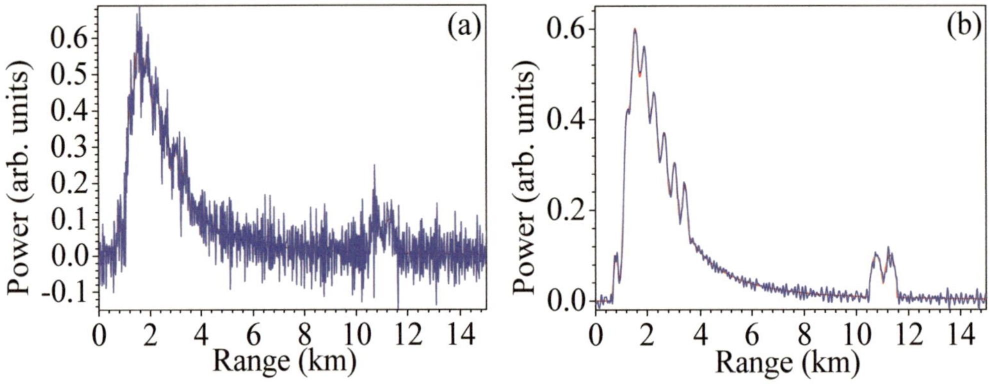

Fig. 5. Profile $P_s(z)$ (red) and the profile restored by use of Fourier deconvolution (blue) in the presence of additive Gaussian correlated and distributed noise with SNR=50 and correlation time $\tau_c=2\Delta t_0$ (a) and $5\Delta t_0$ (b).

The effect of the correlated noise with $\tau_c>\Delta t_0$ (i.e., $\tau_c\sim\tau_q>\Delta t_0$) is gradually lower than that of the white noise [see Eq.(19)]. It is illustrated in Fig.5 where the profiles $P_{sr}(z)$ are shown restored by Fourier deconvolution in the presence of correlated Gaussian noise with $\tau_c=2\Delta t_0$ and $5\Delta t_0$ and SNR=50. As expected, the error magnitude decreases with increasing the correlation time of the noise and at $\tau_c=5\Delta t_0$ the accuracy of the deconvolved lidar profiles is satisfactory even without any filtering applied.

The efficiency of the Fourier deconvolution approach is demonstrated as well in Stoyanov et al., 1996, where data (backscattering power profiles) have been processed, obtained by the National Oceanic and Atmospheric Administration (NOAA) pulsed coherent CO_2 Doppler lidar.

In Fig.6, the profile $P_l(z)$ is shown obtained by convolution of $P_s(z)$ with a rectangular-like sensing laser pulse with $\tau=2$ μs and $\tau_r=0.1$ μs. The recovered by algorithm (31) profiles $P_{sr}(z)$ in the presence of white noise at SNR=50 are represented in Fig.7. As it is seen, the noise influence is strong if no filtering is employed (Fig.7a). At the same time, increasing the computing step [Eq.(34)] up to $\Delta t=4\Delta t_0$ (Fig.7b) or filtering $P_{lm}(z)$ using a smooth monotonic low-pass filter with $4\Delta t_0$-wide window (Fig.7c) lead to comparable substantial reduction of the noise effect at minimum distortion of $P_{sr}(z)$ with respect to $P_s(z)$. The intrinsic noise

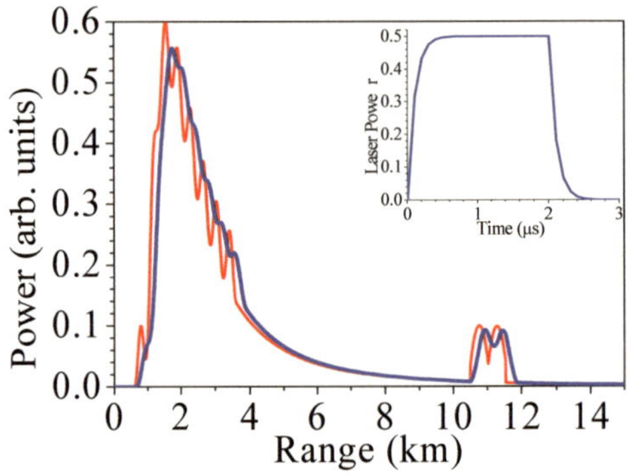

Fig. 6. Short-pulse lidar profile $P_s(z)$ (red) and the corresponding detected lidar return $P_l(z)$ (blue) obtained for the rectangular-like pulse response shape $f(\theta)$ given in the inset.

accumulation with the range is also noticeable. In Fig.8 it is shown that the effect of a correlated noise (with $\tau_c \sim \tau_q > \Delta t_0$) on the retrieval accuracy is considerably lower compared to the effect of white noise. In agreement with the theoretical results [Eq.(33)], the retrieval error decreases with increasing the correlation time of the noise. At $\tau_c = 5\Delta t_0$ the accuracy of the restored profiles is quite acceptable without any filtering performed.

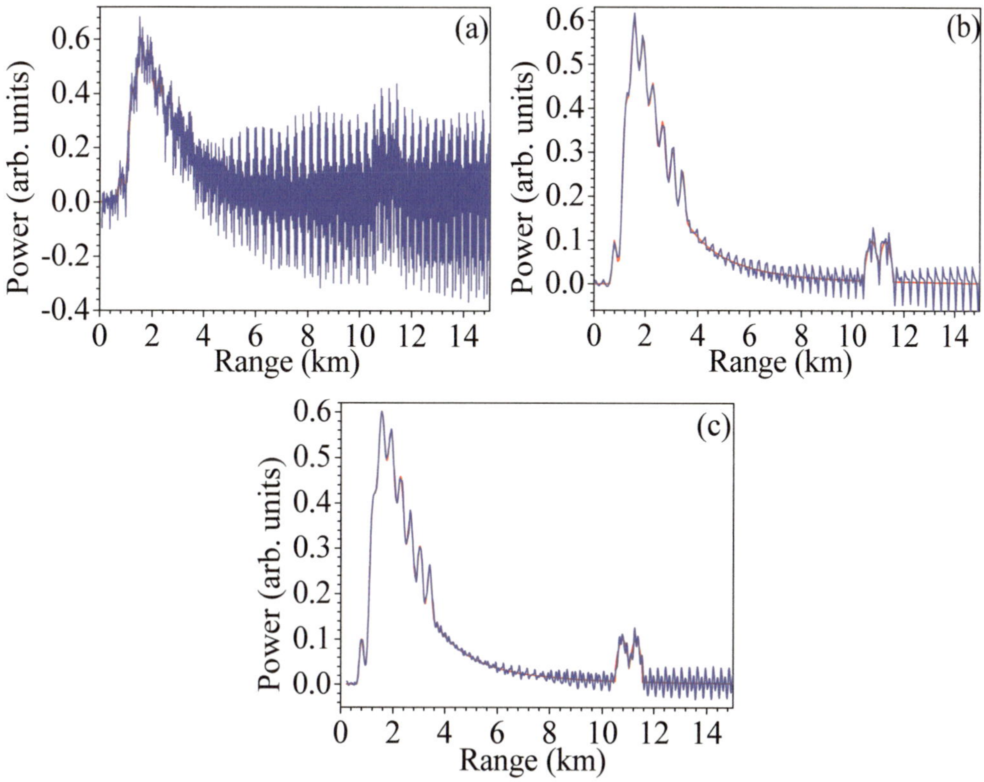

Fig. 7. Profile $P_s(z)$ (red) and the profile restored by use of Fourier deconvolution (blue), in the presence of white Gaussian-distributed noise with SNR=50, at $\Delta t = \Delta t_0$ (a), $\Delta t = 4\Delta t_0$ (b), and when using a smooth monotonic filter with a $4\Delta t_0$-wide window applied to the measured lidar profile (c).

The investigations described in this section show that deconvolution techniques can be successfully used for improving the accuracy and resolution of sensing the atmosphere or other objects by long-pulse elastic direct-detection lidars. At negligibly weak noise a high accuracy in the restoration of the short-pulse lidar profile is achievable at short-enough computing step. Also, the uncertainties in the lidar pulse response function lead to some characteristic retrieval distortions that can be reduced to some extent by using suitable approaches. Even at high initial SNR, a broadband noise, i.e., fast fluctuations with correlation time below the sensing-pulse duration, can cause considerable noise effect such that the retrieved short-pulse lidar profile is fully disguised. In this case, the noise influence can be effectively reduced by using appropriate filtering or choice of the computing step. The filter window or the computing step should exceed the fluctuation correlation time. At

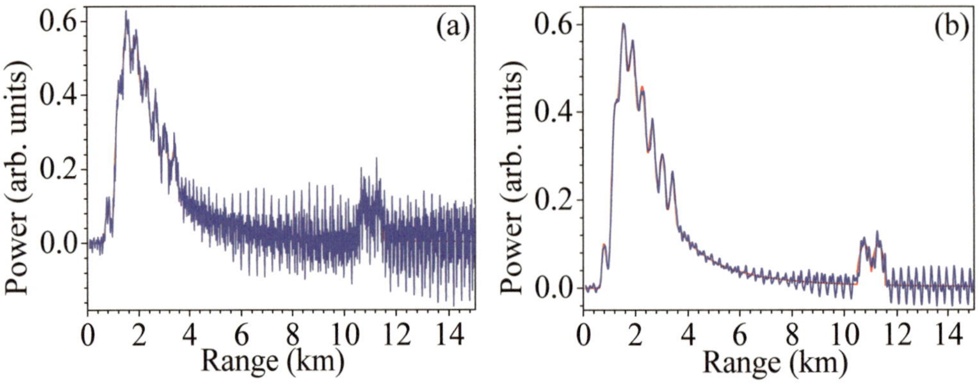

Fig. 8. Profile $P_s(z)$ (red) and the profile restored by use of Fourier deconvolution (blue) in the presence of additive Gaussian correlated and distributed noise with SNR=50 and correlation time $\tau_c=2\Delta t_0$ (a) and $5\Delta t_0$ (b).

the same time, they should be smaller than the least variation scale of the short-pulse lidar profile to avoid essential distortions and lowering of the retrieval resolution. Note as well that the deconvolution algorithm performance decreases the effect of narrow-band noise whose correlation time substantially exceeds the pulse duration. At last, let us mention one more virtue of the deconvolution-based retrieval of the short-pulse lidar profiles. That is, it allows high-resolution sensing of small finite-size objects by longer laser pulses, realizing in this way double-sided linear-strategy optical tomography of such objects.

4. Deconvolution-based improvement of the accuracy of measuring electron temperature profiles in tokamak plasmas by Thomson scattering lidar

The electron temperature T_e and density n_e distributions in the torus are basic characteristics of the tokamak fusion plasma. They are conditioned by the modes of heating and confinement of the high-temperature plasma as well as by the different oscillatory movements of the plasma particles sometimes leading to the appearance of crucial instabilities. Thus, the T_e and n_e profiles are not only important factors of the development and the efficiency of the fusion process but indicators as well of the dynamic plasma state. So far, the most appropriate approach to their simultaneous express determination in a remote contactless way is the Thomson scattering (TS) lidar approach (Salzmann et al., 1988; Kempenaars et al., 2008, 2010). It allows one to obtain the T_e and n_e profiles along a LOS through the torus core. The minimum range resolution interval achievable by the contemporary core TS lidars (Kempenaars et al., 2010) is about 12-15 cm. Such a resolution is relatively good in general, but is insufficient for resolving small-scale inhomogeneities and the edge pedestal areas of T_e and n_e profiles in the so-called high-confinement mode (H-mode) of operation of the tokamak reactors. A way of improving the range resolution of the TS lidars is based on the use of deconvolution techniques for recovering the high-resolution lidar profiles. The deconvolution procedures, however, increase the influence of the noise. Therefore, to achieve acceptable recovered profiles one should apply a final filtering that lowers the sensing resolution to some compromise extent. The statistical modeling is a way to outline some optimal conditions under which the deconvolution techniques lead to satisfactory high-resolution restoration of the T_e profiles (Stoyanov et al., 2009; Dreischuh et al., 2011).

The TS lidar return signal from fussion plasma as well as the plasma light background and other additive noise are convenient to be analyzed on the basis of an equivalent photon counting procedure (Gurdev et al., 2008b). Based on Eqs.(1), (4) and (9), the long-pulse lidar equation in this case, for some say m-th spectral interval [λ_{s1m}, λ_{s2m}], is expressible as

$$N_l(\lambda_{s1m},\lambda_{s2m};z) = N_{lm}(z) = (2/c)\int_0^z dz' f[2(z-z')/c] N_s(\lambda_{s1m},\lambda_{s2m};z), \tag{46}$$

where the maximum-resolved lidar profile N_s is described by the short-pulse lidar equation

$$N_s(\lambda_{s1m},\lambda_{s2m};z) = N_{sm}(z) = (c/2)AN_0(\lambda_i)\int_{\lambda_{s1m}}^{\lambda_{s2m}} d\lambda_s K(\lambda_i,\lambda_s)\Phi(\lambda_i,\lambda_s;z); \tag{47}$$

$K(\lambda_i,\lambda_s)=K_t(\lambda_i)K_t(\lambda_s)K_f(\lambda_s)EQE(\lambda_s)$; $K_t(\lambda_i)$, $K_t(\lambda_s)$, $K_f(\lambda_s)$ and $EQE(\lambda_s)$ are respectively the wavelength-dependent optical transmittance of the plasma-irradiating path, the optical transmittance of the scattered-light collecting path, the receiver filter spectral characteristic, and the effective quantum efficiency of the photon detection accounting for the quantum yield and the Poisson fluctuations of the photoelectron number after the photocathode enhanced in the process of cascade multiplying in the employed microchannel tube; $\Phi(\lambda_i, \lambda_s;z)$ is given by Eq.(2) with $T(\lambda_i,\lambda_s;z) \equiv 1$, $\beta(\lambda_i;z)=\beta(z)=n_e(z)r_0^2$, and

$$\begin{aligned} L[\lambda_s,\lambda_i;z] &= \frac{c}{\sqrt{\pi}\lambda_i v_{th}(z)}\left(1+\frac{15}{16}\frac{v_{th}^2(z)}{c^2}+\frac{105}{512}\frac{v_{th}^4(z)}{c^4}\right)^{-1}\frac{(\lambda_i/\lambda_s)^3}{(1+\lambda_i/\lambda_s)}; \\ &\times\ \exp\left\{-\frac{c^2}{v_{th}^2(z)}\left[(\lambda_i/\lambda_s)^{1/2}+(\lambda_s/\lambda_i)^{1/2}-2\right]\right\} q[\lambda_i,\lambda_s,T_e(z)] \end{aligned} \tag{48}$$

$r_0=e^2/(4\pi\varepsilon_0 m_e c^2)$ is the classical electron radius, e and m_e are respectively the electron charge and rest mass, ε_0 is the dielectric constant of vacuum, $v_{th}(z)=[2k_BT_e(z)/m_e]^{1/2}$ is the rms thermal velocity of the electrons, k_B is the Boltzmann constant, $n_e(z)$ and $T_e(z)$ are respectively the electron density and temperature profiles along the lidar LOS, and $q[\lambda_i,\lambda_s,T_e(z)]$ is the depolarization term accounting for the relativistic depolarization effects on the backscattered radiation. For scattering at 180° the depolarization can be expressed in terms of exponential integral $E_n(p)$ (Naito et al., 1993):

$$q[\lambda_i,\lambda_s,T_e(z)] = 1+2e^p[E_3(p)-3E_5(p)] = 1+\frac{1}{2}\left\{\frac{p^3}{2}-\frac{p^2}{2}-p-1\right\}+p^2\left(1-\frac{p^2}{4}\right)e^pE_1(p) \tag{49}$$

$$p=\frac{m_ec^2}{2k_BT_e(z)}\left(\sqrt{\lambda_s/\lambda_i}+\sqrt{\lambda_i/\lambda_s}\right), \text{ and } E_n(p)=\int_1^\infty \frac{e^{-px}}{x^n}dx.$$

The TS lidar signal is accompanied by the plasma light background that is a serious source of error in the determination of T_e. Its emissivity spectrum per unit solid angle, mainly due to the bremsstrahlung, is given by the expression (Sheffield, 1975; Foord et al., 1982) :

$$\frac{dE}{d\Omega}=\frac{0.95\times10^{-19}}{\lambda\ 4\pi}n_e^2(z)Z_{\rm eff}(z)[k_BT_e(z)]^{-1/2}\exp\left(-\frac{hc}{\lambda k_BT_e(z)}\right)\tilde{g}_{ff}(\lambda,T_e), \tag{50}$$

where $Z_{eff}(z)$ is the effective ion charge, the quantities k_BT_e and hc/λ are in eV, $\exp[-hc/(\lambda k_BT_e)]\approx 1$ and $\tilde{g}_{ff}(\lambda,T_e)$ is the so-called Gaunt factor that depends weakly on T_e and on the radiation wavelength λ, and accounts for the quantum effects, the electron screening of nuclei, etc. (Brusaard & van de Hulst, 1962). For the photoelectron rate characterizing the parasitic background due to plasma light penetrating into the m-th spectral channel we obtain the following expression:

$$N_{\text{bm}}(\lambda_{\text{s}1m},\lambda_{\text{s}2m}) = 6.25\times 10^{-21} A_D \Delta\Omega_D$$
$$\times\int_z dz n_{\text{e}}^2(z)[k_B T_{\text{e}}(z)]^{-1/2} \int_{\lambda_{\text{s}1m}}^{\lambda_{\text{s}2m}} d\lambda_{\text{s}} K_t(\lambda_{\text{s}}) K_f(\lambda_{\text{s}}) EQE(\lambda_{\text{s}}) \lambda_{\text{s}}^{-1} \ln\left[k_B T_{\text{e}}(z)/(13.6h^2c^2/\lambda_{\text{s}}^2)^{1/3}\right], \quad (51)$$

where A_D is the photon detector effective area and $\Delta\Omega_D$ is the solid angle determined by the relative aperture of the receiving optics. In order to take into account additional background light sources, an enhancement factor is included in the simulations.

The center-of-mass wavelength (CMW) approach (Gurdev et al., 2008b; Dreischuh et al., 2009) to the determination of the electron temperature profiles $T_e(z)$ in fusion plasma is based on the unambiguous temperature dependence of the CMW of the relativistic Thomson backscattering spectrum. The TS lidar profiles N_{sm} are measured for M selected spectral intervals $[\lambda_{s1m}, \lambda_{s2m}]$ (m=1,2,…,M) [see Eq.(48)]. The CMW λ_{CM} defined as

$$\lambda_{\text{CM}}(T_{\text{e}};z) = \left(\sum_m \lambda_m N_{sm}(z)\right) \Big/ \left(\sum_m N_{sm}(z)\right) \quad (52)$$

is unambiguous function of the electron temperature (see also Fig.10 below); $\lambda_m=(\lambda_{s1m}+\lambda_{s2m})/2$ is the central wavelength of the m-th interval. Then the temperature is determined on the basis of the inverse function $T_e(\lambda_{CM}, z)$.

The linear error propagation approach leads to the following expression of the rms error δT_e in the determination of T_e on the basis of the dependence $\lambda_{CM}= f(T_e)$ (Gurdev et al., 2008b):

$$\delta T_{\text{e}} = \left|d\ln\lambda_{\text{CM}}(T_{\text{e}})/dT_{\text{e}}\right|^{-1} \left(\sum_m N_{pm}\tau_q\right)^{-1} \left\{\sum_{m=1}^{M} \left(\frac{\lambda_m - \lambda_{\text{CM}}}{\lambda_{\text{CM}}}\right)^2 N_{pm}\tau_q (1 + N_{\text{b}m}/N_{pm})\right\}^{1/2}, \quad (53)$$

where N_{pm} is the convolution of the laser pulse shape and the short-pulse lidar profile. The determinant temporal factor in Eq.(53) is τ_q because it is in practice the signal integration time interval. In case of applying deconvolution techniques for recovering the short-pulse lidar profiles and thus for obtaining more accurate T_e profiles, instead of Eq.(53) we have (Dreischuh et al., 2011)

$$\delta T_{\text{e}} = \left|d\ln\lambda_{\text{CM}}(T_{\text{e}})/dT_{\text{e}}\right|^{-1} \left(\sum_m N_{sm}\tau_\varphi\right)^{-1} \left\{\sum_{m=1}^{M} \left(\frac{\lambda_m - \lambda_{\text{CM}}}{\lambda_{\text{CM}}}\right)^2 N_{sm}\tau_\varphi [1 + \chi(\tau_{\text{s}}/\tau_\varphi) N_{\text{b}m}/N_{sm}]\right\}^{1/2}, \quad (54)$$

where τ_φ is the time-domain filter window and the factor $\chi(\tau_s/\tau_\varphi)$ is an increasing function of the ratio τ_s/τ_φ. This factor is accounting for the fact that the background is initially smoothed (integrated) only by the receiving system response function while the deconvolution is performed using the total lidar response function including the laser pulse shape.

An estimate of the SNR for the m-th spectral channel could be written as follows:

$$SNR_m = \{N_{pm}\tau_q / (1 + N_{bm} / N_{pm})\}^{1/2} \qquad (55)$$

in the case of convolved lidar profiles, and

$$SNR_m = \{N_{sm}\tau_\varphi / (1 + \chi(\tau_s / \tau_\varphi)N_{bm} / N_{sm})\}^{1/2} \qquad (56)$$

in the case of deconvolved lidar profiles.

From Eqs.(53-56) evidently follows that the signal-to-noise ratios SNR_m are the main factor conditioning the statistical retrieval accuracy.

The characteristic parameters of the plasma and the TS lidar used in the simulations are chosen to be close to those of the core TS lidar system on the Joint European Torus (JET) (Casci et al., 2002; Salzmann et al., 1988; Kempenaars et al., 2008, 2010). The sensing laser radiation is assumed to have wavelength λ_i=694 nm and pulse energy $E_0=N_0hc/\lambda_i$ =1 J, and to be injected horizontally along the plasma midplane. The minor radius r of the torus, along the LOS, is supposed to be 1 m. Correspondingly, the plasma is supposed to occupy the region between R=2 m and R=4 m, R being the radial distance from the center of the torus (Casci et al., 2002). Assuming that the LOS coordinate of the center of the torus is z_c, we obtain that $R=z_c- z$. The number of receiving spectrometer channels is chosen to be six. Their absolute spectral responses, including the EQE of the detectors, are also close to those of JET TS core lidar (Kempenaars et al., 2010). In particular, the detectors considered in the simulations are multialkali microchannel plate photomultiplier tubes (MCP-PMTs) with response times of about 650 ps and EQE equal to 0.005 for channel 1 and 0.02 for the other five channels. TS spectrum is observed within the wavelength region from 350 nm to 850 nm. To correct the collection efficiency the values of the solid angle of acceptance given in Kempenaars et al., 2010 are used. They vary from 0.005 sr, at R=2 m, to 0.007 sr at R=4 m. The irradiating and collecting paths optical transmittances assumed are $K_t(\lambda_i)$=0.75 and $K_t(\lambda_s)$=0.25, respectively. The detector's etendue $E=A_D\Delta\Omega_D$ needed for the estimation of the plasma bremsstrahlung photoelectron rate is assumed to have a value of ~0.32 cm^{2}sr. The factor of reducing the plasma bremsstrahlung conditioned by the plasma torus observation pupil is supposed to be 0.3. The effective atomic number of an equivalent plasma ion is chosen to be Z_{eff}=2. The bremsstrahlung background is added multiplied by an enhancement factor of 2 in order to take into account additional background light sources. The temporal sampling interval Δt_0 is supposed to be 200 ps (Δz_0=3 cm spatial interval).

The models of the temperature and density profiles used in the simulations consist of a smooth parabolic component whose parameters are chosen to simulate the real plasma conditions (Dreischuh et al., 2011; see also Figs. 12-14). Additionally, the $T_e(z)$ profile has a multiscale high-resolution component superimposed on the smooth component in order to illustrate the improvement of the retrieval accuracy and resolution depending on the noise level. The central electron density is varied in the range n_e= 2 ÷ 9 x10^{19} m^{-3} to simulate different plasma conditions and SNRs.

The sensing laser pulse shape is chosen to be $s(\theta) = (\theta/ \tau_l^2) \exp(-\theta/ \tau_l)$ for $\theta \geq 0$ and $s(\theta) = 0$ for $\theta < 0$, where τ_l is a time constant. Such a pulse shape can be a good approximation of various

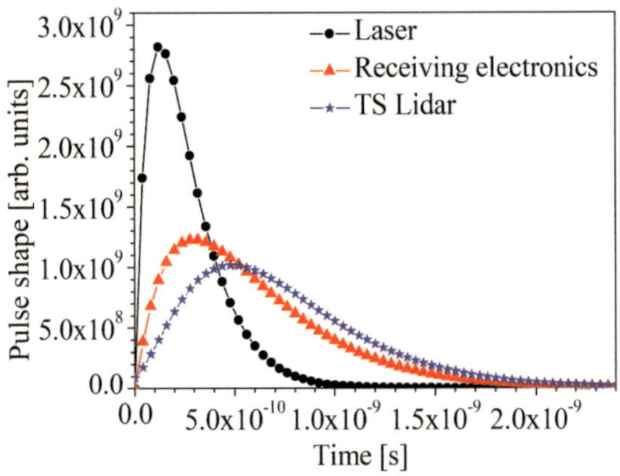

Fig. 9. Models of the laser pulse shape (circles), receiving system response shape (triangles) and the resulting TS lidar system response shape (stars) used in the simulations.

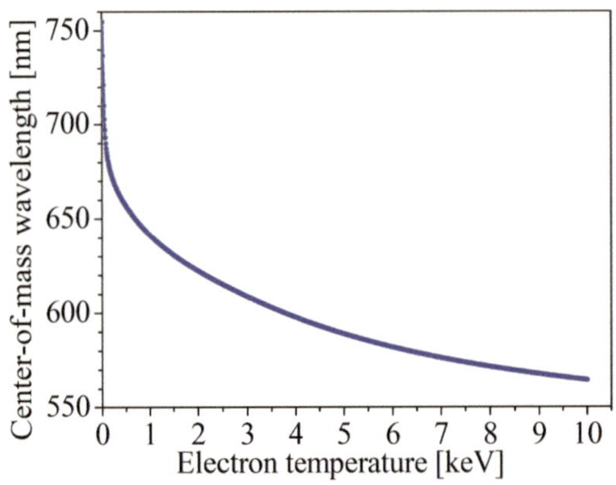

Fig. 10. Reference function $\lambda_{CM}(T_e)$ underlying the CMW approach.

real asymmetric laser pulses (e.g., Dong et al., 2001; Kondoh et al., 2001). The same model is used for the shape of the receiving system response function $q(\theta)$, that is, $q(\theta) = (\theta/\tau_e^2)\exp(-\theta/\tau_e)$ for $\theta \geq 0$, and $q(\theta) = 0$ for $\theta < 0$, where τ_e is another time constant. The Fourier spectrum modulus of the above pulse shapes is equal to $(1+\omega^2\tau_{l,e}^2)^{-1}$, i.e., it has no zeros, which is favorable for applying Fourier-deconvolution algorithm. The values of τ_l and τ_e are chosen so that the effective durations $\tau_s = e\tau_l$ and $\tau_q = e\tau_e$ of $s(\theta)$ and $q(\theta)$ to be respectively about 350 ps (τ_l = 130 ps) and 810 ps (τ_e= 300 ps). Then the effective duration of the resulting system response shape τ_f will be about 1 ns, which corresponds to 15 cm range resolution cell of the TS lidar. The models of the laser pulse shape, the receiving system response shape and the TS lidar system response shape are shown in Fig.9.

The reference function $\lambda_{CM}(T_e)$ is determined on the basis of the temperature dependence of the TS spectrum and is presented in Fig.10 for temperatures up to 10 keV. In the case of long-pulse sensing, when the pulse length exceeds the spatial scale of the temperature inhomogeneities, the temperature information provided by the lidar profiles from the different spectral channels will be distorted. Correspondingly, the recovered temperature profiles will also be distorted with respect to the true ones. The role of the deconvolution here is to reduce, as much as possible at the corresponding noise level, the convolution-due distortions of the recovered T_e profiles.

The Monte-Carlo simulations are performed in the following way. First, the mean values $N_{sm}(z)$ of the TS signal in each spectral channel are determined and then convolved with the laser pulse shape in order to account for the real pulse duration. Next, the mean background photoelectron count rate $N_{bm}(z)$ is evaluated. Then, assuming Poisson statistics of the signal and background photoelectrons within a Δt_0 - long interval and using random - number generator, J realizations of the TS signal $N_{lm}(z)\Delta t_0$ and background $N_{bm}(z)\Delta t_0$ photoelectrons are produced (see Fig11). Further, the receiving system response function is taken into account performing the convolution with it of the profiles of the background and signal count rates in each channel. Assuming an accurate measurement of the mean convolved background count rate, it is subtracted from the corresponding background count rate realizations. Thus, the convolved background count rate fluctuations are obtained. At last, the obtained realizations of the long-pulse lidar profiles including the background fluctuations are deconvolved using the system response function $f(\theta)$. The center of mass wavelength as a function of the coordinate along the LOS is determined according to Eq.(52) on the basis of the deconvolved profiles, and is used together with the reference function $\lambda_{CM}(T_e)$ for obtaining J estimates $\hat{T}_{ej}(z)$ of the electron temperature profile $T_e(z)$, j=1,2,…,J. Then, an estimate $\delta\hat{T}_e(z)$ of the measurement error is obtainable as $\delta\hat{T}_e(z) = \left\{J^{-1}\sum_{j=1}^{J}[\hat{T}_{ej}(z) - T_e(z)]^2\right\}^{1/2}$.

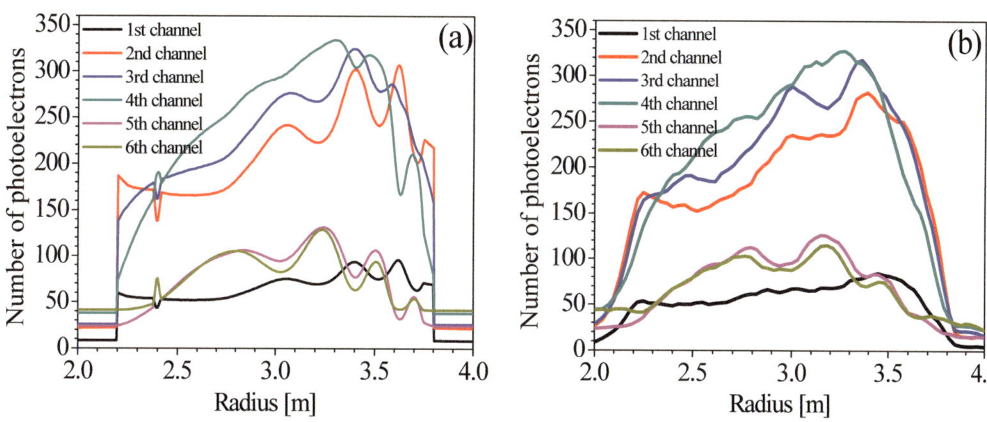

Fig. 11. TS lidar profiles: (a) mean short-pulse lidar profiles including the mean plasma light background, (b) realizations of the measured long-pulse lidar profiles including the background realizations; $n_e = 9\times10^{19}$ m^{-3} .

To simulate correctly the detection of the analog signals, the convolved profiles are calculated almost ideally by a computing step much less than Δt_0. The real ADC step Δt_0 is used when processing further the long-pulse profiles. The obtained mean short-pulse lidar profiles and simulated realizations of the measured long-pulse lidar profiles for the six spectral channels are shown in Figs.11a,b.

As an illustration of the deconvolution effect, the T_e profiles restored in the absence of noise on the basis of the convolved and deconvolved lidar profiles are shown in Fig.12. As it is seen, the direct use of the long-pulse lidar profiles leads to significant distortions in the restored electron temperature profiles. After applying deconvolution techniques to the long-pulse lidar profiles, at negligible noise level the T_e profiles are determined with considerably higher accuracy, and resolution scale of the order of the sampling interval Δz_0.

Because of the strong Poisson fluctuations, some type of low-pass noise filtering is necessary to ensure a satisfactory quality of the restored profiles. However, the filtering procedure lowers the range resolution. The range resolution cell will be already of the order of the width W of the range-domain window of the filter employed. To retain a satisfactory range resolution the value of W should be less than the least variation scale (along the line of sight) of the temperature profile. Then the restored temperature profiles are minimally distorted with respect to the true ones. Different low-pass digital filters are used in the numerical simulations. Results presented below are obtained using filers with $2\Delta z_0$ and $3\Delta z_0$ -wide windows for smoothing the recorded lidar profiles.

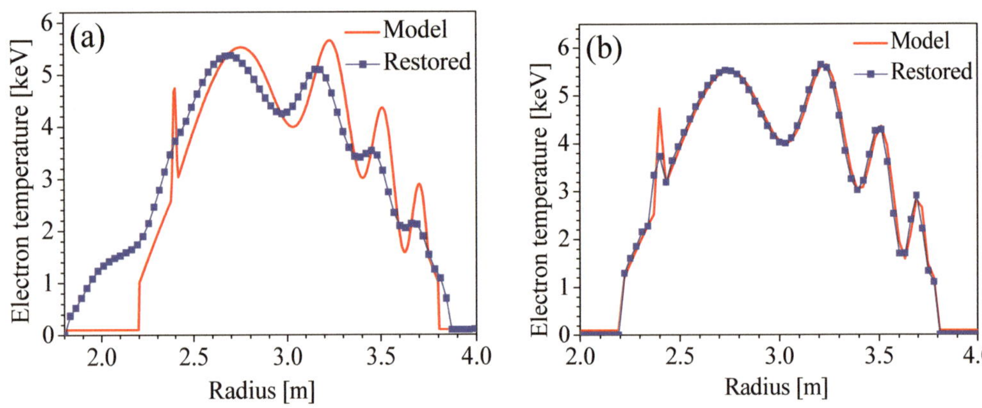

Fig. 12. Electron temperature profiles restored in absence of noise on the basis of the convolved (a) and deconvolved (b) lidar profiles; $n_e = 9\text{x}10^{19}$ m^{-3}.

In Fig.13 the profiles of the electron temperature restored on the basis of the measured convolved and deconvolved lidar profiles for one realization of the Poisson noise are presented. It is well seen that the temperature profiles restored on the basis of convolved lidar profiles (Fig.13a) are essentially distorted with respect to the original model. At the same time, the temperature profile restored on the basis of deconvolved lidar profiles (Fig.13b) is disguised by strongly increased fluctuations. In order to suppress the deconvolution-due increase of the noise, noise controlling filters have been applied (Figs.13c,d) ensuring acceptable accuracy and resolution of the restored electron temperature profiles. It is seen in Fig.13d that even $2\Delta z_0$-wide filter window (corresponding to 6 cm range resolution) ensures good quality of the obtained T_e profile. The theoretical statistical errors presented in these figures are estimated assuming empirically in Eq.(54) that $\chi(\tau_s/\tau_\varphi) = 25$ (Fig.13b), 15 (Fig.13c), and 10 (Fig.13d). When using convolved profiles for determination of T_e (Fig.13a), the factor $\chi(\tau_s/\tau_\varphi)$ is not of importance [Eq.(53)]. In this case, the "empirical" statistical error is estimated by simulations with respect to the temperature profile obtained from the convolved, long-pulse lidar profiles in absence of noise. The same as the above-described is the behavior of the restored profiles in the case of lower electron concentration n_e = 2x10^{19} m^{-3}. Because of the lower SNR in this case, the quality of the restored profiles is somewhat lower compared to the case of n_e=9x10^{19} m^{-3}.

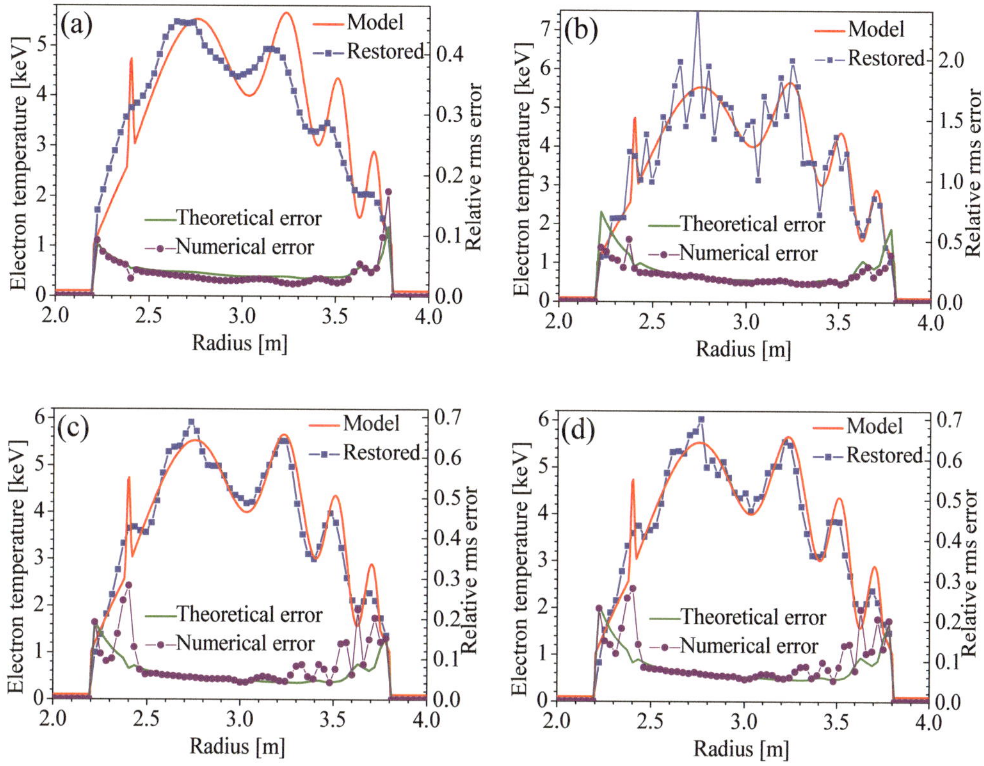

Fig. 13. Electron temperature profiles restored on the basis of the convolved (a) and deconvolved lidar profiles without filtering (b), and on the basis of the deconvolved lidar profiles smoothed by a monotone sharp-cutoff digital filter (with W=3cΔt_0/2) (c) and by a moving average filter (with W=2cΔt_0/2) (d); the right-hand y axis represents the theoretically estimated relative rms errors compared to the numerically obtained ones; n_e = 9x10^{19} m^{-3}.

The results of applying the deconvolution approach in the case of increased sensing pulse energy (E_0=3 J) are shown in Fig.14, where it is seen that the restoration accuracy is higher due to the higher SNR. This allows one to detect reliably smaller-scale inhomogeneities of the finer structure of the electron temperature profiles. In general, the increase of SNR, due for instance to increasing the electron concentration, the sensing pulse energy or the sensitivity of the photodetectors, is determinant for achieving high retrieval accuracy and resolution.

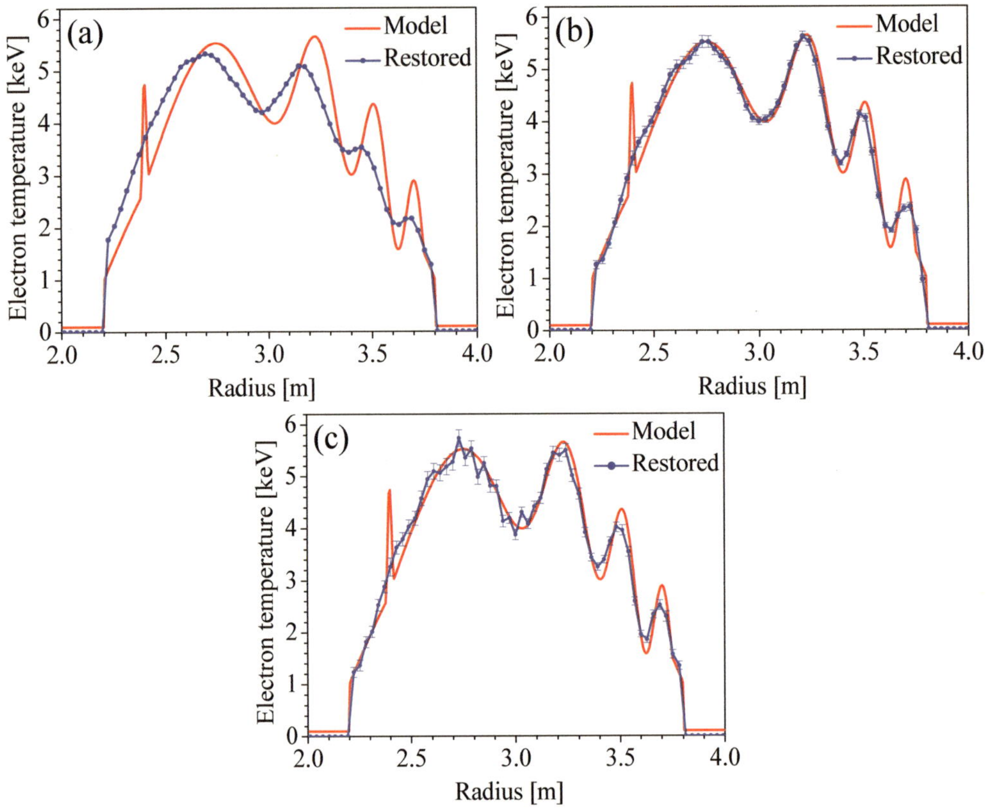

Fig. 14. Electron temperature profiles restored on the basis of the convolved (a) and deconvolved lidar profiles smoothed by a monotone sharp-cutoff digital filter (with W=3c$\Delta t_0/2$) (b) and by a moving average filter (with W=2c$\Delta t_0/2$) (c); n_e=9x10^{19} m^{-3}, E_0= 3 J. The statistical error represented by error bars, (b) and (c), is estimated on the basis of Eq.(54).

5. Conclusions

In the present chapter, the advantages and limitations have been considered of deconvolution techniques for improving the accuracy and resolution of the remote sensing of atmosphere, thermonuclear plasmas, and other objects by lidars of relatively long pulse response function including the laser pulse shape. Analog and photon counting modes of direct signal detection have been concerned. The general Fourier and Volterra deconvolution algorithms have been analyzed as well as some simple and fast special algorithms for the cases of rectangular, rectangular-like and exponentially-shaped pulse response functions. At negligible noise level, a high accuracy of recovering the short-pulse lidar profile is achievable at sufficiently short computing step. Also, by using suitable approaches, in some cases one can reduce the characteristic retrieval distortions due to some pulse response uncertainties. The strong broadband noise effect on the retrieval accuracy and resolution is revealed, including the noise accumulation with the distance of sensing for the recurrence algorithms. The noise influence in this case is shown to be effectively reduced by using appropriate compromise filtering or choice of the computing step. That is, to avoid retrieval distortions, the filter window or the computing step should exceed the noise correlation radius (or time) but be less than the least variation (spatial or temporal) scale of the short-pulse lidar profile. The deconvolution procedures are shown as well to decrease the disturbing effect of narrow-band noise whose correlation time exceeds the pulse duration. Let us also underline the fact that the deconvolution-based retrieval of the short-pulse lidar profiles allows high-resolution sensing of small finite-size objects by longer laser pulses, realizing in this way double-sided linear-strategy optical tomography of such objects.

The investigations performed show as well that Fourier-deconvolution procedures, combined with appropriate low-pass filtering, applied to the measured Thomson scattering lidar profiles lead to several (2-3) times better resolution of recovering electron temperature profiles in fusion plasma, under conditions of plasma light background and amplification-enhanced Poisson noise. The convolution-due systematic errors are essentially corrected for and an acceptable restoration accuracy is achieved allowing one to reveal characteristic inhomogeneities in the distribution of the electron temperature within the plasma torus. It is also shown that, naturally, because of higher signal-to-noise ratio (stronger lidar return) the deconvolution accuracy increases with the increase of the electron concentration and the sensing pulse energy. This means that the deconvolution approach would be especially appropriate for processing data from a new generation of fusion reactors, such as ITER and DEMO, characterized by considerably higher electron concentration and sensing pulse energy compared to these achievable in JET.

6. Acknowledgments

This results described in the chapter was funded partly by the Bulgarian National Science Fund under Projects Ph-447, Ph-1511, and DO 02-107/2009 and the European Communities under the Contract of Association between EURATOM and INRNE (Bulgaria). This work was carried out in part within the framework of the European Fusion Development Agreement. The views and opinions expressed herein do not necessarily reflect those of the European Commission.

7. References

Bahrampour, A.R. & Askari, A.A. (2006). Fourier-Wavelet Regularized Deconvolution (ForWaRD) for Lidar Systems Based on TEA–CO2 laser, *Opt. Commun.*, vol. 257, pp. 97–111.

Brusaard, P.J. & van de Hulst, H.C. (1962). Approximation Formulas for Nonrelativistic Bremsstrahlung and Average Gaunt Factors for a Maxwellian Electron Gas, *Rev. Mod. Phys.*, vol. 34, pp. 507-520.

Casci, F.; Lanzinger, D. & Buceti, G., Eds. (2002). The Joint European Torus - a European Success Story, EFDA Close Support Unit, Culham, Available from http://www.jet.efda.org/wp-content/uploads/jeteuropeansucess.pdf

Dong, J.; Deng, P.; Liu, Y.; Zhang, Y.; Xu, J.; Chen, W & Xie, X. (2001). Passively Q-switched Yb:YAG Laser with Cr4+:YAG as the Saturable Absorber, *Appl. Opt.*, vol. 40, pp. 4303-4307.

Doviak, R.J. & Zrnic, D.S. (1984). *Doppler Radar and Weather Observation,* Academic Press, San Diego, USA.

Dreischuh, T.N.; Gurdev, L.L. & Stoyanov, D.V. (1995). Effect of Pulse-Shape Uncertainty on the Accuracy of Deconvolved Lidar Profiles, *J. Opt. Soc. Am. A*, vol.12, pp.301-306.

Dreischuh, T.N.; Gurdev, L.L. & Stoyanov, D.V. (1996). Lidar Profile Deconvolution Algorithms for Some Rectangular-Like Laser Pulse Shapes, In: *Advances in Atmospheric Remote Sensing with Lidar,* A.Ansmann, R. Neuber, P.Raioux, U. Wandinger, (Eds.), 135-138, Springer Verlag, ISBN 3540618872, Berlin.

Dreischuh, T.N.; Gurdev, L.L.; Stoyanov, D.V.; Protochristov, C.N. & Vankov, O.I. (2007). Application of a Lidar-Type Gamma-Ray Tomography Approach for Detection and Identification of Buried Plastic Landmines, *Proc. SPIE*, vol.6604, paper #660420.

Dreischuh, T.; Gurdev, L.; Stoyanov, D.; Beurskens, M.; Walsh, M. & Capel, A. (2009). Statistical Modeling of the Error in the Determination of the Electron Temperature in JET by a Novel Thomson Scattering LIDAR Approach, In: Proc. 36th EPS Conference on Plasma Physics. Contributed papers, *ECA*, vol.33E, P-2.149, European Physical Society, ISBN:2-914771-61-4, Geneva.

Dreischuh, T.; Gurdev, L. & Stoyanov, D. (2011). Statistical Modeling of Deconvolution Procedures for Improving the Resolution of Measuring Electron Temperature Profiles in Tokamak Plasmas by Thomson Scattering Lidar, *Proc. SPIE*, vol. 7747, paper # 77470T.

Foord, M. E.; Marmar, E. S. & Terry, J. L. (1982). Multichannel Light Detector System for Visible Continuum Measurement on Alcator C, *Rev. Sci. Instrum.*, vol. 53, pp. 1407-1409.

Gurdev, L.L.; Dreischuh, T.N. & Stoyanov, D.V. (1993). Deconvolution Techniques for Improving the Resolution of Long-Pulse Lidars, *J. Opt. Soc. Am. A*, vol.10, No.11, pp. 2296-2306.

Gurdev, L.L.; Dreischuh, T.N. & Stoyanov, D.V. (1996). Deconvolution Algorithms for Improving the Resolution of Exponential-Shape Pulse Lidars", *Proc.SPIE*, vol. 3052, pp. 310-315.

Gurdev, L.L.; Dreischuh, T.N. & Stoyanov, D.V. (1998). Pulse Backscattering Tomography Based on Lidar Principle, *Opt. Commun.*, vol. 151, pp. 339-352.

Gurdev, L.L.; Dreischuh, T.N. & Stoyanov, D.V. (2001). High-Resolution Doppler-Velocity Estimation Techniques for Processing of Coherent Heterodyne Pulsed Lidar Data, *J. Opt. Soc. Am. A*, vol. 18, pp.134-142.

Gurdev, L.L.; Dreischuh, T.N. & Stoyanov, D.V. (2002). High Range Resolution Velocity Estimation Techniques for Coherent Doppler Lidars with Exponentially-Shaped Laser Pulses, *Appl. Opt.*, vol. 41, pp. 1741-1749.

Gurdev, L.L. & Dreischuh, T.N. (2003). High Range Resolution Velocity Estimation Techniques taking into Account the Frequency Chirp in Coherent Doppler Lidars, *Opt. Commun.*, vol. 219, pp.101-116.

Gurdev, L.L.; Stoyanov, D.V.; Dreischuh, T.N.; Protochristov, C.N. & Vankov, O.I. (2007a). Gamma-Ray Backscattering Tomography Approach Based on the Lidar Principle, *IEEE Transactions on Nuclear Science,* vol. 54, No.1, part 2, pp.262-275.

Gurdev, L.L.; Dreischuh, T.N.; Stoyanov, D.V. & Protochristov, C.N. (2007b). Gamma-Ray Lidar (GRAYDAR) in-Depth Sensing of Optically Opaque Media, In: *Nuclear*

Methods for Non-Nuclear Applications, Ch. Stoyanov, (Ed.), Heron Press, ISBN: 978-954-580-209-6, Sofia.

Gurdev, L.L. & Dreischuh, T.N. (2008a). On an approach for improving the range resolution of pulsed coherent Doppler lidars, *J. Mod. Optics*, vol. 55, No.9, pp.1441-1462.

Gurdev, L.L.; Dreischuh, T.N. & Stoyanov, D.V. (2008b). Potential Accuracies of Some New Approaches for Determination by Thomson Scattering Lidar of the Electron Temperature Profiles in Thermonuclear Plasmas, *Proc. SPIE*, vol. 7027, # 702711.

Hannon, S.M. & Thomson, J.A. (1994). Aircraft Wake Vortex Detection and Measurement with Pulsed Solid-State Coherent Laser Radar, *J. Mod. Optics*, vol. 41, pp.2175–2196.

Johnstone, I.; Kerkyacharian, G.; Picard, D., & Raimondo, M (2004). Wavelet Deconvolution in a Periodic Setting, *J. Roy. Stat. Soc. B*, vol. 66, pp. 547–573, with discussion on pp. 627-652.

Kempenaars, M.; Flanagan, J. C.; Giudicotti, L.; Walsh, M. J.; Beurskens, M. & Balboa, I. (2008). Enhancement of the JET Edge LIDAR Thomson Scattering Diagnostic with Ultrafast Detectors, *Rev. Sci. Instrum.*, vol. 79, 10E728.

Kempenaars, M.; Balboa, I.; Beurskens, M.; Flanagan, J. C. & Maslov, M. (2010). The JET Core LIDAR Diagnostic, Int. Conference on Plasma Diagnostics, Pont-à-Mousson, France, 12-16 April 2010, paper P30.

Kondoh, T.; Lee, S.; Hutchinson, D. P. & Richards, R. (2001). Collective Thomson Scattering Using a Pulsed CO2 Laser in JT-60U, *Rev. Sci. Instrum.*, vol. 72, pp. 1143-1146.

Kovalev, V.A. & Eichinger, W.E. (2004). *Elastic Lidar: Theory, Practice, and Analysis Methods*, Wiley, New York, USA.

Marzano, F.S. & Visconti, G., Eds. (2002). *Remote Sensing of Atmosphere and Ocean from Space: Models, Instruments and Techniques*, Kluwer Academic Publishers, Dodrecht.

Measures, R.M. (1984). *Laser Remote Sensing*, Wiley, New York, USA.

Naito, O.; Yoshida, H. & Matoba, T. (1993). Analytic Formula for Fully Relativistic Thomson Scattering Spectrum, *Phys. Fluids B*, vol. 5, pp. 4256-4258.

Park, Y. J.; Dho, S.W. & Kong, H.J. (1997). Deconvolution of Long-Pulse Lidar Signals with Matrix Formulation, *Appl. Opt.*, vol. 36, pp. 5158-5161.

Pensky, M. & Sapatinas, T. (2009). Functional Deconvolution in a Periodic Setting: Uniform Case, *Ann. Statist.*, vol. 37, pp. 73-104.

Pensky, M. & Sapatinas, T. (2010). On Convergence Rates Equivalency and Sampling Strategies in Functional Deconvolution models, *Ann. Statist.*, vol. 38, pp. 1793-1844.

Refaat, T.F.; Ismail, S.; Abedin, M. N.; Spuler, S.M. & Singh, U. N. (2008). Lidar Backscatter Signal Recovery from Phototransistor Systematic Effect by Deconvolution, *Appl. Opt.*, vol. 47, No. 29, pp. 5281-5295.

Rytov, S.M. (1976). *Introduction to Statistical Radiophysics*, vol. 1, Nauka, Moscow.

Salzmann, H. et al., (1988). The LIDAR Thomson Scattering Diagnostic on JET, *Rev. Sci. Instrum.*, vol. 59, pp. 1451-1456.

Sheffield, J. (1975). *Plasma Scattering of Electromagnetic Radiation*, Academic, New York.

Stoyanov, D.V.; Gurdev, L.L. & Dreischuh, T.N. (1996). Reduction of the Pulse Spike Cut Error in Fourier-Deconvolved Lidar Profiles, *Appl. Opt.*, vol.35, pp.4798-4802.

Stoyanov, D.V.; Gurdev, L.L.; Kolarov, G.V. & Vankov, O.I. (2000). Lidar Profiling by Long Rectangular-Like Chopped Laser Pulses, *Opt. Engineering*, vol. 39, pp. 1556-1567.

Stoyanov, D.V.; Dreischuh, T.N.; Vankov, O.I. & Gurdev, L.L. (2004). Measuring the Shape of Randomly Arriving Pulses Shorter than the Acquisition Step, *Meas. Sci. Technol.*, vol. 15, pp.2361-2369.

Stoyanov, D.; Nedkov, I. & Ausloos, M. (2007). Retrieving True Images through fine grid steps for enhancing the resolution beyond the classical limits: theory and simulations, *Journal of Microscopy*, vol. 226, pp. 270–283.

Stoyanov, D.; Dreischuh, T.; Gurdev, L.; Beurskens, M.; Ford, O.; Flanagan, J.; Kempenaars, M.; Balboa, I. & Walsh, M. (2009). Deconvolution of JET Core LIDAR data and Pedestal Detection in Retrieved Electron Temperature and Density Profiles, In: Proc. 36th EPS Conference on Plasma Physics. Contributed papers, *ECA*, vol.33E, P-2.155, European Physical Society, ISBN:2-914771-61-4, Geneva.

Stoyanov, D.; Beurskens, M.; Dreischuh, T.; Gurdev, L.; Ford, O.; Flanagan, J.; Kempenaars, M.; Balboa, I. & Walsh, M. (2010). Resolving the Plasma Electron Temperature Pedestal in JET from Thomson Scattering Core LIDAR Data, In: Proc. 37th EPS Conference on Plasma Physics. Contributed papers, *ECA*, vol.34A, P5.133 European Physical Society, ISBN 2-914771-62-2, Geneva.

Van Trees, H.L. (2001). *Detection, Estimation, and Modulation Theory, Part III, Radar-Sonar Signal Processing and Gaussian Signals in Noise*, Wiley, New York, USA.